Conway Lloyd Morgan

Animal Life and Intelligence

Conway Lloyd Morgan

Animal Life and Intelligence

ISBN/EAN: 9783743332218

Manufactured in Europe, USA, Canada, Australia, Japa

Cover: Foto ©berggeist007 / pixelio.de

Manufactured and distributed by brebook publishing software (www.brebook.com)

Conway Lloyd Morgan

Animal Life and Intelligence

ANIMAL LIFE AND INTELLIGENCE.

BY

C. LLOYD MORGAN, F.G.S.,

PROFESSOR IN AND DEAN OF UNIVERSITY COLLEGE, BRISTOL;
LECTURER AT THE BRISTOL MEDICAL SCHOOL;
PRESIDENT OF THE BRISTOL NATURALISTS' SOCIETY, ETC.

AUTHOR OF
"ANIMAL BIOLOGY," "THE SPRINGS OF CONDUCT," ETC.

BOSTON, U.S.A.:
GINN & COMPANY, PUBLISHERS.
1891.

PREFACE.

There are many books in our language which deal with Animal Intelligence in an anecdotal and conventionally popular manner. There are a few, notably those by Mr. Romanes and Mr. Mivart, which bring adequate knowledge and training to bear on a subject of unusual difficulty. In the following pages I have endeavoured to contribute something (imperfect, as I know full well, but the result of several years' study and thought) to our deeper knowledge of those mental processes which we may fairly infer from the activities of dumb animals.

The consideration of Animal Intelligence, from the scientific and philosophical standpoint, has been my primary aim. But so inextricably intertwined is the subject of Intelligence with the subject of Life, the subject of organic evolution with the subject of mental evolution, so closely are questions of Heredity and Natural Selection interwoven with questions of Habit and Instinct, that I have devoted the first part of this volume to a consideration of Organic Evolution. The great importance and value of Professor Weismann's recent contributions to biological science, and their direct bearing on questions of Instinct, rendered such treatment of my subject, not only advisable, but necessary. Moreover, it seemed to me, and to those whom I consulted in the matter, that a general work on Animal Life and Intelligence, if adequately knit into a connected whole, and based on sound principles of science and of philosophy, would not be unwelcomed by biological students, and by that large and increasing class of readers who, though not professed students, follow with eager interest the development of the doctrine of Evolution.

Incidentally, but only incidentally, matters concerning man, as compared with the dumb animals, have been introduced. It is contended that in man alone, and in no dumb animal, is the rational faculty, as defined in these pages, developed; and it is contended that among human-folk that process of natural selection, which is so potent a factor in the lower reaches of organic life, sinks into comparative insignificance. Man is a creature of ideas and ideals. For him the moral factor becomes one of the very highest importance. He conceives an ideal self which he strives to realize; he conceives an ideal humanity towards which he would raise his fellow-man. He becomes a conscious participator in the evolution of man, in the progress of humanity.

But while we must not be blind to the effects of new and higher factors of progress thus introduced as we rise in the scale of phenomena, we must at the same time remember that biological laws still hold true, though moral considerations and the law of duty may profoundly modify them. The eagle soars aloft apparently in defiance of gravitation; but the law of gravitation still holds good; and no treatment of the mechanism of flight which neglected it would be satisfactory. Moral restraint, a higher standard of comfort, and a perception of the folly and misery of early and improvident marriage may tend to check the rate of growth of population: but the "law of increase" still holds good, as a law of the factors of phenomena; and Malthus did good service to the cause of science when he insisted on its importance. We may guide or lighten the incidence of natural selection through competition; we may in our pity provide an asylum for the unfortunates who are suffering elimination; but we cannot alter a law which, as that of one of the factors of organic phenomena, still obtains, notwithstanding the introduction of other factors.

However profoundly the laws of phenomena may be modified by such introduction of new and higher factors, the older and lower factors are still at work beneath the surface. And he who would adequately grasp the social problems of our time should bring to them a mind prepared by a study of the laws of organic life: for human beings, rational and moral though they may be, are still organisms; and man can in no wise alter or annul those deep-lying facts which nature has throughout the ages been weaving into the tissue of life.

Some parts of this work are necessarily more technical, and therefore more abstruse, than others. This is especially the case with Chapters III., V., and VI.; while, for those unacquainted with philosophical thought, perhaps the last chapter may present difficulties of a different order. With these exceptions, the book will not be beyond the ready comprehension of the general reader of average intelligence.

I have to thank many kind friends for incidental help. Thanks are also due to Professor Flower, who courteously gave permission that some of the exhibits in our great national collection in Cromwell Road might be photographed and reproduced; and to Messrs. Longmans for the use of two or three illustrations from my text-book of "Animal Biology."

C. LLOYD MORGAN.

UNIVERSITY COLLEGE, BRISTOL,
October, 1890.

CONTENTS.

CHAPTER I.

THE NATURE OF ANIMAL LIFE.

	PAGE
The characteristics of animals	2
The relation of animals to food-stuffs	15
,, ,, ,, the atmosphere	15
,, ,, ,, energy	16

CHAPTER II.

THE PROCESS OF LIFE.

Illustration from respiration	21
,, ,, nutrition	25
The utilization of the materials incorporated	27
The analogy of a gas-engine. Explosive metabolism	30

CHAPTER III.

REPRODUCTION AND DEVELOPMENT.

Reproduction in the protozoa	37
Fission in the metazoa	41
The regeneration of lost parts	41
Reproduction by budding	42
Sexual reproduction	42
Illustration of development	51
Parental sacrifice	56
The law of increase	58

CHAPTER IV.

VARIATION AND NATURAL SELECTION.

The law of persistence	61
The occurrence of variations	63
Application of the law of increase	76
Natural selection	77

	PAGE
Elimination and selection	79
Modes of natural elimination illustrated	80
Protective resemblance and mimicry	82
Selection proper illustrated	93
The effects of natural selection	95
Isolation or segregation	99
Its modes, geographical, preferential and physiological	99
Its effects	108
Utility of specific characters	110
Variations in the intensity of the struggle for existence	112
Convergence of characters	117
Modes of adaptation: Progress	119
Evolution and Revolution	120

CHAPTER V.

HEREDITY AND THE ORIGIN OF VARIATIONS.

Heredity in the protozoa	123
Regeneration of lost parts	124
Sexual reproduction and heredity	129
The problem of hen and egg	130
Reproductive continuity	131
Pangenesis	131
Modified pangenesis	134
Continuity of germ-plasm	138
Cellular continuity with differentiation	142
The inheritance or non-inheritance of acquired characters	146
Origin of variations on the latter view	149
Hypothesis of organic combination	150
The extrusion of the second polar cell	153
The protozoan origin of variations	156
How can the body influence the germ?	159
Is there sufficient evidence that it does?	162
Summary and conclusion	175

CHAPTER VI.

ORGANIC EVOLUTION.

The diversity of animal life	177
The evolution theory	181
Natural selection: not to be used as a magic formula	183
Panmixia and disuse	189
Sexual selection or preferential mating	197
Use and disuse	209
The nature of variations	216
The inheritance of variations	223
The origin of variations	231
Summary and conclusion	241

CHAPTER VII.

THE SENSES OF ANIMALS.

	PAGE
The primary object of sensation	243
Organic sensations and the muscular sense	244
Touch	245
The temperature-sense	249
Taste	250
Smell	257
Hearing	261
Sense of rotation or acceleration	269
Sight	273
Restatement of theory of colour-vision	278
Variation in the limits of colour-vision	281
The four types of "visual" organs	293
Problematical senses	294
Permanent possibilities of sensation	298

CHAPTER VIII.

MENTAL PROCESSES IN MAN.

The physiological aspect	302
The psychological aspect	304
Sensations: their localization, etc.	306
Perceptual construction	312
Conceptual analysis	321
Inferences perceptual and conceptual	328
Intelligence and reason	330

CHAPTER IX.

MENTAL PROCESSES IN ANIMALS: THEIR POWERS OF PERCEPTION AND INTELLIGENCE.

The two factors in phenomena	331
The basis in organic evolution	336
Perceptual construction in mammalia	338
Can animals analyze their constructs?	347
The generic difference between the minds of man and brute	350
Perceptual construction in other vertebrates	350
"Understanding" of words	354
Perceptual construction in the invertebrates	356
"The psychic life of micro-organisms"	360
The inferences of animals	361
Intelligent not rational	365
Use of words defined	372
Language and analysis	374

CHAPTER X.

THE FEELINGS OF ANIMALS: THEIR APPETENCES AND EMOTIONS.

	PAGE
Pleasure and pain: their organic limits	379
Their directive value	380
An emotion exemplified	382
Sensitiveness and sensibility	385
The expression of the emotions	385
The postponement of action	385
The three orders of emotion	390
The capacities of animals for pleasure and pain	391
Sense-feelings	393
Some emotions of animals	395
The necessity for caution in interpretation	399
The sense of beauty	407
Can animals be moral?	413
Conclusion	414

CHAPTER XI.

ANIMAL ACTIVITIES: HABIT AND INSTINCT.

The nature of animal activities	415
The outer and inner aspect	417
The inherited organization	419
Habitual activities	420
Instinctive activities	422
Innate capacity	426
Blind prevision	429
Consciousness and instinct	432
Mr. Romanes's treatment of instinct	434
Lapsed intelligence and modern views on heredity	435
Three factors in the origin of instinctive activities	447
The emotional basis of instinct	449
The influence of intelligence on instinct	452
The characteristics of intelligent activities	456
The place of volition	459
Perceptual and conceptual volition	460
Consciousness and consentience	461
Classification of activities	462

CHAPTER XII.

MENTAL EVOLUTION.

Is mind evolved from matter?	464
Kinesis and metakinesis	467
Monistic assumptions	470
The nature of ejects	476

	PAGE
The universe as eject	478
Metakinetic environment of mind	481
Conceptual ideas not subject to natural selection	483
Elimination through incongruity	486
Interneural evolution	490
Interpretations of nature	492
Can fetishism have had a natural genesis?	493
The origin of interneural variations	496
Are acquired variations inherited?	497
Summary and conclusion	501

LIST OF ILLUSTRATIONS.

FIG.		PAGE
	Kentish Plover with Eggs and Young *Frontispiece*	
1.	Spiracles and Air-tubes of Cockroach	3
2.	Gills of Mussel	4
3.	A Cell greatly magnified	11
4.	Amœba	12
5.	Egg-cell and Sperm-cell	13
6.	Diagram of Circulation	23
7.	Protozoa	38
8.	Hydra Virides	43
9.	Aurelia : Life-cycle	45
10.	Liver-Fluke—Embryonic Stages	47
11.	Diagram of Development	51
12.	Wing of Bat (Pipistrelle)	64
13.	Variations of the Noctule	67
14.	,, ,, Long-eared Bat	68
15.	,, ,, Pipistrelle	69
16.	,, ,, Whiskered Bat	70
17.	Variations adjusted to the Standard of the Noctule	73
18.	Caterpillar of a Moth on an Oak Spray	85
19.	Locust resembling a Leaf	86
20.	Mimicry of Bees by Flies	91
21.	Egg and Hen	141
22.	Stag-Beetles	180
23.	Tactile Corpuscules	247
24.	Touch-hair of Insect	248
25.	Taste-buds of Rabbit	250
26.	Antennule of Crayfish	259
27.	Diagram of Ear	263
28.	Tail of Mysis	266
29.	Leg of Grasshopper	266
30.	Diagram of Semicircular Canals	270

FIG.		PAGE
31.	The Human Eye	274
32.	Retina of the Eye	274
33.	Variation in the Limits of Colour-vision	281
34.	Pineal Eye	288
35.	Skull of Melanerpeton	288
36.	Eyes and Eyelets of Bee	289
37.	Eye of Fly	290
38.	Diagram of Mosaic Vision	291
39.	Direction-retina	295
40.	Antennary Structures of Hymenoptera	297

ANIMAL LIFE AND INTELLIGENCE.

CHAPTER I.

THE NATURE OF ANIMAL LIFE.

I ONCE asked a class of school-boys to write down for me in a few words what they considered the chief characteristics of animals. Here are some of the answers—

1. Animals move about, eat, and grow.
2. Animals eat, grow, breathe, feel (at least, most of them do), and sleep.
3. Take a cat, for example. It begins as a kitten; it eats, drinks, plays about, and grows up into a cat, which does much the same, only it is more lazy, and stops growing. At last it grows old and dies. But it may have kittens first.
4. An animal has a head and tail, four legs, and a body. It is a living creature, and not a vegetable.
5. Animals are living creatures, made of flesh and blood.

Combining these statements, we have the following characteristics of animals:—

1. Each has a proper and definite form, at present described as "a head and tail, four legs, and a body."
2. They breathe.
3. They eat and drink.
4. They grow.
5. They also "grow up." The kitten grows up into a cat, which is somewhat different from the kitten.
6. They move about and sleep.

7. They feel—"at least some of them do."
8. They are made of "flesh and blood."
9. They grow old and die.
10. They reproduce their kind. The cat may have kittens.
11. They are living organisms, but "not vegetables."

Now, let us look carefully at these characteristics, all of which were contained in the five answers, and were probably familiar in some such form as this to all the boys, and see if we cannot make them more general and more accurate.

1. *An animal has a definite form.* My school-boy friend described it as a head and tail, four legs, and a body. But it is clear that this description applies only to a very limited number of animals. It will not apply to the butterfly, with its great wings and six legs; nor to the lobster, with its eight legs and large pincer-claws; to the limbless snake and worm, the finned fish, the thousand-legs, the oyster or the snail, the star-fish or the sea-anemone. The animals to which my young friend's description applies form, indeed, but a numerically insignificant proportion of the multitudes which throng the waters and the air, and not by any means a large proportion of those that walk upon the surface of the earth. The description applies only to the backboned vertebrates, and not to nearly all of them.

It is impossible to summarize in a sentence the form-characteristics of animals. The diversities of form are endless. Perhaps the distinguishing feature is the prevalence of curved and rounded contours, which are in striking contrast to the definite crystalline forms of the inorganic kingdom, characterized as these are by plane surfaces and solid angles. We may say, however, that all but the very lowliest animals have each and all a proper and characteristic form of their own, which they have inherited from their immediate ancestors, and which they hand on to their descendants. But this form does not remain constant throughout life. Sometimes the change is slight; in many cases, however, the form alters very markedly during the successive stages of the life of the individual, as is seen in the frog, which begins life as a tadpole, and perhaps even more conspicuously in the butterfly, which passes through a caterpillar and a chrysalis stage. Still, these changes are always the same for the same kind of animal. So that we may say, each animal has a definite form and shape or series of shapes.

2. *Animals breathe.* The essential thing here is that oxygen is taken in by the organism, and carbonic acid gas is produced by the organism. No animal can carry on its life-processes unless certain chemical changes take place in the substance of which it is composed. And for these chemical changes oxygen is essential. The products of these changes, the most familiar of which are carbonic acid gas and urea, must be got rid of by the process of excretion. Respiration and excretion are therefore essential and characteristic life-processes of all animals.

In us, and in all air-breathing vertebrates, there are special organs set apart for respiration and excretion of carbonic acid gas. These are the lungs. A great number of insects also breathe air, but in a different way. They have no lungs, but they respire by means of a number of apertures in their sides, and these open into a system of delicate branching tubes which ramify throughout the body. Many organisms, however, such as fish and lobsters and molluscs, breathe the air dissolved in the water in which they live. The special organs developed for this purpose are the gills. They are freely exposed to the water from which they abstract the air dissolved therein. When the air dissolved in the water is used up, they sicken and die. There can be nothing more cruel than to keep

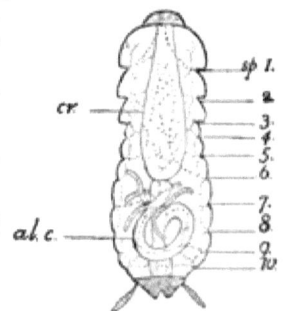

Fig. 1.—Diagram of spiracles and air-tubes (tracheæ) of an insect (cockroach).

The skin, etc., of the back has been removed, and the crop (*cr.*) and alimentary canal (*al.c.*) displayed. The air-tubes are represented by dotted lines. The ten spiracles are numbered to the right of the figure.

aquatic animals in a tank or aquarium in which there is no means of supplying fresh oxygen, either by the action of green vegetation, or by a jet of water carrying down air-bubbles, or in some other way. And then there are a number of animals which have no special organs set apart for breathing. In them respiration is carried on by the general surface of the body. The common earthworm is

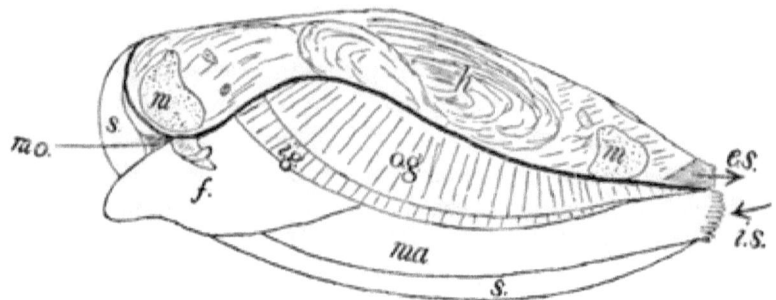

Fig. 2.—Gills of mussel.

o.g., outer gill; *i.g.*, inner gill; *mo.*, mouth; *m.*, muscles for closing shell; *ma.*, mantle; *s.*, shell; *f.*, foot; *h.*, position of heart; *e.s.*, exhalent siphon, whence the water passes out from the gill-chamber; *i.s.*, inhalent siphon, where the water enters.
The left valve of the shell has been removed, and the mantle cut away along the dark line.

one of these; and most microscopic organisms are in the same condition. Still, even if there be no special organs for breathing, the process of respiration must be carried on by all animals.

3. *They eat and drink.* The living substance of an animal's body is consumed during the progress of those chemical changes which are consequent upon respiration; and this substance must, therefore, be made good by taking in the materials out of which fresh life-stuff can be formed. This process is called, in popular language, feeding. But the food taken in is not identical with the life-stuff formed. It has to undergo a number of chemical changes before it can be built into the substance of the organism. In us, and in all the higher animals, there is a complex system of organs set aside for the preparation, digestion, and absorption of the food. But there are certain lowly organisms which can take in food at any portion of their surface, and digest it in any part of their

substance. One of these is the amœba, a minute speck of jelly-like life-stuff, which lives in water, and tucks in a bit of food-material just as it comes. And there are certain degenerate organisms which have taken to a parasitic life, and live within the bodies of other animals. Many of these can absorb the material prepared by their host through the general surface of their simple bodies. But here, again, though there may be no special organs set apart for the preparation, absorption, and digestion of food, the process of feeding is essential to the life of all animals. Stop that process for a sufficient length of time, and they inevitably die.

4. *They grow.* Food, as we have just seen, has to be taken in, digested, and absorbed, in order that the loss of substance due to the chemical changes consequent on respiration may be made good. But where the digestion and absorption are in excess of that requisite for this purpose, we have the phenomenon of growth.

What are the characteristics of this growth? We cannot, perhaps, describe it better than by saying (1) that it is organic, that is to say, a growth of the various organs of the animal in due proportion; (2) that it takes place, not merely by the addition of new material (for a crystal grows by the addition of new material, layer upon layer), but by the incorporation of that new material into the very substance of the old; and (3) that the material incorporated during growth differs from the material absorbed from without, which has undergone a preparatory chemical transformation within the animal during digestion. The growth of an animal is thus dependent upon the continued absorption of new material from without, and its transformation into the substance of the body.

The animal is, in fact, a centre of continual waste and repair, of nicely balanced constructive and destructive processes. These are the invariable concomitants of life. Only so long as the constructive processes outbalance the destructive processes does growth continue. During the greater part of a healthy man's life, for example, the two processes, waste and repair, are in equilibrium. In old age, waste slowly but surely gains the mastery; and at death the balanced process ceases, decomposition sets in, and the elements of the body are scattered to the winds or returned to mother earth.

There are generally limits of growth which are not exceeded by any individuals of each particular kind of animal. But these limits are somewhat variable among the individuals of each kind. There are big men and little men, cart-horses and ponies, bloodhounds and lap-dogs. Wild animals, however, when fully grown, do not vary so much in size. The period of growth is also variable. Many of the lower backboned animals probably grow during the whole of life, but those which suckle their young generally cease growing after a fraction (in us from one-fourth to one-fifth) of the allotted span of life is past.

5. But animals not only grow—*they also " grow up."* The kitten grows up into a cat, which is somewhat different from the kitten. We speak of this growing up of an animal as its *development*. The proportion of the various parts and organs progressively alter. The relative lengths of the arms and legs, and the relative size of the head, are not the same in the infant as in the man or woman. Or, take a more marked case. In early spring there is plenty of frog-spawn in the ponds. A number of blackish specks of the size of mustard seeds are embedded in a jelly-like mass. They are frogs' eggs. They seem unorganized. But watch them, and the organization will gradually appear. The egg will be hatched, and give rise to a little fish-like organism. This will by degrees grow into a tadpole, with a powerful swimming tail and rounded head and body, but with no obvious neck between them. Legs will appear. The tail will shrink in size and be gradually drawn into the body. The tadpole will have developed into a minute frog.

There are many of the lower animals which go through a not less wonderful, if not more wonderful, metamorphosis.

The butterfly or the silkworm moth, beginning life as a caterpillar and changing into a chrysalis, from which the perfect insect emerges, is a familiar instance. And hosts of the marine invertebrates have larval forms which have but little resemblance to their adult parents.

Such a series of changes as is undergone by the frog is called *metamorphosis*, which essentially consists in the temporary development of certain provisional embryonic organs (such as gills and a powerful swimming tail) and the appearance of adult organs (such as lungs and legs) to take their place. In metamorphosis these changes occur during the free life of the organism. But beneath the eggshell of birds and within the womb of mammals scarcely less wonderful changes are slowly but surely effected, though they are hidden from our view. There is no metamorphosis during the free life of the organism, but there is a prenatal *transformation*. The little embryo of a bird or mammal has no gills like the tadpole (though it has for a while gill-slits, pointing unmistakably to its fishy ancestry), but it has a temporary provisional breathing organ, called the allantois, pending the full development and functional use of its lungs.

All the higher animals, in fact—the dog, the chick, the serpent, the frog, the fish, the lobster, the butterfly, the worm, the star-fish, the mollusc, it matters not which we select—take their origin from an apparently unorganized egg. They all, therefore, pass during their growth from a comparatively simple condition to a comparatively complex condition by a process of change which is called development. But there are certain lowly forms, consisting throughout life of little more than specks of jelly-like life-stuff, in which such development, if it occurs at all, is not conspicuous.

6. *They move about and sleep.* This is true of our familiar domestic pets. The dog and the cat, after periods of restless activity, curl themselves up and sleep. The canary that has all day been hopping about its cage, or perhaps been allowed the freedom of the dining-room, tucks

its head under its wing and goes to sleep. The cattle in the meadows, the sheep in the pastures, the horses in the stables, the birds in the groves, all show alternating periods of activity and repose. But is this true of all animals? Do all animals "move about and sleep"? The sedentary oyster does not move about from place to place; the barnacle and the coral polyp are fixed for the greater part of life; and whether these animals sleep or not it is very difficult to say. We must make our statement more comprehensive and more accurate.

If we throw it into the following form, it will be more satisfactory: Animals exhibit certain activities; and periods of activity alternate with periods of repose.

I shall have more to say hereafter concerning the activities of animals. Here I shall only say a few words concerning the alternating periods of repose. No organism can continue in ceaseless activity unbroken by any intervening periods of rest. Nor can the organs within an organism, however continuous their activity may appear, work on indefinitely and unrestfully. The heart is apparently restless in its activity. But in every five minutes of the continued action of the great force-pump (ventricle) of the heart, two only are occupied in the efforts of contraction and work, while three are devoted to relaxation and repose. What we call sleep may be regarded as the repose of the higher brain-centres after the activity of the day's work—a repose in which the voluntary muscles share.

The necessity for rest and repose will be readily understood. We have seen that the organism is a centre of waste and repair, of nicely balanced destructive and reconstructive processes. Now, activity is accompanied by waste and destruction. But it is clear that these processes, by which the substance of the body and its organs is used up, cannot go on for an indefinite period. There must intervene periods of reconstruction and recuperation. Hence the necessity of rest and repose alternating with the periods of more or less prolonged activity.

7. *They feel*—"*at least some of them do.*" The quali-

fication was a wise one, for in truth, as we shall hereafter see, we know very little about the feelings of the lower organisms. The one animal of whose feelings I know anything definite and at first hand, is myself. Of course, I believe in the feelings of others; but when we come to very lowly organisms, we really do not know whether they have feelings or not, or, if they do, to what extent they feel.

Shall we leave this altogether out of account? Or can we throw it into some form which is more general and less hypothetical? This, at any rate, we know—that all animals, even the lowest, are sensitive to touches, sights, or sounds. It is a matter of common observation that their activities are generally set agoing under the influence of such suggestions from without. Perhaps it will be objected that there is no difference between feeling and being sensitive. But I am using the word "sensitive" in a general sense—in that sense in which the photographer uses it when he speaks of a sensitive plate, or the chemist when he speaks of a sensitive test. When I say that animals are sensitive, I mean that they answer to touches, or sounds, or other impressions (what are called stimuli) coming from without. They may feel or not; many of them undoubtedly do. But that is another aspect of the sensitiveness. Using the term, then, with this meaning, we may say, without qualification, that all animals are more or less sensitive to external influences.

8. *They are made of "flesh and blood."* Here we have allusion to the materials of which the animal body is composed. It is obviously a loose and unsatisfactory statement as it stands. An American is said to have described the difference between vertebrates and insects by saying that the former are composed of flesh and bone, and the latter of skin and squash. But even if we amend the statement that animals are made of "flesh and blood" by the addition of the words, "or of skin and squash," we shall hardly have a sufficiently satisfactory statement of the composition of the animal body.

The essential constituent of animal (as indeed also of vegetable) tissues is protoplasm. This is a nearly colourless, jelly-like substance, composed of carbon, hydrogen, nitrogen, and oxygen, with some sulphur and phosphorus, and often, if not always, some iron; and it is permeated by water. Protoplasm, together with certain substances, such as bony and horny matter, which it has the power of producing, constitutes the entire structure of simple organisms, and is built up into the organs of the bodies of higher animals. Moreover, in these organs it is not arranged as a continuous mass of substance, but is distributed in minute separate fragments, or corpuscles, only visible under the microscope, called cells. These cells are of very various shapes—spherical, discoidal, polyhedral, columnar, cubical, flattened, spindle-shaped, elongated, and stellate.

A great deal of attention has been devoted of late years to the minute structure of cells, and the great improvements in microscopical powers and appliances have enabled investigators to ascertain a number of exceedingly interesting and important facts. The external surface of a cell is sometimes, but not always in the case of animals, bounded by a film or membrane. Within this membrane the substance of the cell is made up of a network of very delicate fibres (the *plasmogen*), enclosing a more fluid material (the *plasm*); and this network seems to be the essential living substance. In the midst of the cell is a small round or oval body, called the nucleus, which is surrounded by a very delicate membrane. In this nucleus there is also a network of delicate plasmogen fibres, enclosing a more fluid plasm material. At certain times the network takes the form of a coiled filament or set of filaments, and these arrange themselves in the form of rosettes and stars. In the meshwork of the net or in the coils of the filament there may be one or more small bodies (nucleoli), which probably have some special significance in the life of the cell. These cells multiply or give birth to new cells by dividing into two, and this process is often accompanied by special changes in the nucleus (which also divides) and by the arrangement of its network or filaments into the rosettes and stars before alluded to.

Instead, therefore, of the somewhat vague statement that animals are made of flesh and blood, we may now say that the living substance of which animals are composed is a complex material called protoplasm; that organisms are formed either of single cells or of a number of related cells, together with certain life-products of these cells; and

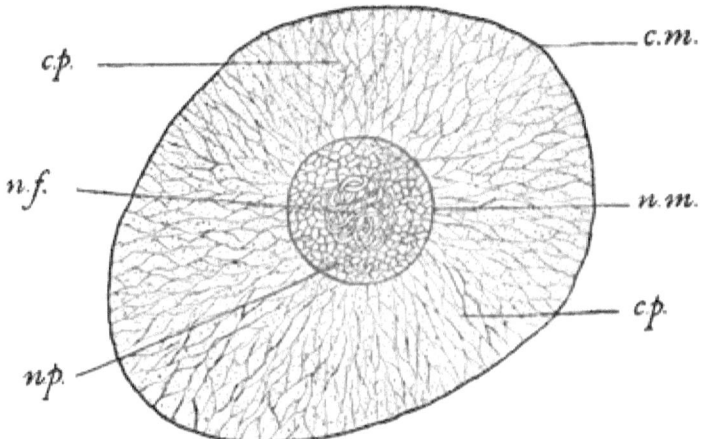

Fig. 3.—A cell, greatly magnified.

c.m., cell-membrane; c.p., cell-protoplasm; n.m., nuclear membrane; n.p., nuclear protoplasm; n.f., coiled nuclear filament.

that each cell, small as it is, has a definite and wonderful minute structure revealed by the microscope.

9. *Animals grow old and die.* This is a familar observation. Apart from the fact that they are often killed by accident, by the teeth or claws of an enemy, or by disease, animals, like human beings, in course of time become less active and less vigorous; the vital forces gradually fail, and eventually the flame of life, which has for some time been burning dimmer and dimmer, flickers out and dies. But is this true of all animals? Can we say that death— as distinct from being killed—is the natural heritage of every creature that lives?

One of the simplest living creatures is the amœba. It consists of a speck of nucleated protoplasm, no larger than a small pin's head. Simple as it is, all the essential life-processes are duly performed. It is a centre of waste and repair; it is sensitive and responsive to a stimulus; respiration and nutrition are effected in a simple and primitive fashion. It is, moreover, reproductive. First the nucleus and then the protoplasm of the cell divide, and in place of one amœba there are two. And these two are, so far as we can tell, exactly alike. There is no saying which is mother and which is daughter; and, so far as we can see at present, there is no reason why either should die. It is conceivable that amœbæ never die, though they may be killed in immense numbers. Hence it has been plausibly

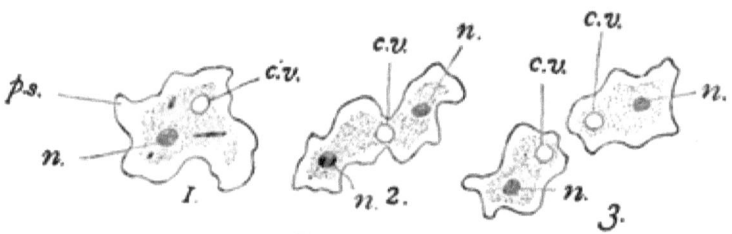

Fig. 4.—Amœba.

1. An amœba, showing the inner and outer substance (endosarc and ectosarc); a pseudopodium, *p.s.*; the nucleus, *n.*; and the contractile vesicle, *c.v.* 2. An amœba dividing into two. 3. The division just effected.

maintained that the primitive living cell is by nature deathless; that death is not the heritage of all living things; that death is indeed an acquisition, painful indeed to the individual, but, since it leaves the stage free for the younger and more vigorous individuals, conducive to the general good.

In face of this opinion, therefore, we cannot say that all animals grow old and die; but we may still say that all animals, with the possible exception of some of the lowest and simplest, exhibit, after a longer or a shorter time, a waning of the vital energies which sooner or later ends in death.

10. *Animals reproduce their kind.* We have just seen

the nature of reproduction in the simple unicellular amœba. The reproduction of the constituent cells in the complex multicellular organism, during its natural growth or to make good the inevitable loss consequent on the wear and tear of life, is of the same character.

When we come to the higher organisms, reproduction is effected by the separation of special cells called egg-cells, or ova, from a special organ called the ovary; and these, in a great number of cases, will not develop into a new organism unless they be fertilized by the union with them in each case of another cell—the sperm-cell—produced by a different individual. The separate parents are called male

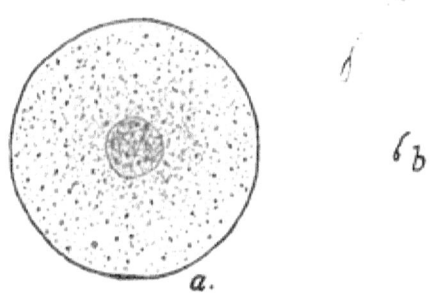

Fig. 5.—Egg-cell and sperm-cell.
a, ovum or egg; *b*, spermatozoon or sperm.

and female, and reproduction of this kind is said to be sexual.

The wonderful thing about this process is the power of the fertilized ovum, produced by the union of two minute cells from different parents, to develop into the likeness of these parents. This likeness, however, though it extends to minute particulars, is not absolute. The offspring is not exactly like either parent, nor does it present a precise mean between the characters of the two parents. There is always some amount of individual variability, the effects of which, as we shall hereafter see, are of wide importance. We are wont to say that these phenomena, the transmission of parental characteristics, together with a margin of difference, are due to heredity with variation. But this merely names the facts. How the special reproductive cells have acquired the secret of developing along special lines, and reproducing, with a margin of variability, the likeness of the organisms which produced them, is a matter concerning which we can at present only make more or less plausible guesses.

Scarcely less wonderful is the power which separated bits of certain organisms, such as the green freshwater hydra of our ponds, possess of growing up into the complete organism. Cut a hydra into half a dozen fragments, and each fragment will become a perfect hydra. Reproduction of this kind is said to be asexual.

We shall have, in later chapters, to discuss more fully some of the phenomena of reproduction and heredity. For the present, it is sufficient to say that animals reproduce their kind by the detachment of a portion of the substance of their own bodies, which portion, in the case of the higher animals, undergoes a series of successive developmental changes constituting its life-history, the special nature of which is determined by inheritance, and the result of which is a new organism in all essential respects similar to the parent or parents.

11. *Animals are living organisms, and "not vegetables."* The first part of this final statement merely sums up the characteristics of living animals which have gone before. But the latter part introduces us to the fact that there are other living organisms than those we call animals, namely, those which belong to the vegetable kingdom.

It might, at first sight, be thought a very easy matter to distinguish between animals and plants. There is no chance, for example, of mistaking to which kingdom an oak tree or a lion, a cabbage or a butterfly, belongs. But when we come down to the simpler organisms, those whose bodies are constituted by a single cell, the matter is by no means so easy. There are, indeed, lowly creatures which are hovering on the boundary-line between the two kingdoms. We need not discuss the nature of these boundary forms. It is sufficient to state that unicellular plants are spoken of as *protophyta*, and unicellular animals as *protozoa*, the whole group of unicellular organisms being classed together as *protista*. The animals whose bodies are formed of many cells in which there is a differentiation of structure and a specialization of function, are called *metazoa*, and the multicellular plants *metaphyta*. The relations of these groups may be thus expressed—

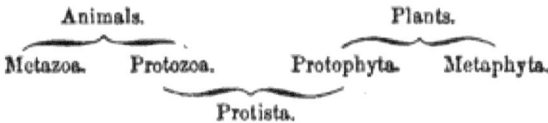

There are three matters with regard to the life-process of animals and plants concerning which a few words must be said. These are (1) their relation to food-stuffs; (2) their relation to the atmosphere; (3) their relation to energy, or the power of doing work.

With regard to the first matter, that of food-relation, the essential fact seems to be the dependence of animals on plants. Plants can manufacture protoplasm out of its constituents if presented to them in suitable inorganic form scattered through earth and air and water. Hence the peculiar features of their form, the branching and spreading nature of those parts which are exposed to the air, and the far-reaching ramifications of those parts which are implanted in the earth. Hence, too, the flattened leaves, with their large available surface. Animals are unable to manufacture protoplasm in this way. They are, sooner or later, dependent for food on plant-products. It is true that the carnivora eat animal food, but the animals they eat are directly or indirectly consumers of vegetable products. Plants are nature's primary producers of organic material. Animals utilize these products and carry them to higher developments.

In relation to the atmosphere, animals require a very much larger quantity of oxygen than do plants. This, during the respiratory process, combines with carbon so

as to form carbonic acid gas; and the atmosphere would be gradually drained of its oxygen and flooded with carbonic acid gas were it not that plants, through their green colouring matter (**chlorophyll**), under the influence of light, have the power of decomposing the carbonic acid gas, seizing on the carbon and building it into their tissues, and setting free the oxygen. Thus are animals and green plants complementary elements in the scheme of nature.*
The animal eats the carbon elaborated by the plant into organic products (starch and others), and breathes the oxygen which the plant sets free after it has abstracted the carbon. In the animal's body the carbon and oxygen recombine; its varied activities are thus kept going; and the resultant carbonic acid gas is breathed forth, to be again separated by green, growing plants into carbonaceous food-stuff and vitalizing oxygen. It must be remembered, however, that vegetable protoplasm, like animal protoplasm, respires by the absorption of oxygen and the formation of carbonic acid gas. But in green plants this process is outbalanced by the characteristic action of the chlorophyll, by which carbonic acid gas is decomposed.

Lastly, we have to consider the relations of animals and plants to energy. Energy is defined as the power of doing work, and it is classified by physicists under two modes— potential energy, or energy of position; and kinetic energy, or energy of motion. The muscles of my arm contain a store of potential energy. Suppose I pull up the weight

---

* An interesting problem concerning the atmosphere is suggested by certain geological facts. In our buried coal-seams and other carbonaceous deposits a great quantity of carbon, for the most part abstracted from the atmosphere, has been stored away. Still greater quantities of carbon are imprisoned in the substance of our limestones, which contain, when pure, 44 per cent. of this element. A large quantity of oxygen has also been taken from the atmosphere to combine with other elements during their oxidation. The question is—Was the atmosphere, in the geological past, more richly laden with carbonic acid gas, of which some has entered into combination with lime to form limestone, while some has been decomposed by plants, the carbon being buried as coal, and the oxygen as products of oxidation? Or, has the atmosphere been furnished with continuous fresh supplies of carbonic acid gas?

of an old-fashioned eight-day clock. Some of the potential energy of my arm is converted into the potential energy of the weight; that is, the raised weight is now in a position of advantage, and capable of doing work. It has energy of position, or potential energy. If the chain breaks, down falls the weight, and exhibits the energy of motion. But, under ordinary circumstances, this potential energy is utilized in giving a succession of little pushes to the pendulum to keep up its swing, and in overcoming the friction of the works. Again, the energy of an electric current may be utilized in decomposing water, and tearing asunder the oxygen and hydrogen of which it is composed. The oxygen and hydrogen now have potential energy, and, if they be allowed to combine, this will manifest itself as the light and heat of the explosion. These examples will serve to illustrate the nature of the changes which energy undergoes. These are of the nature of transferences of energy from one body to another, and of transformations from one mode or manifestation to another. The most important point that has been established during this century with regard to energy is that, throughout all its transferences and transformations, it can be neither created nor destroyed. But there is another point of great importance. Transformations of energy take place more readily in certain directions than in others. And there is always a tendency for energy to pass from the higher or more readily transformable to the lower or less readily transformable forms. When, for example, energy has passed to the low kinetic form of the uniformly distributed molecular motion of heat, it is exceedingly difficult, or practically impossible, to transform it into a higher and more available form.

Now, both animals and plants are centres of the transformation of energy; and in them energy, notwithstanding that it is being raised to a high position of potentiality, is constantly tending to be degraded to lower forms. Hence the necessity of some source from which fresh stores of available energy may be constantly supplied. Such a source is solar radiance. This it is which gives the succession of little pushes which keeps the pendulum of life a-swinging. And it is the green plants which, through their chlorophyll, are in the best position to utilize the solar energy. They utilize it in building up, from the necessary constituents diffused through the atmosphere and the soil, complex forms of organic material, of which the first visible product seems to be starch; and these not only contain large stores of potential energy, but are capable, when combined with oxygen, of containing yet larger stores. The animal, taking into its body these complex materials, and elaborating them together with oxygen into yet more complex and more unstable compounds, then, during its vital activity, makes organized use of the transformation of the potential energy thus stored into lower forms of energy. Thus there go on side by side, in both animals and plants, a building up or synthesis of complex and unstable chemical compounds, accompanied by a storage of potential energy, and a breaking down or analysis of these compounds into lower and simpler forms, accompanied by a setting free of kinetic energy. But in the plant, synthetic changes and storage of energy are in excess, while in the animal, analytic changes and the setting free of kinetic energy are more marked. Hence the variety and volume of animal activities.

The building up of complex organic substances with abundance of stored energy may be roughly likened to the building up, by the child with his wooden bricks, of houses and towers and pyramids. The more complex they become the more unstable they are, until a touch will shatter the edifice and liberate the stored-up energy of position acquired by the bricks. Thus, under the influence of solar energy, do plants build up their bricks of hydrogen, carbon, and oxygen into complex molecular edifices. Animals take advantage of the structures so elaborated, modify them, add to them, and build yet more complex molecular edifices. These, at the touch of the appropriate stimulus, topple over and break down—not, indeed, into the elemental bricks, but into simpler molecular forms, and these again in later stages into yet simpler forms, which are then got rid of or excreted from the body. Meanwhile the destructive fall of the molecular edifice is accompanied by the liberation of energy—as heat, maintaining the warmth of the body; as visible or hidden movements, in locomotion, for example, and the heart-beat; and sometimes as electrical energy (in electric fishes); as light (in phosphorescent animals and the glow-worm), or as sound. It is this abundant liberation of energy, giving rise to many and complex activities, which is one of the distinguishing features of animals as compared with plants.

We have now, I trust, extended somewhat and rendered somewhat more exact our common and familiar knowledge of the nature of animal life. In the next chapter we will endeavour to extend it still further by a consideration of the process of life.

CHAPTER II.

THE PROCESS OF LIFE.

In the foregoing chapter, on "The Nature of Animal Life," we have seen that animals breathe, feed, grow, are sensitive, exhibit various activities, and reproduce their kind. These may be regarded as primary life-processes, in virtue of which the animal characterized by them is a living creature. We have now to consider some of these life-processes—the sum of which we may term the process of life—a little more fully and closely.

The substance that exhibits these life-processes is protoplasm, which exists in minute separate masses termed cells. It seems probable, however, that these cells, separate as they seem, are in some cases united to each other by minute protoplasmic filaments. In the higher animals the cells in different parts of the body take on different forms and perform different functions. Like cells with like functions are also aggregated together into tissues. Thus the surfaces of the body, external and internal, are bounded by or lined with epithelial tissue; the bones and framework of the body are composed of skeletal tissue; nervous tissue goes to form the brain and nerves; contractile tissue is found in the muscles; while the blood and lymph form a peculiar nutritive tissue. The organs of the body are distinct parts performing definite functions, such as the heart, stomach, or liver. An organ may be composed of several tissues. Thus the heart has contractile tissue in its muscular walls, epithelial tissue lining its cavities, and skeletal tissue forming its framework. Still, notwithstanding their aggregation into tissues and organs, it remains true that the body of one of the higher animals is composed of cells, together with certain cell-products, horny, calcareous, or other. The simplest animals, called protozoa, are, however, unicellular, each organism being constituted by a single cell.

We must notice that, even during periods of apparent inactivity—for example, during sleep—many life-processes are still in activity, though the vigour of action may be somewhat reduced. When we are fast asleep, respiration, the heart-beat,* and the onward propulsion of food through the alimentary canal, are still going on. Even at rest, the living animal is a *going* machine. In some cases, however, as during the hibernating sleep of the dormouse or the bear, the vital activities fall to the lowest possible ebb. Moreover, in some cases, the life-processes may be temporarily arrested, but again taken up when the special conditions giving rise to the temporary arrest are removed. Frogs, for example, have been frozen, but have resumed their life-activities when subsequently thawed.

Let us take the function of respiration as a starting-point in further exemplification of the nature of the life-processes of animals.

The organs of respiration, in ourselves and all the mammalia, are the lungs, which lie in the thoracic cavity of the chest, the walls of which are bounded by the ribs and breast-bone, its floor being formed of a muscular and movable partition, the diaphragm, which separates it from the stomach and other alimentary viscera in the abdominal region. The lungs fit closely, on either side of the heart, in this thoracic cavity; and when the size of this cavity is altered by movements of the ribs and diaphragm, air is either sucked into or expelled from the lungs through the windpipe, which communicates with the exterior through the mouth or nostrils. It is unnecessary to describe

---

* It has before been noticed that the organs themselves have their periods of rest. The rhythm of rest and repose in the heart is not that of the activity and sleep of the organism, but that of the contraction and relaxation of the organ itself.

the minute structure of the lungs; suffice it to say that, in the mammal, they contain a vast number of tubes, all communicating eventually with the windpipe, and terminating in little expanded sacs or bags. Around these little sacs courses the blood in a network of minute capillary vessels, the walls of which are so thin and delicate that the fluid they contain is only separated from the gas within the sacs by a film of organic tissue.

The blood is a colourless fluid, containing a great number of round red blood-discs, which, from their minute size and vast numbers, seem to stain it red. They may be likened to a fleet of little boats, each capable of being laden with a freight of oxygen gas, while the stream in which they float is saturated with carbonic acid gas. This latter escapes into the air-sacs as the fluid courses through the delicate capillary tubes.

Whither goes the oxygen? Whence comes the carbonic acid gas? The answer to these questions is found by following the course of the blood-circulation. The propulsion of the blood throughout the body is effected by the heart, an organ consisting, in mammals, of two receivers (auricles) into which blood is poured, and two powerful force-pumps (ventricles), supplied with blood from the receivers and driving it through great arteries to various parts of the body. There are valves between the receivers and the force-pumps and at the commencement of the great arterial vessels, which ensure the passage of the blood in the right direction. The two receivers lie side by side; the two force-pumps form a single muscular mass; and all four are bound up into one organ; but there is, during adult life, no direct communication between the right and left receivers or the right and left force-pumps.

Let us now follow the purified stream, with its oxygen-laden blood-discs, as it leaves the capillary tubes of the lungs. It generally collects, augmented by blood from other similar vessels, into large veins, which pour their contents into the left receiver. Thence it passes on into the left force-pump, by which it is propelled, through a great arterial vessel and the numerous branches it gives off, to the head and brain, to the body and limbs, to the abdominal viscera; in short, to all parts of the body except the lungs. In all the parts thus supplied, the vessels at length break up into a delicate capillary network, so that the blood-fluid is separated from the tissue-cells only by the delicate organic film of the capillary walls. Then the blood begins to re-collect into larger and larger veins. But a change has taken place; the blood-discs have delivered up to the tissues their freight of oxygen; the stream in which they float has been charged with carbonic acid gas. The veins leading from various parts of the body converge upon the heart and pour their contents into the right receiver; thence the blood passes into the right force-pump, by which it is propelled, by arteries, to the lungs. There the blood-discs are again laden with oxygen, the stream is again purified of its carbonic acid gas, and the blood proceeds on its course, to renew the cycle of its circulation.

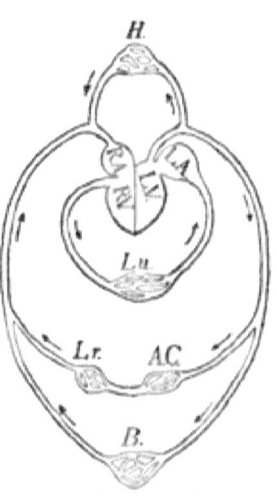

Fig. 6.—Diagram of circulation.

L.A., left auricle of the heart; *L.V.*, left ventricle; *H.*, capillary plexus of the head; *B.*, capillary plexus of the body; *A.C.*, alimentary canal; *Lr.*, liver; *R.A.*, right auricle of the heart; *R.V.*, right ventricle; *Lu.*, lungs.

Now, if we study the process of respiration and that of circulation, with which it is so closely associated, in other forms of life, we shall find many differences in detail. In the bird, for example, the mechanism of respiration is different. There is no diaphragm, and the lungs are scarcely distensible. There are, however, large air-sacs in the abdomen, in the thoracic region, in the fork of the merry-thought, and elsewhere. These are distensible, and to reach them the air has to pass through the lungs, and as it thus passes through the delicate tubes of the lungs, it supplies the blood with oxygen and takes away carbonic acid gas. In the frog there is no diaphragm, and there are no ribs. The lungs are hollow **sacs** with honey-combed **sides,** and they **are inflated from the mouth,** which is used as **a force-pump for this purpose.** In the fish there are no lungs, respiration being effected by **means of gills. In** these organs **the** blood is separated **from the** water which **passes** over **them** (being gulped **in by the** mouth **and** forced out between **the** gill-covers) by **only a thin** organic film, so that it can **take up the** oxygen dissolved in the water, and give up to the water the carbonic acid it contains. In fishes, too, we have only one receiver **and** one force-pump, the blood passing through the gills **on its** way to the various **parts** of the body. **In** the lobster, again, there are gills, **but the mechanism** by which the water is **drawn** over **them is quite different, and** the blood passes **through them on its way to the heart, after** passing **through the various organs of the body, not on its** way **from the heart, as in vertebrate fishes.** The blood, too, has no red **blood-discs. In the air-breathing** insects the mechanism is, again, altogether **different. The** air, which obtains access to the body by **spiracles in** the sides (see Fig. 1, p. 3), is distributed by delicate **and** beautiful tubes to all parts of the organs; so that the oxygen is supplied to **the** tissues directly, **and not through** the intervention of **a** blood-stream. **In the earthworm, on** the other hand, **there is** a distributing blood-stream, but there **is no mechanism for** introducing the air within the body; **while in some of the** lowliest forms of life there is neither any **introduction of air within** the body nor any distribution **by means of a circulating fluid.** Beginning, therefore, **with the** surface **of the body** simply absorbent **of** oxygen, we have the concentration of the absorbent parts **in** special regions, and an increase **in the** absorbent **surface,** either **(1) by** the **pushing** out of processes **into the** surrounding **medium, as in** gills; **or (2)** by the **formation** of internal cavities, tubes, **or branching passages, as in** lungs and the tracheal air-system of insects.

What, then, is the essential nature of the respiratory

process thus so differently manifested? Clearly the supply of oxygen to the cellular tissue-elements, and, generally closely associated with this, the getting rid of carbonic acid gas.

Let us now glance at the life-processes which minister to nutrition, beginning, as before, with the mode in which these processes are effected in ourselves.

The alimentary canal is a long tube running through the body from the mouth to the vent. In the abdominal region it is coiled upon itself, so that its great length may be conveniently packed away. Opening into this tube are the ducts of certain glands, which secrete fluids which aid in the digestion of the food. Into the mouth there open the ducts of the salivary glands, which secrete the saliva; in the stomach there are a vast number of minute gastric glands; in the intestine, besides some minute tubular glands, there are the ducts of the large liver (which secretes the bile) and the pancreas, or sweetbread. Since, with the exception of the openings of these ducts, the alimentary canal is a closed tube, its contents, though lying within the body, are in a sense outside it, just as the fuel in a tubular boiler, though within the boiler, is really outside it. The organic problem, therefore, is how to get the nutritive materials through the walls of the tube and thus into the body.

At an ordinary meal we are in the habit of consuming a certain amount of meat, with some fat, together with bread and potatoes, and perhaps some peas or beans and a little salt. This is followed by, say, milky rice-pudding, with which we take some sugar; and a cheese course may, perhaps, be added. The whole is washed down with water more or less medicated with other fluid materials. Grouping these substances, there are (1) water and salts, including calcium phosphate in the milk; (2) meat, peas, milk, and cheese, all of which contain albuminous or allied materials; (3) bread, potatoes, and rice, which contain starchy matters; and here we may place the sugar; (4) fat, associated with the meat or contained in the cream of the milk. Now, of all the materials thus consumed, only **the water, salts, and** sugar are **capable, in** their unaltered condition, of passing through the lining membrane of the alimentary canal, and thus of **entering the** body. The albuminous materials, **the** starchy matter, **and the fat**—that is to say, the main elements of the food—are, **in their** raw state, absolutely useless for nutritive purposes.

The preparation of the food begins **in the mouth.** The saliva here acts upon **some of the starchy matter, and** converts it into a kind of **sugar, which** *can* pass **through** the lining membrane of the alimentary canal, and thus enter the body. The fats and albuminous matters here remain unaltered, though they are torn to pieces by the **mastication effected by** the teeth. In the stomach the **albuminous constituents of** the **meat** are attacked by **the gastric juice and** converted into peptones; and in this **new condition they, too, can soak** through **the** lining membrane of the alimentary **canal, and thus can** enter the body. In the stomach all **action on starch is** arrested; but **in** the intestine, through **the effect of** a ferment contained in the pancreatic juice, this action is resumed, and the **rest** of **the** starch is converted **into** absorbable sugar. Another principle contained **in pancreatic juice** takes effect on the albuminous matters, and **converts them into** absorbable **peptones. The** pancreatic juice **also acts** on the fats, converting **them** into an emulsion, that **is to** say, causing them to break up into exceedingly minute globules, like the butter globules **in** milk. It furthermore **contains a ferment** which splits up the fats into fatty acids **and glycerine;** and these fatty acids, with **an alkaline** carbonate contained in small quantities in **pancreatic juice,** form soluble soaps, **which** further **aid in emulsifying fats. The bile** also aids **in emulsifying fats.**

The effect, then, **of the various digestive** fluids upon **the** food is to convert **the starch,** albuminous material, **and** fat into sugar, peptones, glycerine, and soap, and thus render them capable of passing through the lining membrane of the **canal into** the body.

The materials thus absorbed are either taken up into the blood-stream or pass into a separate system of vessels called lacteals. All the blood which comes away from the alimentary canal passes into the liver, and there undergoes a good deal of elaboration in that great chemical laboratory of the body. The fluid in the lacteals passes through lymphatic glands, in which it too undergoes some elaboration before it passes into the blood-stream by a large vessel or duct.

Thus the blood, which we have seen to be enriched with oxygen in the lungs, is also enriched with prepared nutritive material through the processes of digestion and absorption in the alimentary organs and elaboration in the liver and lymphatic glands.

Here let us again notice that the details of the process of nutrition vary very much in different forms of life. In some mammals the organs of digestion are specially fitted to deal with a flesh diet; in others they are suited for a diet of herbs. In the graminivorous birds the grain is swallowed whole, and pounded up in the gizzard. The leech swallows nothing but blood. The earthworm pours out a secretion on the leaves, by which they are partially digested before they enter the body. Many parasitic organisms have no digestive canal, the nutritive juices of their host being absorbed by the general external surface of the body. But the essential life-process is in all cases the same—the absorption of nutritive matter to be supplied to the cell or cells of which the organism is built up.

Thus in the mammal the blood, enriched with oxygen in the lungs, and enriched also with nutritive fluids, is brought, in the course of its circulation, into direct or indirect contact with all the myriads of living cells in the body.

In the first place, the material thus supplied is utilized for and ministers to the growth of the organs and tissues. This growth is effected by the multiplication of the constituent cells. The cells themselves have a very limited power of growth. But, especially in the early stages of the life of the organism, when well supplied with nutriment, the cells multiply rapidly, by a process of fission, or the division of each cell into two daughter cells. The first part of the cell to divide is the nucleus, the protoplasmic network of which shows, during the process, curious and interesting arrangements and groupings of the fibres. When the nucleus has divided, the surrounding protoplasm is constricted, and separates into two portions, each of which contains a daughter nucleus.

In addition to the multiplication of cells, there is the formation, especially during periods of growth, of certain products of cell-life and cell-activity. Bone, for example, is a more or less permanent product of the activity of certain specialized cells.

There is, perhaps, no more wonderful instance of rapid and vigorous growth than the formation of the antlers of deer. These splendid weapons and adornments are shed and renewed every year. In the spring, when they are growing, they are covered over with a dark skin provided with short, fine, close-set hair, and technically termed "the velvet." If you lay your hand on the growing antler, you will feel that it is hot with the nutrient blood that is coursing beneath it. It is, too, exceedingly sensitive and tender. An army of tens of thousands of busy living cells is at work beneath that velvet surface, building the bony antlers, preparing for the battles of autumn. Each minute cell knows its work, and does it for the general good—so perfectly is the body knit into an organic whole. It takes up from the nutrient blood the special materials it requires; out of them it elaborates the crude bone-stuff, at first soft as wax, but ere long to become as hard as stone; and then, having done its work, having added its special morsel to the fabric of the antler, it remains embedded and immured, buried beneath the bone-products of its successors or descendants. No hive of bees is busier or more replete with active life than the antler of a stag as it grows beneath the soft, warm velvet. And thus are built up in the course of a few weeks those splendid "beams," with their "tynes" and "snags," which, in the case of the wapiti, even in the confinement of our Zoological Gardens, may reach a weight of thirty-two pounds, and which, in the freedom of the Rocky Mountains, may reach such a size that a man may walk, without stooping, beneath the archway made by setting up upon their points the shed antlers. When the antler has reached its full size, a circular ridge makes its appearance at a short distance from the base. This is the "burr," which divides the antler into a short "pedicel" next the skull, and the "beam" with its branches above. The circulation in the blood-vessels of the beam now begins to languish, and the velvet dies and peels off, leaving the hard, dead, bony substance exposed. Then is the time for fighting, when the stags challenge each other to single combat, while the hinds stand timidly by. But when the period of battle is over, and the wars and loves of the year are past, the bone beneath the burr begins to be eaten away and absorbed, through the activity of certain large bone-eating cells, and, the base of attachment being thus weakened, the beautiful antlers are shed; the scarred surface skins over and heals, and only the hair-covered pedicel of the antler is left.*

Not only are there these more or less permanent products of cell-activity which are built up into the framework of the body; there are other products of a less enduring, but, in the case of some of them, not less useful character. The secretions, for example, which, as we have seen, minister in such an important manner to nutrition, are of this class. The salivary fluids, the gastric juice, the pancreatic products, and the bile,—all of these are products of cell-life and cell-activity. And then there are certain products of cell-life which must be cast out from the body as soon as possible. These are got rid of in the excretions, of which the carbonic acid gas expelled in the lungs and the waste-products eliminated through the kidneys are examples. They are the ultimate organic

* From a popular article of the author's on "Horns and Antlers," in *Atalanta*.

products of the combustion that takes place in the muscular, nervous, and other tissues.

The animal organism has sometimes been likened to a steam-engine, in which the food is the fuel which enters into combustion with the oxygen taken in through the lungs. It may be worth while to modify and modernize this analogy—always remembering, however, that it is an analogy, and that it must not be pushed too far.

In the ordinary steam-engine the fuel is placed in the fire-box, to which the oxygen of the air gains access; the heat produced by the combustion converts the water in the boiler into steam, which is made to act upon the piston, and thus set the machinery in motion. But there is another kind of engine, now extensively used, which works on a different principle. In the gas-engine the fuel is gaseous, and it can thus be introduced in a state of intimate mixture with the oxygen with which it is to unite in combustion. This is a great advantage. The two can unite rapidly and explosively. In gunpowder the same end is effected by mixing the carbon and sulphur with nitre, which contains the oxygen necessary for their explosive combustion. And this is carried still further in dynamite and gun-cotton, where the elements necessary for explosive combustion are not merely mechanically mixed, but are chemically combined in a highly unstable compound.

But in the gas-engine, not only is the fuel and the oxygen thus intimately mixed, but the controlled explosions and the resulting condensation are caused to act directly on the piston, and not through the intervention of water in a boiler. Whereas, therefore, in the steam-engine the combustion is to some extent external to the working of the machine, in the gas-engine it is to a large extent internal and direct.

Now, instead of likening the organism as a whole to a steam-engine, it is more satisfactory to liken each cell to a gas-engine. We have seen that the cell-substance around the nucleus is composed of a network of protoplasm, the plasmogen, enclosing within its meshes a more fluid material, the plasm. It is probable that this more fluid material is an explosive, elaborated through the vital activity of the protoplasmic network. During the period of repose which intervenes between periods of activity, the protoplasmic network is busy in construction, taking from the blood-discs oxygen, and from the blood-fluid carbonaceous and nitrogenous materials, and knitting these together into relatively unstable explosive compounds. These explosive compounds are like the mixed air and gas of the gas-engine. A rested muscle may be likened to a complex and well-organized battery of gas-engines. On the stimulus supplied through a nerve-channel a series of co-ordinated explosions takes place: the gas-engines are set to work; the muscular fibres contract; the products of the explosions (one of which is carbonic acid gas) are taken up and hurried away by the blood-stream; and the protoplasm sets to work to form a fresh supply of explosive material. Long before the invention of the gas-engine, long before gun-cotton or dynamite were dreamt of, long before some Chinese or other inventor first mixed the ingredients of gunpowder, organic nature had utilized the principle of controlled explosions in the protoplasmic cell.

Certain cells are, however, more delicately explosive than others. Those, for example, on or near the external surface of the body—those, that is to say, which constitute the end organs of the special senses—contain explosive material which may be fired by a touch, a sound, an odour, the contact with a sapid fluid or a ray of light. The effects of the explosions in these delicate cells, reinforced in certain neighbouring nerve-knots (ganglionic cells), are transmitted down the nerves as along a fired train of gunpowder, and thus reach that wonderful aggregation of organized and co-ordinated explosive cells, the brain. Here it is again reinforced and directed (who, at present, can say how?) along fresh nerve-channels to muscles, or glands, or other organized groups of explosives. And in the brain, somehow associated with the explosion of its cells, consciousness and the mind-element emerges; of which we need only notice here that it belongs to a *wholly different order of being* from the physical activities and products with which we are at present concerned.

No analogies between mechanical contrivances and organic processes can be pushed very far. To liken the organic cell to a gas-engine is better than to liken the organism to a steam-engine, because it serves to indicate the fact that the fuel does not simply combine with the oxygen in combustion, but that an unstable or explosive combination of "fuel" and oxygen is first formed; and again, because the effect of this is direct, and not through the intervention of any substance to which the combustion merely supplies the necessary heat. But beyond the fact that a kind of explosive is formed which, like a fulminating compound, can be fired by a touch, there is no very close analogy to be drawn. Nor must we press the explosion analogy too far. The essential thing would seem to be this—which, perhaps, the analogy may have served to lead up to—that the vital protoplasmic network of the cell has the power of building up complex and unstable chemical compounds, which are probably stored in the plasm within the spaces between the threads of the network; and that these unstable compounds, under the influence of a stimulus (or, possibly, sometimes spontaneously) break down into simpler and more stable compounds.* In the case of muscle-cells, this latter change is accompanied by an alteration in length of the fibres and consequent movements in the organism, the products of the disruptive change being useless or harmful, and being, therefore, got rid of as soon as possible. But very

---

* It will be well here to introduce the technical terms for these changes. The general term for chemical actions occurring in the tissues of a living creature is *metabolism;* where the change is of such a nature that complex and unstable compounds are built up and stored for a while, it is called *anabolism;* where complex unstable compounds break up into less complex and relatively stable compounds, the term *katabolism* is applied. We shall speak of anabolic changes as ***constructive;*** katabolic, as *disruptive,* or sometimes, ***explosive.***

frequently the products of explosive activity are made use of. In the case of bone-cells, one of the products of disruption is of permanent use to the organism, and constitutes the solid framework of the skeleton. In the case of the secreting cells of the salivary and other digestive glands, one of the disruptive products is of temporary value for the preparation of the food. It is exceedingly probable that these useful products of disruption, permanent or temporary, took their origin in waste products for which natural selection has found a use, and which have been, through natural selection, rendered more and more efficacious. This, however, is a question we are not at present in a position to discuss.

In the busy hive of cells which constitutes what we call the animal body, there is thus ceaseless activity. During periods of apparent rest the protogen filaments of the cell-net are engaged in constructive work, building up fresh supplies of complex and unstable materials, which, during periods of apparent activity, break up into simpler and more stable substances, some of which are useful to the organism while others must be got rid of as soon as possible. From another point of view, the cells during apparent rest are storing up energy which is utilized by the organism during its periods of activity. The storing up of available energy may be likened to the winding up of a watch or clock; it is during apparent rest that the cell is winding itself up; and thus we have the apparent paradox that the cell is most active and doing most work when it is at rest. During the repose of an organ, in fact, the cells are busily working in preparation for the manifestation of energetic action that is to follow. Just as the brilliant display of intellectual activity in a great orator is the result of the silent work of a lifetime, so is the physical manifestation of muscular power the result of the silent preparatory work of the muscle-cells.*

One point to be specially noted is the varied activity

* I do not mean, of course, to imply that there is no reconstruction during activity, but that it is then distinctly outbalanced by disruptive changes.

of the cells. While they are all working for the general good of the organism, they are divided into companies, each with a distinct and definite kind of work. This is known as the physiological division of labour. It is accompanied by a morphological differentiation of structure. By the form of a cell, therefore, we can generally recognize the kind of work it has to perform. The unstable compounds produced by the various cells must also be different, though not much is known at present on this subject. The unstable compound which forms bone and that which forms the salivary ferment, the unstable matter elaborated by nerve-cells and that built up by muscle-cells, are in all probability different in their chemical nature. Whether the formative plasmogen from which these different substances originate is in all cases the same or in different cases different, we do not know.

It may, perhaps, seem strange that the products of cellular life should be reached by the roundabout process of first producing a very complex substance out of which is then formed a less complex substance, useful for permanent purposes, as in bone, or temporary purposes, as in the digestive fluids. It seems a waste of power to build up substances unnecessarily complex and stored with an unnecessarily abundant supply of energy. Still, though we do not know that this course is adopted in all cases, there is no doubt that it is adopted in a great number of instances. And the reason probably is that by this method the organs are enabled to act under the influence of stimuli. They are thus like charged batteries ready to discharge under the influence of the slightest organic touch. In this way, too, is afforded a means by which the organ is not dependent only upon the products of the immediate activity of the protoplasm at the time of action, but can utilize the store laid up during a considerable preceding period.

Sufficient has now been said to illustrate the nature of the process of life. The fact that I wish to stand out clearly is that the animal body is stored with large quantities of available energy resident in highly complex and unstable chemical compounds, elaborated by the constructive energy of the formative protoplasm of its constituent cells. These unstable compounds, eminently explosive according to our analogy, are built up of materials derived from two different sources—from the nutritive matter (containing carbon, hydrogen, and nitrogen) absorbed in the digestive organs, and from oxygen taken up from the air in the lungs. The cells thus become charged with energy that can be set free on the application of the appropriate stimulus, which may be likened to the spark that fires the explosive.

Let us note, in conclusion, that it is through the blood-system, ramifying to all parts of the body, and the nerve-system, the ramifications of which are not less perfect, that the larger and higher organisms are knit together into an organic whole. The former carries to the cell the raw materials for the elaboration of its explosive products, and, after the explosions, carries off the waste products which result therefrom. The nerve-fibres carry the stimuli by which the explosive is fired, while the central nervous system organizes, co-ordinates, and controls the explosions, and directs the process of reconstruction of the explosive compounds.

CHAPTER III.

REPRODUCTION AND DEVELOPMENT.

We have now to turn to a fresh aspect of animal life, that of reproduction; and it will be well to connect this process as closely as possible with the process of life in general, of which it is a direct outcome.

It will be remembered that, in the last chapter, it was shown that the essential feature in the process of life is the absorption by living protoplasm of oxygen on the one hand and nutritive matter on the other hand, and the kneading of these together, in subtle metabolism, into unstable compounds, which we likened to explosives. This is the first, or constructive, stage of the life-process. Thereupon follows the second, or disruptive, stage. The unstable compounds break down into more stable products,— they explode, according to our analogy; and accompanying the explosions are manifestations of motor activity—of heat, sometimes of light and electrical phenomena. But in the economy of nature the products of explosion are often utilized, and in the division of labour among cells the explosions of some of them are directed specially to the production of substances which shall be of permanent or temporary use—for digestion, as in the products of the salivary, gastric, and intestinal glands; for support, as in bone, cartilage, and skeletal tissue generally; or as a store of nutriment, in fat or yolk. The constructive products of protoplasmic activity seem for the most part to be lodged in the spaces between the network of formative protoplasm. The disruptive products—those of them, that is to say, which are of temporary or permanent value to the organism—accumulate either within the cell, sometimes at one pole, sometimes at the centre, as in the case of the yolk of eggs, or around the cell, as in the case of cartilage or bone.

Apart from and either preceding or accompanying these phenomena, is the growth or increase of the formative protoplasm itself; concerning which the point to be here observed is that it is not indefinite, but limited. This was first clearly enunciated by Herbert Spencer, and may be called Spencer's law. In simplest expression it may thus be stated: *Volume tends to outrun surface*. Take a cube measuring one inch in the side; its volume is one cubic inch, its surface six square inches. Eight such cubes will have a surface of (6×8) forty-eight square inches. But let these eight be built into a larger cube, two inches in the side, and it will be found that the surface exposed is now only twenty-four square inches. While the volume has been increased eight times, the surface has been increased only four times. With increase of size, volume tends to outrun surface. But in the organic cell the nutritive material and oxygen are absorbed at the surface, while the explosive changes occur throughout its mass. Increase of size, therefore, cannot be carried beyond certain limits, for the relatively diminished surface is unable to supply the relatively augmented mass with material for elaboration into unstable compounds. Hence the cell divides to afford the same mass increased surface. This process of cell-division is called fission, and in some cases cleavage.

We will now proceed to pass in review the phenomena of reproduction and development in animals.

Attention has already been drawn to the difference between those lowly organisms, each of which is composed of a single cell—the protozoa, as they are termed—and those higher organisms, called metazoa, in which there are many cells with varied functions. Confining our attention at first to the former group of unicellular animals, we find considerable diversities of form and habit, from the relatively large, sluggish, parasitic *Gregarina*, to the active slipper-animalcule, or *Paramœcium*, or the beautiful, stalked bell-animalcule, or *Vorticella*; and from the small, slow-moving amœba to the minute, intensely active monad. In many cases reproduction is by simple fission, as in the amœba, where the nucleus first undergoes division; and then the whole organism splits into two parts, each with its own nucleus. In other cases, also numerous, the

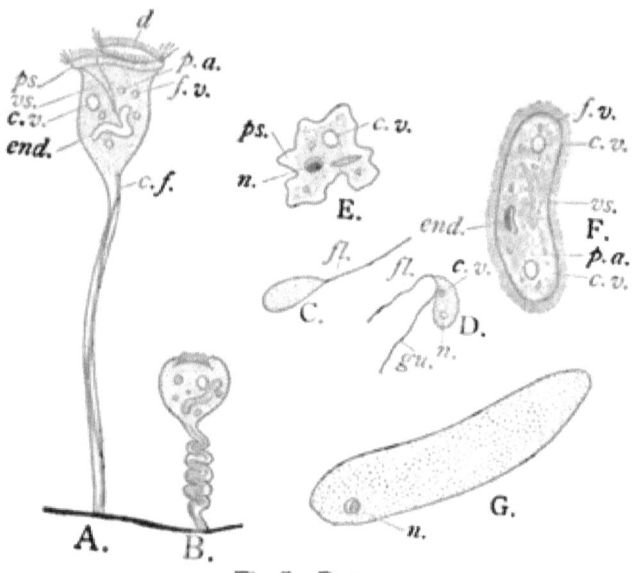

Fig. 7.—Protozoa.

A, vorticella extended. B, the same contracted. C, D, monads. E, amœba. F, Paramœcium. G, Gregarina. c.f., contractile fibre; c.v., contractile vesicle; d., disc; end., endoplast; f.v., food-vacuole; fl., flagellum; gu., gubernaculum; n., nucleus; p.a., potential anus; ps., (in A) peristome, (in E) pseudopodium; vs., vestibule.

organism passes into a quiescent state, and becomes surrounded with a more or less toughened cyst. The nucleus then disappears, and the contents of the cyst break up into a number of small bodies or spores. Eventually the cyst bursts, and the spores swarm forth. In the case of some active protozoa the minute creatures that swarm forth are more or less like the parent; but in the more sluggish kinds the minute forms are more active than the parent. Thus in the case of the gregarina, the minute spore-products are like small amœbæ; while in other instances the embryos, if so we may call them, have a whip-like cilium like the monads.

Very frequently, however, there is, in the protozoa, a further process, which would seem to be intimately associated with fission or the formation of spores, as the case may be. This is known as conjugation. Among monads, for example, two individuals may meet together, conjugate, and completely fuse the one into the other. A triangular cyst results. After a while, the cyst bursts, and an apparently homogeneous fluid escapes. The highest powers of the microscope fail to disclose in it any germ of life; and there, at first sight, would seem to be an end of the matter. But wait and watch; and there will appear in the field of the microscope, suddenly and as if by magic, countless minute points, which prolonged watching shows to be growing. And when they have further grown, each distinct point is seen to be a monad.

In the slipper-animalcule, conjugation is temporary. But during the temporary fusion of the two individuals important changes are said to occur. In these infusorians there is, beside the nucleus, a smaller body, the paranucleus. This, in the case of conjugating paramœcia, appears to divide into two portions, of which one is mutually exchanged. Thus when two slipper-animalcules are in conjugation, the paranucleus of each breaks into two parts, a and b, of which a is retained and b handed over in exchange. The old a and the new b then unite, and each paramœcium goes on its separate way. M. Maupas, who has lately reinvestigated this matter, considers, as the result of his observations on another infusorian (*Stylonichia*), that without conjugation these organisms become exhausted, and multiplication by fission comes to a standstill. If this be so, conjugation is, in these organisms, necessary for the continuance of the race. But Richard Hertwig has recently shown that this is, at any rate, not universally true.

In the bell-animalcule, fission takes place in such a manner as to divide the bell into two equal portions. Thus there are two bells to one stalk. But the fate of the two is not the same. One remains attached to the stalk, and expands into a complete vorticella. The other remains pear-shaped, and develops round the posterior region of the body a girdle of powerful vibratile cilia, by the lashing of which the animalcule tears itself away from the parent stem, and swims off through the water. After a short active existence, it settles down in a convenient spot, adhering by its posterior extremity. The hinder girdle of cilia is lost or absorbed, a stalk is rapidly developed, and the organism expands into a perfect vorticella.

In some cases, however, the fission is of a different character, with different results. It may be very unequal, so that a minute, free-swimming animalcule is disengaged; or minute animalcules may result by repetition of division. In either case the minute form conjugates with an ordinary vorticella, its smaller mass being completely merged in the larger volume of its mate.

There are, of course, many variations in detail in the modes of protozoan reproduction; but we may say that, omitting such details, reproduction is either by simple fission or by spore-formation; and that these processes are in some cases associated with, and perhaps dependent on, the temporary or permanent union of two individuals in conjugation.

It is essential to notice that the results of fission or of spore-formation separate, each going on its own way. Hence such development as we find in the protozoa results from differentiations within the limits of the single cell. Thus the bell-animalcule has a well-defined and constant form; a definite arrangement of cilia round the rim and in the vestibule by which food finds entrance to the body. The outer layer of the body forms a transparent cuticle, beneath which is a so-called "myophan" layer, continuous with a contractile thread in the stalk. Within the substance of the body is a pulsating cavity, or contractile vesicle, and a nucleus. Such is the nature of the differentiation which may go on within the protozoan cell.

When we pass to the metazoa, we find that the method

of differentiation is different. These organisms are composed of many cells; and instead of the parts of the cell differentiating in several directions, the several cells differentiate each in its own special direction. This is known as the physiological division of labour. The cells merge their individuality in the general good of the organism. Each, so to speak, cultivates some special protoplasmic activity, and neglects everything else in the attainment of this end. The adult metazoan, therefore, consists of a number of cells which have diverged in several, sometimes many, directions.

In some of the lower metazoans, reproduction may be effected by fission. Thus the fresh-water hydra is said to divide into two parts, each of which grows up into a perfect hydra. It is very doubtful, however, whether this takes place normally in natural life. But there is no doubt that if a hydra be artificially divided into a number of special pieces, each will grow up into a perfect organism, so long as each piece has fair samples of the different cells which constitute the body-wall. Sponges and sea-anemones may also be divided and subdivided, each part having the power of reproducing the parts that are thus cut away. When a worm is cut in half by the gardener's spade, the head end grows a new tail; and it is even stated that a worm not only survived the removal of the first five rings, including the brain, mouth, and pharynx, but within fifty-eight days had completely regenerated these parts.

Higher up in the scale of metazoan life, animals have the power of regenerating lost limbs. The lobster that has lost a claw reproduces a new one in its stead. A snail will reproduce an amputated "horn," or tentacle, many times in succession, reproducing in each case the eye, with its lens and retina. Even a lizard will regenerate a lost tail or a portion of a leg. In higher forms, regeneration is restricted to the healing of wounds and the mending of broken bones.

Closely connected with this process of regeneration of lost parts is the widely prevalent process of reproduction by budding. The cut stump of the amputated tentacle of the hydra or the snail buds forth a new organ. But in the hydra, during the summer months, under normal circumstances, a bud may make its appearance and give rise to a new individual, which will become detached from the parent, to lead a separate existence. In other organisms allied to the hydra the buds may remain in attachment, and a colony will result. This, too, is the result of budding in many of the sponges. In some worms, too, budding may occur. In the fresh-water worm (*Chætogaster limnæi*) the animal, as we ordinarily see it, is a train of individuals, one budded off behind the other—the first fully developed, those behind it in various stages of development. The individuals finally separate by transverse division. Another more lowly worm (*Microstomum lineare*, a Turbellarian) may bud off in similar fashion a chain of ten or fifteen individuals. In these cases budding is not far removed from fission.

Now, in the case of reproduction by budding, as in the hydra, a new individual is produced from some group of cells in the parent organism. From this it is but a step—a step, however, of the utmost importance—to the production of a new individual from a single cell from the tissues of the parental organism. Such a reproductive cell is called an egg-cell, or ovum. In the great majority of cases, to enable the ovum to develop into a new individual, it is necessary that the egg-cell should conjugate or fuse with a minute, active sperm-cell, generally derived from a different parent. This process of fusion of germinal cells is called fertilization (see Fig. 5, p. 13).

In sponges, the cells which become ova or sperms lie scattered in the mid-layer between the ciliated layers which line the cavities and spaces of the organism. Sometimes the individual sponge produces only ova; sometimes only sperms; sometimes both, but at different periods. The cells which become ova increase in size, are passive, and rich in reserve material elaborated by their protoplasm. The cells which become sperms divide again and again, and thus produce minute active bodies, adance with restless motion. These opposite tendencies are repeated and emphasized throughout the animal kingdom—ova relatively large, passive, and accumulative of reserve material; sperms minute, active, and the result of repeated fission. The active sperm, when it unites with the ovum, imports into it a tendency to fission, or cleavage; but the resulting cells do not part and scatter—they remain associated together, and in mutual union give rise to a new sponge.

In the hydra, generally near the foot or base of attachment, a rounded swelling often makes its appearance in

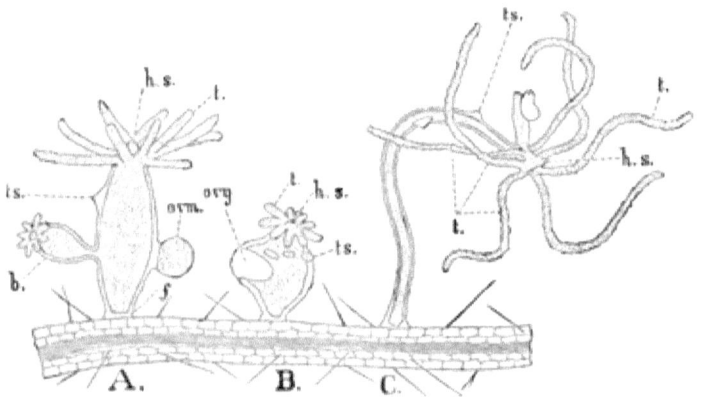

Fig. 8.—Hydra viridis.

A, hydra half retracted, with a bud and an ovum attached to the shrunken ovary; B, a small hydra firmly retracted; C, a hydra fully extended. b., bud; f., foot; h.s., hypostome; ovm., ovum; ovy., ovary; t., tentacles; ts, testis.

autumn. Within this swelling one central cell increases enormously at the expense of the others. It becomes an ovum. Eventually it bursts through the swelling, but remains attached for a time. Rarely in the same hydra, more frequently in another, one or two swellings may be seen higher up, beneath the circle of tentacles. Within these, instead of the single ovum may be seen a swarm of sperms, minute and highly active. When these are discharged, one may fuse with and fertilize an ovum, occasionally in the same, but more frequently in another individual, with the result that it develops into a new hydra. Here there are definite organs—an ovary and a testis—producing the ova or the sperms. But they are indefinite and not permanent in position.

In higher forms of life the organs which are set apart for the production of ova or sperms become definite in position and definite in structure. Occasionally, as in the snail, the same organ produces both sperms and ova, but then generally in separate parts of its structure. The two products also ripen at different times. Not infrequently, as in the earthworm, each individual has both testes and ovaries, and thus produces both ova and sperms, but from different organs. The ova of one animal are, however, fertilized by sperms from another. But in the higher invertebrates and vertebrates there is a sex-differentiation among the individuals, the adult males being possessed of testes only and producing sperms, the adult females possessed of ovaries only and producing ova. There are also, in many cases, accessory structures for ensuring that the ova shall be fertilized by sperms, while sexual appetences are developed to further the same end. But however the matter may thus be complicated, the essential feature is the same—the union of a sluggish, passive cell, more or less laden with nutritive matter, with a minute active cell with an hereditary tendency to fission.*

It is not, however, necessary in all cases that fertilization of the ovum should take place. The plant-lice, or *Aphides* of our rose trees, may produce generation after

---

* Professor Geddes and Mr. J. Arthur Thomson, in their interesting work on "The Evolution of Sex," regard the ovum in especial, and the female in general, as preponderatingly anabolic (see note, p. 32); while the sperm in especial, and the male in general, are on their view preponderatingly katabolic. Regarding, as I do, the food-yolk as a katabolic product, I cannot altogether follow them. The differentiation seems to me to have taken place along divergent lines of katabolism. In the ovum, katabolism has given rise to storage products; in the sperm, to motor activities associated with a tendency to fission. The contrast is not between anabolic and katabolic tendencies, but between storage katabolism and motor katabolism. Nor do I think that "the essentially katabolic male-cell brings to the ovum a supply of characteristic waste products, or katastates, which stimulate the latter to division" (*l.c.*, p. 162). I believe that it brings an inherited tendency to fission, and thus reintroduces into the fertilized ovum the tendency which, as ovum, it had renounced in favour of storage katabolism.

generation, and their offspring in turn reproduce in like manner, without any union or fusion of ovum or sperm. The same is true of the little water-fleas, or *Daphnids;* while in some kinds of rotifers fertilization is said never to occur. It is a curious and interesting fact, which seems now to be established beyond question, that drone bees are developed from unfertilized ova, the fertilized ova producing either queens or workers, according to the nature of the food with which the grubs are supplied. Where, as in the case of aphids and daphnids, fertilization occasionally takes place, it would seem that lowered temperature and diminished food-supply are the determining conditions. Fertilization, therefore, generally takes place in the autumn ; the fertilized ovum living on in a quiescent state during the winter, and developing with the warmth of the succeeding spring. In the artificial summer of a greenhouse, reproduction may continue for three or four years without the occurrence of any fertilization.

Mention may here be made of some peculiarly modified modes of reproduction among the metazoa. The aurelia is a well-known and tolerably common jelly-fish. These

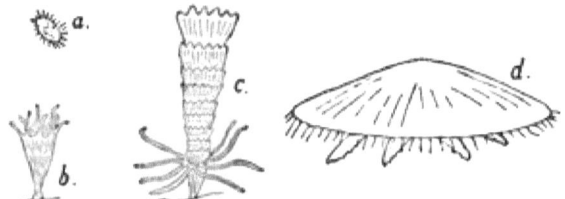

Fig. 9.—Aurelia: Life-cycle.

a, embryo; *b*, *Hydra tuba*; *c*, *Hydra tuba*, with medusoid segments; *d*, medusa separated to lead free existence.

produce ova, which are duly fertilized by sperms from a different individual. A minute, free-swimming embryo develops from the ovum, which settles down and becomes a little polyp-like organism, the *Hydra tuba*. As growth proceeds, this divides or segments into a number of separable, but at first connected, parts. As these attain their full development, first one and then another is detached from the free end, floats off, and becomes a medusoid aurelia. Thus the fertilized ovum of aurelia develops, not into one, but into a number of medusæ,* passing through the *Hydra tuba* condition as an intermediate stage.

Many of the hydroid zoophytes, forming colonies of hydra-like organisms, give rise in the warm months to medusoid jelly-fish, capable of producing ova and sperms. Fertilization takes place; and the fertilized ova develop into little hydras, which produce, by budding, new colonies. In these new colonies, again, the parts which are to become ovaries or testes float off, and ripen their products in free-swimming, medusoid organisms. Such a rhythm between development from ova and development by budding is spoken of as an alternation of generations.

The fresh-water sponge (*Spongilla*) exhibits an analogous rhythm. The ova are fertilized by sperms from a different short-lived individual. They develop into sponges which have no power of producing ova or sperms. But on the approach of winter in Europe, and of the dry season in India, a number of cells collect and group themselves into a so-called gemmule. Round this is formed a sort of crust beset with spicules, which, in some cases, have the form of two toothed discs united by an axial shaft. When these gemmules have thus been formed, the sponge dies; but the gemmules live on in a quiescent state during the winter or the dry season, and with the advent of spring develop into sponges, male or female. These have the power of producing sperms or ova, but no power of producing gemmules. The power of producing ova, and that of producing gemmules, thus alternates in rhythmic fashion.

But one more example of these modified forms of reproduction can here be cited (from the author's text-book on "Animal Biology"). The liver-fluke is a parasitic organism, found in the liver of sheep. Here it reaches sexual maturity, each individual producing many thousands of eggs, which pass with the bile into the alimentary canal of the host, and are distributed over the fields with the

* On the other hand, three ova of the crustacean *Apus* are said to coalesce to form the single ovum from which one embryo develops.

excreta. Here, in damp places, pools, and ditches, free and active embryos are hatched out of the eggs. Each embryo (Fig. 10, C., much enlarged) is covered with cilia, except at the anterior end, which is provided with a head-

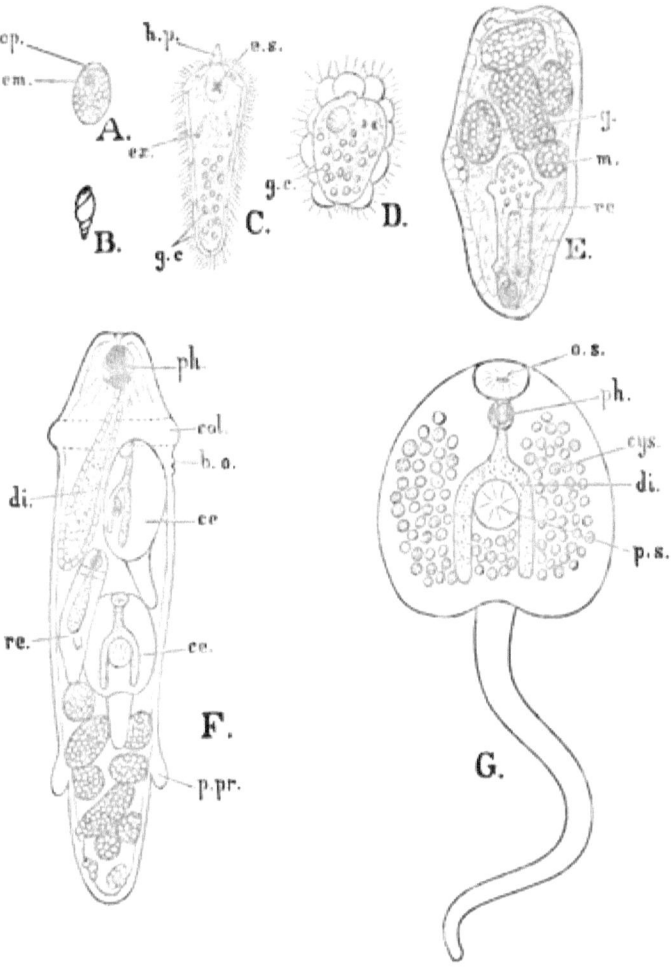

Fig. 10.—Liver-fluke: Embryonic stages. (After A. P. Thomas.)

A. Ovum: *em.*, embryo; *op.*, operculum. B. *Limnœus truncatulus* (natural size). C. Free embryo: *e.s.*, eye-spot; *ex.*, excretory vessel; *g.c.*, germinal cells; *h.p.*, head-papilla. D. Embryo preparing to become a sporocyst: *g.c.*, germinal cells. E. Sporocyst: *g.*, gastrula; *m.*, morula; *re.*, redia. F. Redia: *b.o.*, birth-opening; *ce.*, cercaria; *col.*, collar; *di.*, digestive sac; *ph.*, pharynx; *p.pr.*, posterior processes; *re.*, daughter redia. G. Cercaria: *cys.*, cystogenous organ; *di.*, digestive sac; *o.s.*, oral sucker; *p.s.*, posterior sucker; *ph.*, pharynx.

papilla (*h.p.*). When the embryo comes in contact with any object, it, as a rule, pauses for a moment, and then darts off again. But if that object be the minute water-snail, *Limnæus truncatulus* (Fig. 10, B., natural size), instead of darting off, the embryo bores its way into the tissues until it reaches the pulmonary chamber, or more rarely the body-cavity. Here its activity ceases. It passes into a quiescent state, and is now known as a *sporocyst* (Fig. 10, E.). The active embryo has degenerated into a mere brood-sac, in which the next generation is to be produced. For within the sporocyst special cells undergo division, and become converted into embryos of a new type, which are known as *rediæ* (F.), and which, so soon as they are sufficiently developed, break through the wall of the sporocyst. They then increase rapidly in size, and browse on the digestive gland of the water-snail (known as the *intermediate host*), to which congenial spot they have in the mean time migrated. The series of developmental changes is even yet not complete. For within the rediæ (besides, at times, daughter rediæ) embryos of yet another type are produced by a process of cell-division. These are known as *cercariæ* (Fig. 10, G.). Each has a long tail, by means of which it can swim freely in water. It leaves the intermediate host, and, after leading a short, active life, becomes encysted on blades of grass. The cyst is formed by a special larval organ, and is glistening snowy white. Within the cyst lies the transparent embryonic liver-fluke, which has lost its tail in the process of encystment.

The last chapter in this life-history is that in which the sheep crops the blade of grass on which the parasite lies encysted; whereupon the cyst is dissolved in the stomach of the host, the little liver-fluke becomes active, passes through the bile-duct into the liver of the sheep, and there, growing rapidly, reaches sexual maturity, and lays its thousands of eggs, from each of which a fresh cycle may take its origin. The sequence of phenomena is characterized by discontinuity of development. Instead of the embryo growing up continuously into the adult, with only the atrophy of provisional organs (*e.g.* the gills and tail of the tadpole, or embryo frog), it produces germs from which the adult is developed. Not merely provisional organs, but provisional organisms, undergo atrophy. In the case of the liver-fluke there are two such provisional organisms, the embryo sporocyst and the redia.

We may summarize the life-cycle thus—

1. *Ovum* laid in liver of sheep, passes with bile into intestine, and thence out with the excreta.

2. *Free ciliated embryo*, in water or on damp earth, passes into pulmonary cavity of *Limnæus truncatulus*, and develops into

3. *Sporocyst*, in which secondary embryos are developed, known as

4. *Rediæ*, which pass into the digestive glands of *Limnæus*, and within which, besides daughter rediæ, there are developed tertiary embryos, or

5. *Cercariæ*, which pass out of the intermediate host and become

6. *Encysted* on blades of grass, which are eaten by sheep. The cyst dissolves, and the young flukes pass into the liver of their host, each developing into

7. A *liver-fluke*, sexual, but hermaphrodite.

Here, again, we notice that one fertilized ovum gives rise to not one, but a number of liver-flukes.

We must now pass on to consider the growth and development of organisms. Simple growth results from the multiplication of similar cells. As the child, for example, grows, the framework of the body and the several organs increase in size by continuous cell-multiplication. Development is differential growth; and this may be seen either in the organs or parts of an organism or in the cells themselves. As the child grows up into a man, there is a progressive change in his relative proportions. The head becomes relatively smaller, the hind limbs relatively longer, and there are changes in the proportional size of other organs.

In the development of the embryo from the ovum, the differentiation is of a deeper and more fundamental character. Cells at first similar become progressively dissimilar, and out of a primitively homogeneous mass of cells is developed a heterogeneous system of different but mutually related tissues.

This view of development is, however, the outcome of comparatively modern investigation and perfected microscopical appliances. The older view was that development in all cases is nothing more than differential growth, that there is no differentiation of primitively similar into ultimately different parts. Within the fertilized ovum of the horse or bird lay, it was supposed, in all perfection of structure, a miniature racer or chick, the parts all there, but too minute to be visible. All that was required was that each part should grow in due proportion. Those who held this view, however, divided into two schools. The one believed that the miniature organism was contained within the ovum, the function of the sperm being merely to stimulate its subsequent developmental growth. The other held that the sperm was the miniature organism, the ovum merely affording the food-material necessary for its developmental growth. In either case, this unfolding of the invisible organic bud was the *evolution* of the older writers on organic life. More than this. As Messrs. Geddes and Thomson remind us,* "the germ was more than a marvellous bud-like miniature of the adult. It necessarily included, in its turn, the next generation, and this the next—in short, all future generations. Germ within germ, in ever smaller miniature, after the fashion of an infinite juggler's box, was the corollary logically appended to this theory of preformation and unfolding."

Modern embryology has completely negatived any such view as that of preformation, and as completely established that the evolution is not the unfolding of a miniature germ, but the growth and differentiation of primitively similar cell-elements. In different animals, as might be expected, the manner and course of development are different. We may here illustrate it by a very generalized and so to speak diagrammatic description of the development of a primitive vertebrate.

* "The Evolution of Sex," p. 84.

The ovum before fertilization is a simple spherical cell, without any large amount of nutritive material in the form of food-yolk (*A.*). It contains a nucleus. Previous to fertilization, however, in many forms of life, portions of the nucleus, amounting to three parts of its mass, are got rid of in little "polar cells" budded off from the ovum. The import of this process we shall have to consider in connection with the subject of heredity. The sperm is also

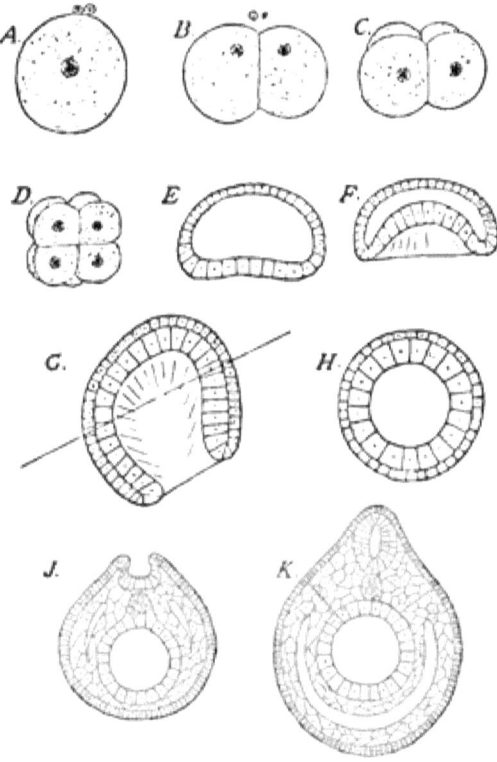

Fig. 11.—Diagram of development.

See text. The fine line across *G.* indicates the plane of section shown in *H.*

a nucleated cell; and on its entrance into the ovum there are for a short time two nuclei—the female nucleus proper to the ovum, and the male nucleus introduced by the sperm. These two unite and fuse to form a joint nucleus. Thus the fertilized ovum starts with a perfect blending of the nuclear elements from two cells produced by different parents.

Then sets in what is known as the segmentation or cleavage of the ovum. First the nucleus and then the cell itself divides into two equal halves (*B*.), each of these shortly afterwards again dividing into two. We may call the points of intersection of these two planes of division the "poles," and the planes "vertical planes." We thus have four cells produced by two vertical planes (*C*.). The next plane of division is equatorial, midway between the poles. By this plane the four cells are subdivided into eight (*D*.). Then follow two more vertical planes intermediate between the first two. By them the eight cells are divided into sixteen. These are succeeded by two more horizontal planes midway between the equator and the poles. Thus we get thirty-two cells. So the process continues until, by fresh vertical and horizontal planes of division, the ovum is divided into a great number of cells.

But meanwhile a cavity has formed in the midst of the ovum. This makes its appearance at about the eight-cell stage, the eight cells not quite meeting in the centre of the ovum. The central cavity so formed is thus surrounded by a single layer of cells, and it remains as a single layer throughout the process of segmentation, so that there results a hollow vesicle composed of a membrane constituted by a single layer of cells (*E*.).

The cells on one side of the vesicle are rather larger than the others, and the next step in the process is the apparent pushing in of this part of the hollow sphere; just as one might take a hollow squash indiarubber ball, and push in one side so as to form a hollow, two-layered cup (*F*.). The vesicle, then, is converted into a cup, the mouth of which gradually closes in and becomes smaller, while the cup itself elongates (*G*.).* Thus a hollow, two-layered, stumpy, worm-like embryo is produced, the outer layer of

---

* In some forms of life the opening of the cup marks the position of the future mouth; in others, of the future vent. In yet others it elongates into a slit, occupying the whole length of the embryo; the middle part of the slit closes up, and the opening at the far ends mark the position, the one of the future mouth, the other of the future vent.

which may be ciliated, so that by the lashing of these cilia it is enabled to swim freely in the water. The inner cavity is the primitive digestive cavity.

A cross-section through the middle of the embryo at this stage will show this central cavity surrounded by a two-layered body-wall (*II.*). A little later the following changes take place (*J. K.*): Along a definite line on the surface of the embryo, marking the region of the back, the outer layer becomes thickened; the edges of the thickened band so produced rise up on either side, so as to give rise to a median groove between them; and then, overarching and closing over the groove, convert it into a tube. This tube is called the neural tube, because it gives rise to the central nervous system. In the region of the head it expands; and from its walls, by the growth and differentiation of the cells, there is formed—in the region of the head, the brain, and along the back, the spinal cord. Immediately beneath it there is formed a rod of cells, derived from the inner layer. This rod, which is called the notochord, is the primitive axial support of the body. Around it eventually is formed the vertebral column, the arches of the vertebræ embracing and protecting the spinal cord.

Meanwhile there has appeared between the two primitive body-layers a third or middle layer.* The cells of which it is composed arise from the inner layer, or from the lips of the primitive cup when the outer and inner layer pass the one into the other. This middle layer at first forms a more or less continuous sheet of cells between the inner and the outer layers. But ere long it splits into two sheets, of which one remains adherent to the inner layer and one to the outer layer. The former becomes the muscular part of the intestinal or digestive tube, the latter the lining of the body-wall. The space between the two is known as the body-cavity. Beneath the throat the heart is fashioned out of this middle layer.

Very frequently—that is to say, in many animals—the

---

* In technical language, the outer layer of cells is called the *epiblast*, the inner layer the *hypoblast*, and the mid-layer between them the *mesoblast*.

opening by which the primitive digestive tube communicated with the exterior has during these changes closed up, so that the digestive cavity does not any longer communicate in any way with the exterior. This is remedied by the formation of a special depression or pit at the front end for the mouth, and a similar pit at the hinder end.* These pits then open into the canal, and communications with the exterior are thus established. The lungs and liver are formed as special outgrowths from the digestive tube. The ovaries or testes make their appearance *at a very early period* as ridges of the middle layer projecting into the body-cavity. For some time it is impossible to say whether they will produce sperms or ova; and it is said that in many cases they pass through a stage in which one portion has the special sperm-producing, and another the special ovum-producing, structure. But eventually one or other prevails, and the organs become either ovaries or testes.

Thus from the outer layer of the primitive embryo is produced the outer skin, together with the hairs, scales, or feathers which it carries; from it also is produced the nervous system, and the end-organs of the special senses. From the inner layer is formed the digestive lining of the alimentary tube and the glands connected therewith; from it also the primitive axial support of the body. But this primitive support gives place to the vertebral column formed round the notochord; and this is of mid-layer origin. Out of the middle layer are fashioned the muscles and framework of the body; out of it, too, the heart and reproductive organs. The tissues of many of the organs are cunningly woven out of cells from all three layers. The lens of the eye, for example, is a little piece of the outer layer pinched off and rendered transparent. The retina of that organ is an outgrowth from the brain, which,

* In technical language, the opening by which the primitive digestive cavity (or *mesenteron*) communicates with the exterior is called the *blastopore*. When this closes, the new opening for the mouth is called the *stomodæum*; that for the vent, the *proctodæum*.

as we have seen, was itself developed from the outer layer. But round the retina and the lens there is woven from the middle layer the tough capsule of the eye and the circular curtain or iris. The lining cells of the digestive tube are cells of the inner layer, but the muscular and elastic coats are of middle-layer origin. The lining cells of the salivary glands arise from the outer layer where it is pushed in to form the mouth-pit; but the supporting framework of the glands is derived from the cells of the middle layer.

Enough has now been said to give some idea of the manner in which the different tissues and organs of the organism are elaborated by the gradual differentiation of the initially homogeneous ovum. The cells into which the fertilized egg segments are at first all alike; then comes the divergence between those which are pushed in to line the hollow of the cup, and those which form its outer layer. Thereafter follows the differentiation of a special band of outer cells to form the nervous system, and a special rod, derived from the inner cells, to form the primitive axial support. And when the middle layer has come into existence, its cells group themselves and differentiate along special lines to form gristle or bone, blood or muscle.

The description above given is a very generalized and diagrammatic description. There are various ways in which complexity is introduced into the developmental process. The store of nutritive material present in the egg, for example, profoundly modifies the segmentation so that where, as in the case of birds' eggs, there is a large amount of food-yolk, not all the ovum, but only a little patch on its surface, undergoes segmentation. In this little patch the embryo is formed. Break open an egg upon which a hen has been sitting for five or six days, and you will see the little embryo chick lying on the surface of the yolk. The large mass of yolk to which it is attached is simply a store of food-material from which the growing chick may draw its supplies.

For it is clear that the growing and developing embryo must obtain, in some way and from some source, the food-stuff for its nutrition. And this is effected, among different animals, in one of three ways. Either the embryo becomes at a very early stage a little, active, voracious, free-swimming larva, obtaining for itself in these early days of life its own living; as is the case, for example, with the oyster or the star-fish. Or the egg from which it is developed contains a large store of food-yolk, on which it can draw without stint; as is the case with birds. Or else the embryo becomes attached to the maternal organism in such a way that it can draw on her for all the nutriment which it may require; as is the case with the higher mammals.

In both these latter cases the food-material is drawn from the maternal organism, and is the result of parental sacrifice; but in different ways. In the case of the bird, the protoplasm of the ovum has acquired the power of storing up the by-products of its vital activity. The ovum of such an animal seems at first sight a standing contradiction to the statement, made some pages back, that the cell cannot grow to any great extent without undergoing division or fission; and this because volume tends to outrun surface. For the yolk of a bird's egg is a single cell, and is often of large size. But when we come to examine carefully these exceptional cases of very large cells—for what we call the yolk of an egg is, I repeat, composed of a single cell—we find that the formative protoplasm is arranged as a thin patch on one side of the yolk in the case of the bird's egg, or as a thin pellicle surrounding the yolk in the case of that of the lobster or the insect. All the rest is a product of protoplasmic life stowed away beneath the patch or within the pellicle. And this stored material is relatively stable and inert, not undergoing those vital disruptive changes which are characteristic of living formative protoplasm. The mass of formative protoplasm, even in the large eggs of birds, is not very great, and is so arranged as to offer a relatively extensive surface. All the rest, the main mass of the visible egg-yolk, is the stored product of a specialized activity of the formative protoplasm. But all this material is of parental origin—is elaborated from the nutriment absorbed and digested by the mother.

Thus we see, in the higher types of life, parental sacrifice, fosterage, and protection. For in the case of mammals and many birds, especially those which are born in a callow, half-fledged condition, even when the connection of mother and offspring is severed, or the supplies of food-yolk are exhausted, and the young are born or hatched, there is still a more or less prolonged period during which the weakly offspring are nourished by milk, by a secretion from the crop ("pigeon's milk"), or by food-stuff brought with assiduous care by the parents. There is a longer or shorter period of fosterage and protection—longer in the case of man than in that of any of the lower animals—ere the offspring are fitted to fend for themselves in life's struggle.

And accompanying this parental sacrifice, first in supplying food for embryonic development, and then in affording fosterage and protection during the early stages of growth, there is, as might well be supposed, a reduction in the number of ova produced and of young brought forth or hatched. Many of the lower organisms lay hundreds of thousands of eggs, each of which produces a living active embryo. The condor has but two downy fledglings in a year; the gannet lays annually but a single egg; while the elephant, in the hundred years of its life, brings forth but half a dozen young.

We shall have to consider by what means these opposite tendencies (a tendency to produce enormous numbers of tender, ill-equipped embryos, and a tendency to produce few well-equipped offspring) have been emphasized. The point now to be noted is that every organism, even the slowest breeder that exists, produces more young than are sufficient to keep up the numbers of the species. If every pair of organisms gave birth to a similar pair, and if this pair survived to do likewise, the number of individuals in the species would have no tendency either to increase or to diminish. But, as a matter of fact, animals actually do produce from three or four times to hundreds or even thousands of times as many new individuals as are necessary in this way to keep the numbers constant. This is the *law of increase*. It may be thus stated: *The number of individuals in every race or species of animals is tending to increase.* Practically this is only a tendency. By war, by struggle, by competition, by the preying of animals upon each other, by the stress of external circumstances, the numbers are thinned down, so that, though the births are many, the deaths are many also, and the survivals few. In the case of those species the numbers of which are remaining constant, out of the total number born only two survive to procreate their kind. We may judge, then, of the amount of extermination that goes on among those animals which produce embryos by the thousand or even the hundred thousand. The effects of this enormous death-rate on the progress of the race or species we shall have to consider in the next chapter, when the question of the differentiation of species is before us.

There is one form of differentiation, however, which we may glance at before closing this chapter—the differentiation of sex. We are not in a position to discuss the ultimate causes of sex-differentiation, but we may here note the proximate causes as they seem to be indicated in certain cases.

Among honey-bees there are males (drones), fertile females (queens), and imperfect or infertile females (workers). It has now been shown, beyond question, that the eggs from which drones develop are not fertilized. The presence or absence of fertilization in this case determines the sex. During the nuptial flight, a special reservoir, possessed by the queen bee, is stored with sperms in sufficient number to last her egg-laying life. It is in her power either to fertilize the eggs as they are laid or to withhold fertilization. If the nuptial flight is prevented, and the reservoir is never stored with sperms, she is incapable of laying anything but drone eggs. The cells in which drones are developed are somewhat smaller than those for ordinary workers; but what may be the nature of the stimulus that prompts the queen to withhold fertilization we at present do not know. The difference between the fertile queen and the unfertile worker seems to be entirely a matter of nutrition. If all the queen-embryos should die, the workers will tear down the partitions so as to throw three ordinary worker-cells into one; they will destroy two of the embryos, and will feed the third on highly nutritious and stimulating diet; with the result that the ovaries and accessory parts are fully developed, and the grub that would have become an infertile worker becomes a fertile queen. And one of the most interesting points about this change, thus wrought by a stimulating diet, is that not only are the reproductive powers thus stimulated, but the whole organism is modified. Size, general structure, sense-organs, habits, instincts, and character are all changed with the development of the power of laying eggs. The organism is a connected whole, and you cannot modify one part without deeply influencing all parts. This is the *law of correlated variation*.

Herr Yung has made some interesting experiments on tadpoles. Under normal circumstances, the relation of females to males is about 57 to 43. But when the tadpoles were well fed on beef, the proportion of females to males rose so as to become 78 to 32; and on the highly nutritious flesh of frogs the proportion became 92 to 8. A highly nutritious diet and plenty of it caused a very large preponderance of females.

Mrs. Treat, in America, found that if caterpillars were half-starved before entering upon the chrysalis state, the proportion of males was much increased; while, if they were supplied with abundant nutritious food, the proportion of female insects was thereby largely increased. The same law is said to hold good for mammals. Favourable vital conditions are associated with the birth of females; unfavourable, with that of males. Herr Ploss attempts to show that, among human folk, in hard times there are more boys born; in good times, more girls.

On the whole, we may say that there is some evidence to show that in certain cases favourable conditions of temperature, and especially nutrition, tend to increase the number of females. We have seen that many animals pass through a stage where the reproductive organs are not yet differentiated into male and female, while in some there is a temporary stage where the outer parts of the organ produce ova and the inner parts sperms. We have also seen that the ova are cells where storage is in excess; the sperms are cells in which fission is in excess. Favourable nutritive conditions may, therefore, not incomprehensibly lead to the formation of well-stored ova; unfavourable nutritive conditions, on the other hand, to the formation of highly subdivided sperms. By correlated variation,* the ova-bearing or sperm-bearing individuals then develop into the often widely different males and females.

* We have seen that when volume tends to outrun surface, fission may take place, whereby the same volume has increased surface. But in unfavourable nutritive conditions, the same surface which had before been sufficient for nutrition may become, under the less favourable circumstances, insufficient, and fission may again take place to give a larger absorbent surface. Hence, possibly, the connection between insufficient nutriment and highly subdivided sperms.

CHAPTER IV.

VARIATION AND NATURAL SELECTION.

EVERYTHING, so far as in it lies, said Benedict Spinoza, tends to persist in its own being. This is *the law of persistence.* It forms the basis of Newton's First Law of Motion, which enunciates that, if a body be at rest, it will remain so unless acted on by some external force; or, if it be in motion, it will continue to move in the same straight line and at a uniform velocity unless it is acted on by some external force. Practically every known body is thus affected by external forces; but the law of persistence is not thereby disproved. It only states what would happen under certain exceptional or perhaps impossible circumstances. To those ignorant of scientific procedure, it seems unsatisfactory, if not ridiculous, to formulate laws of things, not as they are, but as they might be. Many well-meaning but not very well-informed people thus wholly misunderstand and mistake the value of certain laws of political economy, because in those laws (which are generalized statements of fact under narrowed and rigid conditions, and do not pretend to be inculcated as rules of conduct) benevolence, sentiment, even moral and religious duty, are intentionally excluded. These laws state that men, under motives arising out of the pursuit of wealth, will act in such and such a way, unless benevolence, sentiment, duty, or some other motive, lead them to act otherwise. Such laws, which hold good, not for phenomena in their entirety, but for certain isolated groups of facts under narrowed conditions, are called *laws of the factors* of phenomena. And since the complexity of phenomena is such that it is difficult for the human mind to grasp all the interlacing threads of causation at a single glance, men of science have endeavoured to isolate their several strands, and, applying the principle of analysis, without which reasoning is impossible, to separate out the factors and determine their laws. In this chapter we have to consider some of the factors of organic progress, and endeavour to determine their laws.

The law of heredity may be regarded as that of persistence exemplified in a series of organic generations. When, as in the amœba and some other protozoa, reproduction is by simple fission, two quite similar organisms being thus produced, there would seem to be no reason why (modifications by surrounding circumstances being disregarded) hereditary persistence should not continue indefinitely. Where, however, reproduction is effected by the detachment of a single cell from a many-celled organism, hereditary persistence * will be complete only on the condition that this reproductive cell is in some way in direct continuity with the cells of the parent organism or the cell from which that parent organism itself developed. And where, in the higher animals, two cells from two somewhat different parents coalesce to give origin to a new individual, the phenomena of hereditary persistence are still further complicated by the blending of characters handed on in the ovum and the sperm; still further complication being, perhaps, produced by the emergence in the offspring of characters latent in the parent, but derived from an earlier ancestor. And if characters acquired by the parents in the course of their individual life be handed on to the offspring, yet further complication will be thus introduced.

It is no matter for surprise, therefore, that, notwith-

---

* Samuel Butler in England, and Ewald Hering in Prague, have ingeniously likened this hereditary persistence to "organic memory." What are ordinarily called memory, habit, instinct, and embryonic reconstruction, are all referable to the memory of organic matter. The analogy, if used with due caution, is a helpful one, what we call memory being the psychical aspect (under certain special organic and neural conditions) of what under the physical aspect we call persistence.

standing the law of hereditary persistence, variations should occur in the offspring of animals. At the same time, it must be remembered that the occurrence of variations is not and cannot be the result of mere chance; but that all such variations are determined by some internal or external influences, and are thus legitimate and important subjects of biological investigation. In the next chapter we shall consider at some length the phenomena of heredity and the origin of variations. Here we will accept them without further discussion, and consider some of their consequences. But even here, without discussing their origin, we must establish the fact that variations do actually occur.

Variations may be of many kinds and in different directions. In colour, in size, in the relative development of different parts, in complexity, in habits, and in mental endowments, organisms or their organs may vary. Observers of mammals, of birds, and of insects are well aware that colour is a variable characteristic. But these colour-variations are not readily described and tabulated. In the matter of size the case is different. In Mr. Wallace's recent work on "Darwinism" a number of observations on size-variations are collected and tabulated. As this is a point of great importance, I propose to illustrate it somewhat fully from some observations I have recently made of the wing-bones of bats. In carrying out these observations and making the necessary measurements, I have had the advantage of the kind co-operation of my friend Mr. Henry Charbonnier, of Clifton, an able and enthusiastic naturalist.*

The nature of the bat's wing will be understood by the aid of the accompanying figure (Fig. 12). In the fore limb the arm-bone, or humerus, is followed by an elongated bone composed of the radius and ulna. At the outer end of the radius is a small, freely projecting digit, which carries a claw. This answers to the thumb. Then follow four long, slender bones, which answer to the bones in the

* I have also to thank Mr. Edward Wilson for kindly giving me the measurements of three or four bats in the Bristol Museum.

palm of our hand. They are the metacarpals, and are numbered II., III., IV., and V. in the tabulated figures in which the observations are recorded. The metacarpals of the second and third digits run tolerably close together, and form the firm support of the anterior margin of the wing.

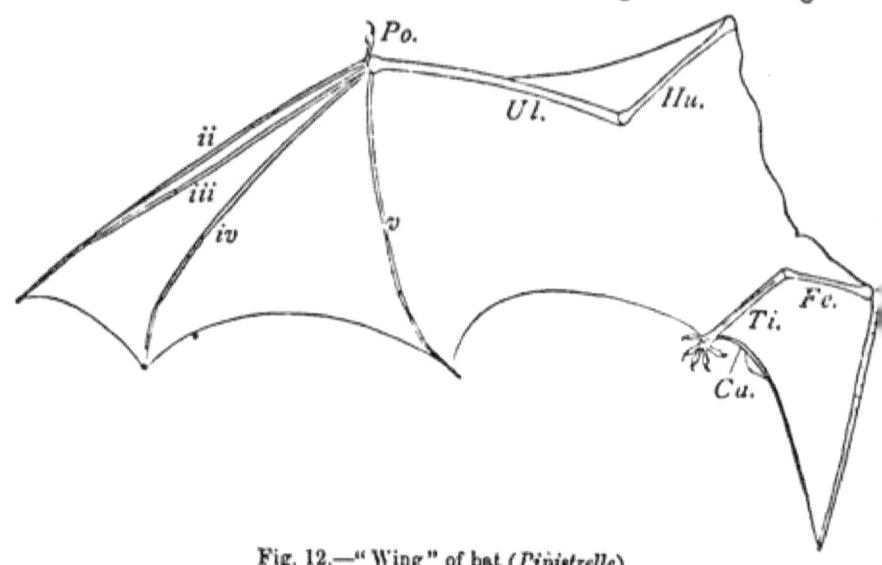

Fig. 12.—" Wing" of bat (*Pipistrelle*).

Hu., humerus, or arm-bone; *Ul.*, conjoined radius and ulna, a bone in the forearm; *Po.*, pollex, answering to our thumb; II., III., IV., V., second, third, fourth, and fifth digits of the manus, or hand. The figures are placed near the metacarpals, or palm-bones. These are followed by the phalanges. *Fe.*, femur or thigh-bone; *Ti.*, tibia, the chief bone of the shank. The digits of the pes, or foot, are short and bear claws. *Ca.*, calcar.

Those of the third and fourth make a considerable angle with these and with each other, and form the stays of the mid part of the wing. Beyond the metacarpals are the smaller joints or phalanges of the digits, two or three to each digit. The third digit forms the anterior point or apex of the wing. The fourth and fifth digits form secondary points behind this. Between these points the wing is scalloped into bays.

From the point of the fifth or last digit the leathery wing membrane sweeps back to the ankle. The bones of the hind limb are the femur, or thigh-bone, and the tibia (with a slender, imperfectly developed fibula). There are five toes, which bear long claws. From the ankle there runs backward a long, bony and gristly spur, which serves to support the membrane which stretches from the ankle to the tip (or near the tip) of the tail.

Thus the wing of the bat consists of a membrane stretched on the expanded or spread fingers of the hand, and sweeping from the point of the little finger to the ankle. Behind the ankle there is a membrane reaching to the tip of the tail. This forms a sort of net in which some bats, at any rate, as I have myself observed, can catch insects.

I have selected the wing of the bat to exemplify variation, (1) because the bones are readily measured even in dried specimens; (2) because they form the mutually related parts of a single organ; and (3) because they offer facilities for the comparison of variations, not only among the individuals of a single species, but also among several distinct species.

The method employed has been as follows: The several bones have been carefully measured in millimetres,* and all the bones tabulated for each species. Such tables of figures are here given in a condensed form for three species of bats.

BAT-MEASUREMENTS (IN MILLIMETRES).

	Radius and Ulna.	Pollex.	2ND DIGIT. Metacarpal.	THIRD DIGIT. Metacarpal.	Phalange 1.	Phalanges 2, 3.	FOURTH DIGIT. Metacarpal.	Phalange 1.	Phalanges 2, 3.	FIFTH DIGIT. Metacarpal.	Phalange 1.	Phalanges 2, 3.	Tibia.	
Hairy-armed bat (*Vesperugo leisleri*).	41	6·5	38	40	16	19	38	14	7	32	8	7	16	♂
	41	6	38	40	16	19	39	15·5	7	33	8	6·5	16	,,
	41	6	39	40	16	18	39	16	6·5	33	8	7	16	,,
	41·5	5	39	40·5	17	20	39	16	7	33	8	7	15	♀
	40	6	39	37	15·5	18	37	14·5	7	32	8	6·5	15	♀
	41	5·5	38·5	39	16·5	20	39	15	7·5	33	8	7·5	17	,,
	41	6	39	40	15·5	20·5	39	15·5	7	33	8	7	16	,,
Horseshoe bat (*Rhinolophus ferri-equinum*).	51	5	39	36	19	29	40	11	18	40	13	15	22	♂
	54	5	40	36	18	32	40	11	19	40	14	16	23	♀
	52	5	39	36	18	31	39	10	19	40	13	14	23	,,
	54	5	39	36	18	32	40	11	17	40	13	13	25	,,
	46	5	36	34	16	29	36	10	19	36	13	17	22	?
Lesser horseshoe bat (*Rhinolophus hipposideros*).	34	4	25	23	12	17	26	6·5	12	26	9	13	17	♂
	37	3	26	24	13	20	28	8	13	24	9	14	17	,,
	35	3	26	24·5	13	17	27	7	12	26	10	12	15	,,

It would be troublesome to the reader to pick out the

* A millimetre is about 1/25 of an inch, or more exactly ·03937 inch.

meaning from these figures. I have, therefore, plotted in the measurements for four other species of bats in tabular form (Figs. 13, 14, 15, 16).

Fig. 13, for example, deals with the common large noctule bat, which may often be seen flying high up on summer evenings. Now, the mean length of the radius and ulna in eleven individuals was 51·5 millimetres. Suppose all the eleven bats had this bone (for the two bones form practically one piece) of exactly the same length. There would then be no variation. We may express this supposed uniformity by the straight horizontal line running across the part of the figure dealing with the radius and ulna. Practically the eleven bats measured did not have this bone of the same length; in some of them it was longer, in others it was shorter than the mean. Let us run through the eleven bats (which are represented by the numbers at the head of the table) with regard to this bone. The first fell below the average by a millimetre and a half, the length being fifty millimetres. This is expressed in the table by placing a dot or point three quarters of a division below the mean line. Each division on the table represents two millimetres, or, in other words, the distance between any two horizontal lines stands for two millimetres measured. Half a division, therefore, is equivalent to one measured millimetre; a quarter of a division to half a millimetre. The measurements are all made to the nearest half-millimetre. The second bat fell short of the mean by one millimetre. The bone measured 50·5 millimetres. The third exceeded the mean by a millimetre and a half; the fourth, by three millimetres and a half. The fifth was a millimetre and a half above the mean; and the sixth and seventh were both half a millimetre over the mean. The eighth fell short by half a millimetre; the ninth and tenth by a millimetre and a half; and the eleventh by two millimetres and a half. The points have been connected together by lines, so as to give a curve of variation for this bone.

The other curves in these four tables are drawn in exactly

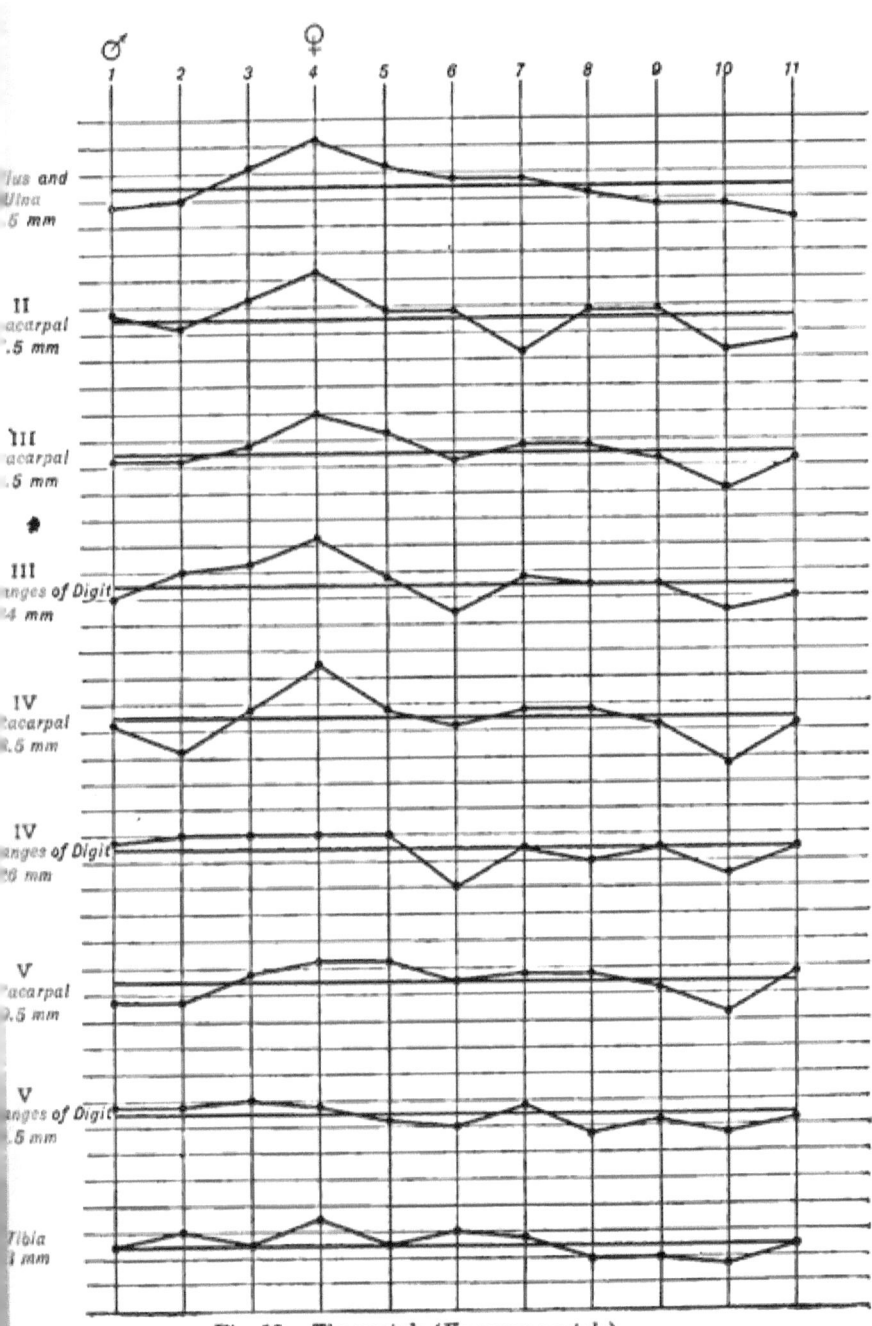

Fig. 13.—The noctule (*Vesperugo noctula*).

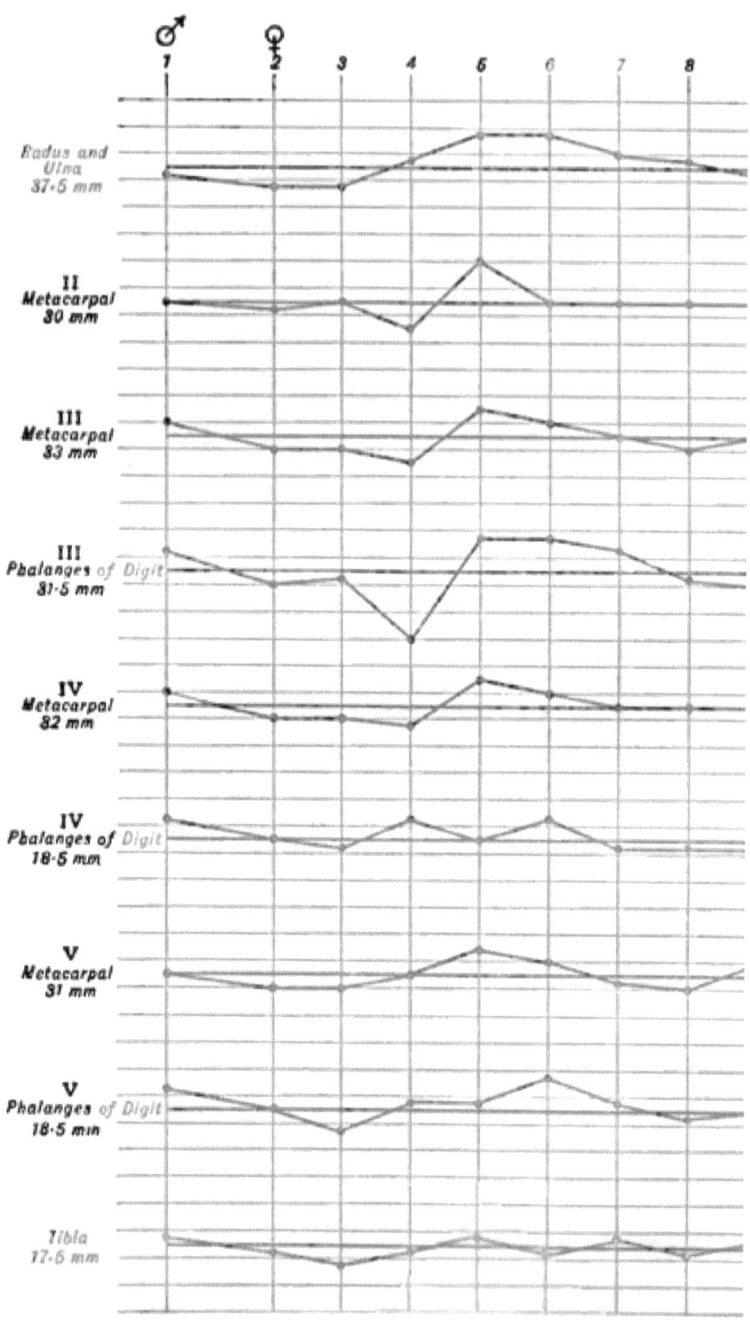

Fig. 14.—The long-eared bat (*Plecotus auritus*).

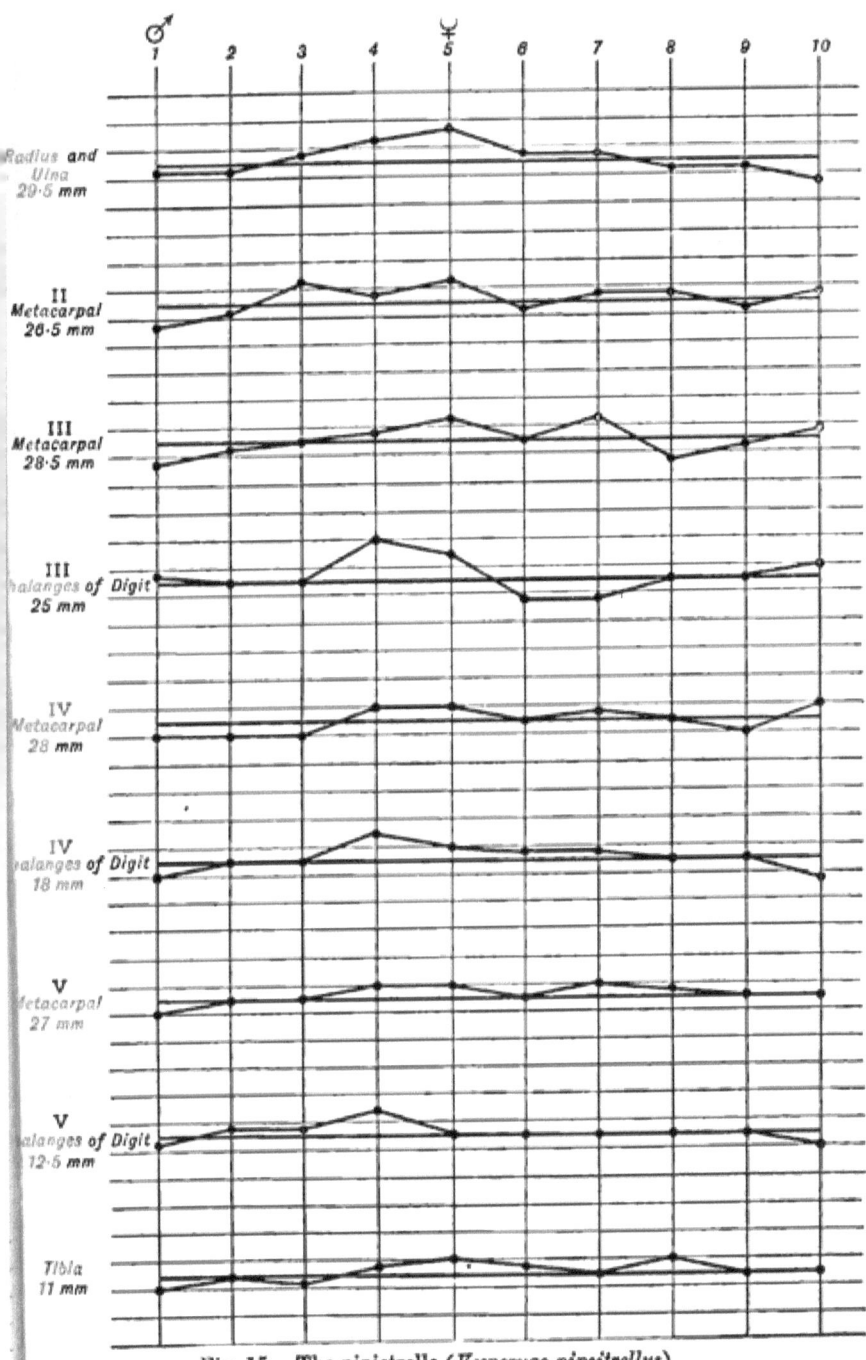

Fig. 15.—The pipistrelle (*Vesperugo pipsitrellus*).

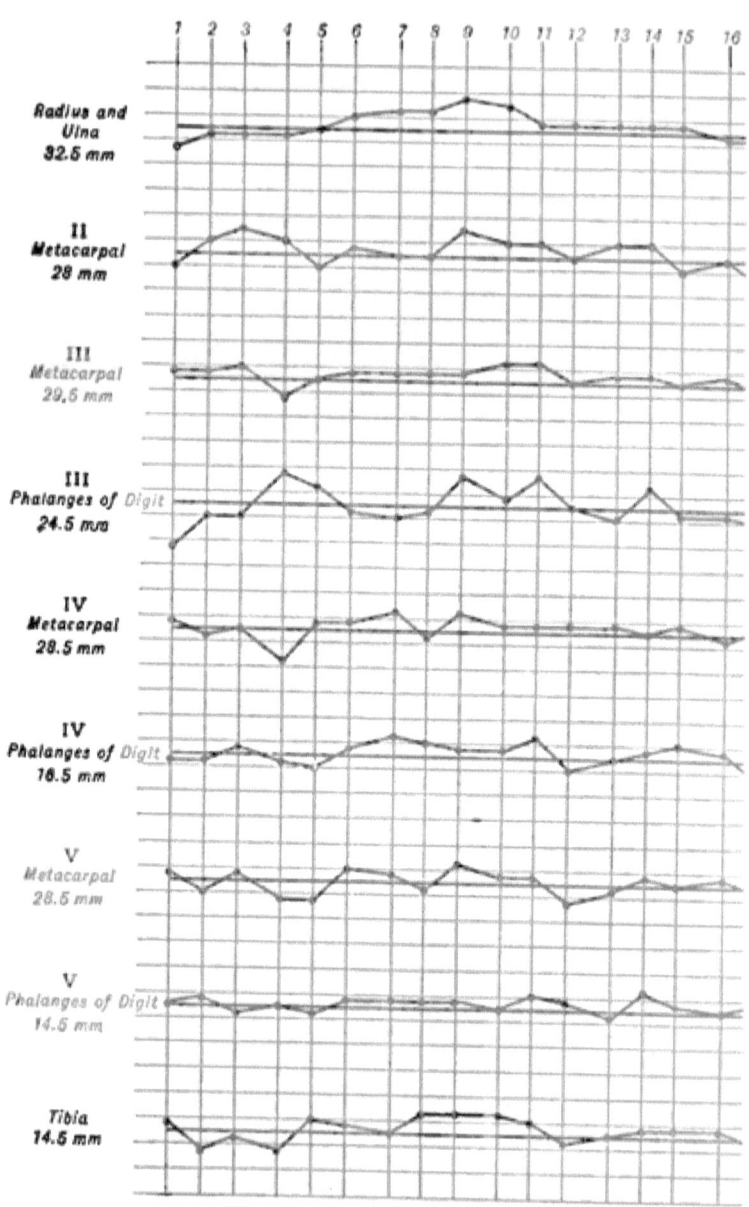

Fig. 16.—The whiskered bat (*Vespertilio mystacinus*).

the same way. The mean length is stated; and the amount by which a bone in any bat exceeds or falls short of the mean can be seen and readily estimated by means of the horizontal lines of the table. Any one can reconvert the tables into figures representing our actual measurements.

Now, it may be said that, since some bats run larger than others, such variation is only to be expected. That is true. But if the bones of the wing all varied equally, *all the curves would be similar.* That is clearly not the case. The second metacarpal is the same length in 5 and 6. But the third metacarpal is two millimetres shorter in 6 than in 5. In 10 the radius and ulna are *longer* than in 11; but the second metacarpal is *shorter* in 10 than in 11. A simple inspection of the table as a whole will show that there is a good deal of *independent* variation among the bones.

The amount of variation is itself variable, and in some cases is not inconsiderable. In the long-eared bats 4 and 5 in Fig. 14, the phalanges of the third digit measured 26·5 millimetres in 4, and 34 millimetres in 5—a difference of more than 28 per cent. This is unusually large, and it is possible that there may have been some slight error in the measurements.* A difference of 10 or 12 per cent. is, however, not uncommon.

In any case, the observations here tabulated show (1) that variations of not inconsiderable amount occur among the related bones of the bat's wing; and (2) that these variations are to a considerable extent independent of each other.

So far we have compared a series of individuals of the same species of bat, each table in Figs. 13-16 dealing with a distinct species. Let us now compare the different species with each other. To effect such a comparison, we must take some one bone as our standard, and we must level up our bats for the purposes of tabulation. I have selected the radius and ulna as the standard. In both the

---

* In nearly all cases the measurements were checked by comparing the two wings. In one or two instances there were differences of as much as two or three millimetres between the bones of the two sides of the body, but in most cases they exactly corresponded.

noctule and the greater horseshoe bats the mean length of this bone is 51·5 millimetres. The bones of each of the other bats have been multiplied by such a number as will bring them up to the level of size in these two species. Mr. Galton, in his investigations on the variations of human stature, had to take into consideration the fact that men are normally taller than women. He found, however, that the relation of man to woman, so far as height is concerned, is represented by the proportion 108 to 100. By multiplying female measurements by 1·08, they were brought up to the male standard, and could be used for purposes of comparison. In the same way, by multiplying in each case by the appropriate number, I have brought all the species in the table (Fig. 17) up to the standard of the noctule. When so multiplied, the radius and ulna (selected as the standard of comparison) has the same length in all the species, and is hence represented by the horizontal line in the table.

Compared with this as a standard, the mean length of the second metacarpal in the seven species is forty-three millimetres; that of the third metacarpal, forty-four millimetres; and so on. The amount by which each species exceeds or falls short of the mean is shown on the table, and the points are joined up as before. Here, again, the table gives the actual measurements in each case. For example, if the mean length of the third metacarpal of the greater horseshoe bat be required, it is seen by the table to fall short of the mean by four horizontal divisions and a quarter, that is to say, by eight millimetres and a half. The length is therefore $(44 - 8\frac{1}{2})$ 35·5 millimetres.

Now, it will be seen from the table that the variation in the mean length of the bones in different species is much greater than the individual variations in the members of the same species. The table also brings out in an interesting way the variation in the general character of the wing. The noctule, for example, is especially strong in the development of the second and third metacarpals, the phalanges of the third digit being also a little above the

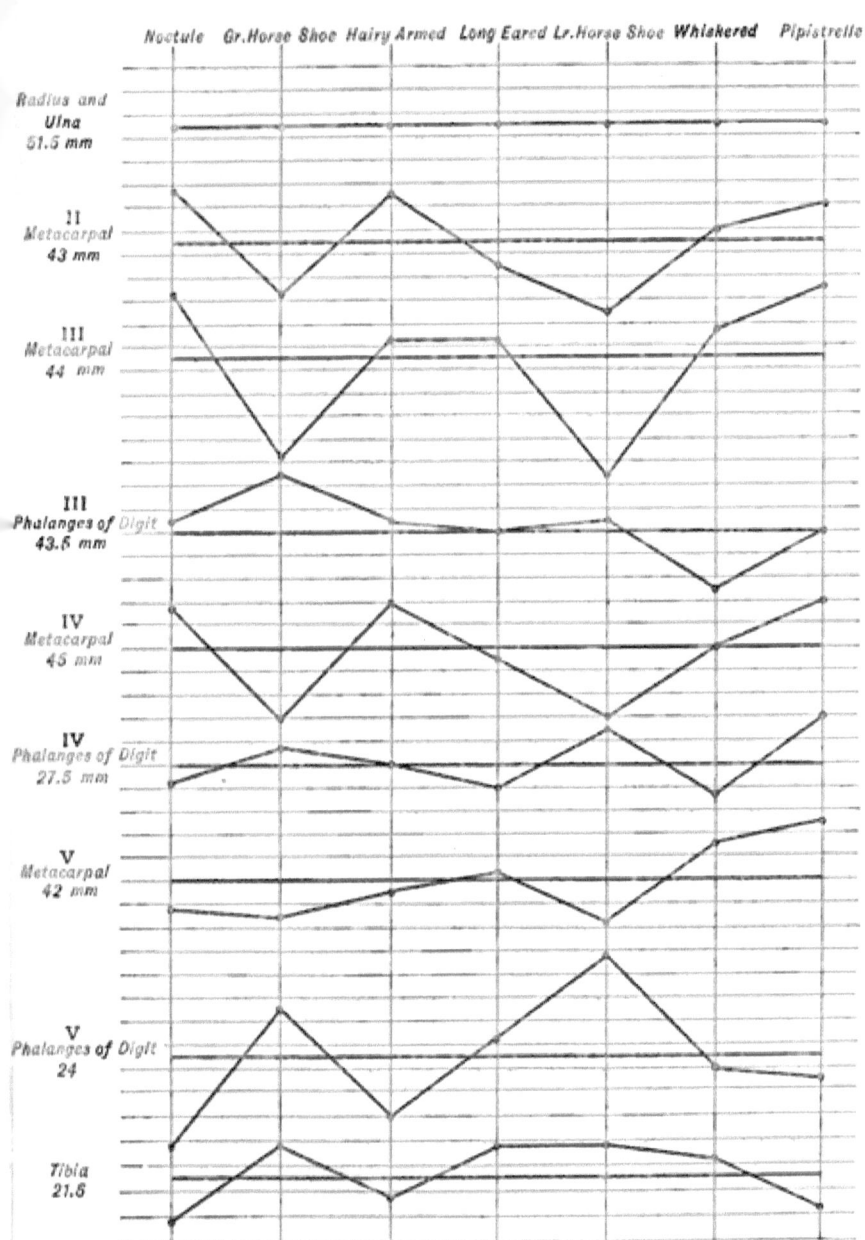

Fig. 17.—Variations adjusted to the standard of the noctule.

average. Reference to the figure of the bat's wing on p. 64 will show that these excellences give length to the wing. It fails, however, in the metacarpal and phalanges of the fifth digit, and in the length of the hind leg as represented by the tibia. On consulting the figure of the wing, it is seen that these are the bones which give breadth to the wing. Here the noctule fails. Its wing is, therefore, long and narrow. It is a swallow among bats.

On the other hand, the horseshoe bats fail conspicuously in the second and third metacarpals, though they make up somewhat in the corresponding digits. On the whole, the wing is deficient in length. But the phalanges of the fourth and fifth digits, and the length of the hind limb represented by the tibia, give a corresponding increase of breadth. The wing is, therefore, relatively short and broad. The long-eared bat, again, has the third metacarpal and its digits somewhat above the mean, and therefore a somewhat more than average length. But it has the fifth metacarpal with its digit and also the tibia decidedly above the mean, and therefore more than average breadth. Without possessing the great length of the noctule's wing, or the great breadth of that of the horseshoe, it still has a more than average length and breadth.

The total wing-areas are very variable, the females having generally an advantage over the males. I do not feel that our measurements are sufficiently accurate to justify tabulation. Taking, however, the radius and ulna as the standard for bringing the various species up to the same level, the greater horseshoe seems to have decidedly the largest wing-area; the noctule stands next; then come the lesser horseshoe and the long-eared bat; somewhat lower stands the hairy-armed bat; while the pipistrelle and the whiskered bat (both small species) stand lowest.*

Sufficient has now been said in illustration of the fact

---

* We are anxious to extend our observations and to compare series of bats from different localities. If any of my readers should feel disposed to help us, by sending specimens (*with the locality duly indicated*) to Mr. H. Charbonnier, 7, The Triangle South, Clifton, Bristol, we shall be grateful.

that variations in the lengths of the bones in the bat's wing do actually occur in the various individuals of one species; that the variations are independent; and that the different species and genera have the character of the wing determined by emphasizing, so to speak, variations in special directions. I make no apology for having treated the matter at some length. Those who do not care for details will judiciously exercise their right of skipping.

As before mentioned, Mr. Wallace has collected and tabulated other observations on size and length variations. And in addition to such variations, there are the numerous colour-variations that do not admit of being so readily tabulated. Mr. Cockerell tells us that among snail-shells, taking variations of banding alone, he knows of 252 varieties of *Helix nemoralis* and 128 of *H. hortensis*.*

That variations do occur under nature is thus unquestionable. And it is clear that all variations necessarily fall under one of three categories. Either they are of advantage to the organism in which they occur; or they are disadvantageous; or they are neutral, neither advantageous nor disadvantageous to the animal in its course through life.

We must next revert to the fact to which attention was drawn in the last chapter, that every species is tending, through natural generation, to increase in numbers. Even in the case of the slow-breeding elephant, the numbers tend to increase threefold in each generation; for a single pair of elephants give birth to three pairs of young. In many animals the tendency is to increase ten, twenty, or thirtyfold in every generation; while among fishes, amphibians, and great numbers of the lower organisms, the tendency is to multiply by a hundredfold, a thousandfold, or even in some cases ten thousandfold. But, as before noticed, this is only a tendency. The law of increase is a law of one factor in life's phenomena, the reproductive factor. In any

---

* *Nature*, vol. xli. p. 393. The variation in molluscs is often considerable. In one of the bays in the basement hall of the Natural History Museum is a series showing the variation in size, form, and sculpturing of *Paludomus loricatus*, which is found in the streams of Ceylon. These varieties have in former times been named as ten distinct species!

area, the conditions of which are not undergoing change, the numbers of the species which constitute its fauna remain tolerably constant. They are not actually increasing in geometrical progression. There is literally no room for such increase. The large birth-rate of the constituent species is accompanied by a proportionate death-rate, or else the tendency is kept in check by the prevention of certain individuals from mating and bearing young.*

Now, the high death-rate is, to a large extent among the lower organisms and in a less degree among higher animals, the result of indiscriminate destruction. When the ant-bear swallows a tongue-load of ants, when the Greenland whale engulfs some hundreds of thousands of fry at a gulp, when the bear or the badger destroys whole nests of bees,—in such cases there is wholesale and indiscriminate destruction. Those which are thus destroyed are nowise either better or worse than those which escape. At the edge of a coral reef minute, active, free-swimming coral embryos are set free in immense numbers. Presently they settle down for life. Some settle on a muddy bottom, others in too great a depth of water. These are destroyed. The few which take up a favourable position survive. But they are no better than their less fortunate neighbours. The destruction is indiscriminate. So, too, among fishes

---

* More observations and fuller knowledge on this latter point and on the relative numbers of the sexes in different species are much to be desired. It is clear that the number of offspring mainly depends upon the number of females. But if it be true that good times and favourable conditions lead to an increased production of females, while hard times and unfavourable conditions lead to a relative increase of males, then it is evident that good times will lead to a more rapid increase and hard times to a less rapid increase of the species. Suppose, for example, in a particular district food and other conditions were especially favourable for frogs. Among the well-nourished tadpoles there would be a preponderance of females. In the next generation the many females would produce abundant offspring (for one male may fertilize the ova laid by several females). There would be a greater number of tadpoles to compete for the same amount of nutriment. They would be less nourished. There would be less females; and in the succeeding generation a diminished number of tadpoles. Thus to some extent a balance between the number of tadpoles and the amount of available nutrition would be maintained. These conclusions are, perhaps, too theoretical to be of much value, while the tendency here indicated would be but one factor among many.

and the many marine forms which produce a great number of fertilized eggs giving rise to embryos that are from an early period free-swimming and self-supporting. Such embryos are decimated by a destruction which is quite indiscriminate. And again, to take but one more example, the liver-fluke, whose life-history was sketched in the last chapter, produces its tens or hundreds of thousands of ova. But the chances are enormously against their completing their life-cycle. If the conditions of temperature and moisture are not favourable, the embryo is not hatched or soon dies; even if it emerges, no further development takes place unless it chances to come in contact with a particular and not very common kind of water-snail. When it emerges from the intermediate host and settles on a blade of grass, it must still await the chance of that blade being eaten by a sheep or goat. It is said that the chances are eight millions to one against it, and for the most part its preservation is due to no special excellence of its own. The destruction is to a large extent, though not entirely, indiscriminate.

Even making all due allowance, however, for this indiscriminate destruction—which is to a large extent avoided by those higher creatures which foster their young—there remain more individuals than suffice to keep up the normal numbers of the species. Among these there arises a struggle for existence, and hence what Darwin named *natural selection*.

"How will the struggle for existence"—I quote, with some omissions, the words of Darwin—"act in regard to variation? Can the principle of selection, which is so potent in the hands of man, apply under nature? I think that we shall see that it can act most efficiently. Let the endless number of slight variations and individual differences be borne in mind; as well as the strength of the hereditary tendency. Let it also be borne in mind how infinitely complex and close-fitting are the mutual relations of all organic beings to each other and to their physical conditions of life; and consequently what infinitely varied diversities of structure might be of use to each being under changing conditions of life. Can it, then, be thought improbable, seeing that variations useful to man have undoubtedly occurred, that other variations, useful in some way to each being in the great and complex battle of life, should occur in the course of many successive generations? If such do occur, can we doubt (remembering that many more individuals are born than can possibly survive) that individuals having any advantage, however slight, over others, would have the best chance of surviving and of procreating their kind? On the other hand, we may feel sure that any variation in the least degree injurious would be rigidly destroyed. This preservation of favourable individual differences and variations, and the destruction of those which are injurious, I have called Natural Selection, or the Survival of the Fittest. Variations neither useful nor injurious would not be affected by natural selection, and would be left either a fluctuating element, or would ultimately become fixed, owing to the nature of the organism and the nature of the conditions." *

"The principle of selection," says Darwin, elsewhere, "may conveniently be divided into three kinds. *Methodical selection* is that which guides a man who systematically endeavours to modify a breed according to some predetermined standard. *Unconscious selection* is that which follows from men naturally preserving the most valued and destroying the less valued individuals, without any thought of altering the breed. Lastly, we have *Natural selection*, which implies that the individuals which are best fitted for the complex and in the course of ages changing conditions to which they are exposed, generally survive and procreate their kind." †

I venture to think that there is a more logical division than this. A man who is dealing with animals or plants under domestication may proceed by one of two well-contrasted methods. He may either select the most satisfac-

---

* "Origin of Species," pp. 62, 63.
† "Animals and Plants under Domestication," vol. ii. p. 177.

tory individuals or he may reject the most unsatisfactory. We may term the former process *selection*, the latter *elimination*. Suppose that a gardener is dealing with a bed of geraniums. He may either pick out first the best, then the second best, then the third, and so on, until he has selected as many as he wishes to preserve. Or, on the other hand, he may weed out first the worst, then in succession other unsatisfactory stocks, until, by eliminating the failures, he has a residue of sufficiently satisfactory flowers. Now, I think it is clear that, even if the ultimate result is the same (if, that is to say, he selects the twenty best, or eliminates all but the twenty best), the method of procedure is in the two cases different. Selection is applied at one end of the scale, elimination at the other. There is a difference in method in picking out the wheat-grains (like a sparrow) and scattering the chaff by the wind.

Under nature both methods are operative, but in very different degrees. Although the insect may select the brightest flowers, or the hen-bird the gaudiest or most tuneful mate, the survival of the fittest under nature is in the main the net result of the slow and gradual process of the elimination of the unfit.* The best-adapted are not, save in exceptional cases, selected; but the ill-adapted are weeded out and eliminated. And this distinction seems to me of sufficient importance to justify my suggestion that *natural selection* be subdivided under two heads—*natural elimination*, of widespread occurrence throughout the animal world; and *selection proper*, involving the element of individual or special choice.

The term "natural elimination" for the major factor serves definitely to connect the natural process with that struggle for existence out of which it arises. The struggle for existence is indeed the reaction of the organic world called forth by the action of natural elimination. Organisms are tending to increase in geometrical ratio. There is not

---

* I may here draw attention to the fact that the bats whose wing-bone measurements were given above are those which have so far survived and escaped such elimination as is now in progress.

room or subsistence for the many born. The tendency is therefore held in check by elimination, involving the struggle for existence. And the factors of elimination are three: first, elimination through the action of surrounding physical or climatic conditions, under which head we may take such forms of disease as are not due to living agency; secondly, elimination by enemies, including parasites and zymotic diseases; and thirdly, elimination by competition. It will be convenient to give some illustrative examples of each of these.

Elimination through the action of surrounding physical conditions, taken generally, deals with the very groundwork or basis of animal life. There are certain elementary mechanical conditions which must be fulfilled by every organism however situated. Any animal which fails to fulfil these conditions will be speedily eliminated. There are also local conditions which must be adequately met. Certain tropical animals, if transferred to temperate or sub-Arctic regions, are unable to meet the requirements of the new climatic conditions, and rapidly or gradually die. Fishes which live under the great pressure of the deep sea are killed by the expansion of the gases in their tissues when they are brought to the surface. Many fresh-water animals are killed if the lake in which they live be invaded by the waters of the sea. If the water in which corals live be too muddy, too cold, or too fresh—near the mouth of a great river on the Australian coast, for example—they will die off. During the changes of climate which preceded and followed the oncoming of the glacial epoch, there must have been much elimination of this order. Even under less abnormal conditions, the principle is operative. Darwin tells us that in the winter of 1854–5 four-fifths of the birds in his grounds perished from the severity of the weather, and we cannot but suppose that those who were thus eliminated were less able than others to cope with or stand the effects of the inclement climatic conditions. My colleague, Mr. G. Munro Smith, informs me that, in cultivating microbes, certain forms, such as *Bacillus violaceus* and *Micrococcus prodigiosus*, remain in the field during cold weather when other less hardy microbes have perished. The insects of Madeira may fairly be regarded as affording another instance. The ground-loving forms allied to insects of normally slow and heavy flight have in Madeira become wingless or lost all power of flight. Those which attempted to fly have been swept out to sea by the winds, and have thus perished; those which varied in the direction of diminished powers of flight have survived this eliminating process. On the other hand, among flower-frequenting forms and those whose habits of life necessitate flight, the Madeira insects have stronger wings than their mainland allies. Here, since flight could not be abandoned without a complete change of life-habit, since all must fly, those with weaker powers on the wing have been eliminated, leaving those with stronger flight to survive and procreate their kind.[*] In Kerguelen Island Mr. Eaton has found that all the insects are incapable of flight, and most of them in a more or less wingless condition.[†] Mr. Wallace regards the reduction in the size of the wing in the Isle of Man variety of the small tortoiseshell butterfly as due to the gradual elimination of larger-winged individuals.[‡] These are cases of elimination through the direct action of surrounding physical conditions. Even among civilized human folk, this form of elimination is still occasionally operative—in military campaigns, for example (where the mortality from hardships is often as great as the mortality from shot or steel), in Arctic expeditions, and in arduous travels. But in early times and among savages it must be a more important factor.

Elimination by enemies needs somewhat fuller exemplification. Battle within battle must, throughout nature, as Darwin says, be continually recurring with varying success. The stronger devour the weaker, and wage war with each other over the prey. In the battle among coordinates the weaker are eliminated, the stronger prevail.

---

[*] "Origin of Species," p. 109.   [†] "Darwinism," p. 106.
[‡] Ibid. p. 106.

When the weaker are preyed upon by the stronger and a fair fight is out of the question, the slow and heavy succumb, the agile and swift escape; stupidity means elimination, cunning, survival; to be conspicuous, unless it be for some nasty or deleterious quality, is inevitably to court death: the sober-hued stand at an advantage. In these cases, if there be true selection at work, it is the selection of certain individuals, the plumpest and most toothsome to wit, for destruction, not for survival.

This mode of elimination has been a factor in the development of protective resemblance and so-called mimicry, and we may conveniently illustrate it by reference to these qualities. If the hue of a creature varies in the direction of resemblance to the normal surroundings, it will render the animal less conspicuous, and therefore less liable to be eliminated by enemies. This is well seen in the larvæ or caterpillars of many of our butterflies and moths. It is not easy to distinguish the caterpillar of the clouded yellow, so closely does its colour assimilate to the clover leaves on which it feeds, nor that of the Lulworth skipper on blades of grass. I would beg every visitor to the Natural History Museum at South Kensington to look through the drawers containing our British butterflies and moths and their larvæ, in the further room on the basement, behind the inspiring statue of Charles Darwin. Half an hour's inspection will serve to bring home the fact of protective resemblance better than many words.

It may,' however, be remarked that not all the caterpillars exhibit protective resemblance; and it may be asked—How have some of these conspicuous larvæ, that of the magpie moth, for example, escaped elimination? What is sauce for the Lulworth goose should be sauce for the magpie gander. How is it that these gaudy and variable caterpillars, cream-coloured with orange and black markings, have escaped speedy destruction? Because they are so nasty. No bird, or lizard, or frog, or spider would touch them. They can therefore afford to be bright-coloured. Nay, their very gaudiness is an advantage, and saves them from being the subject of unpleasant experiments in the matter. Other caterpillars, like the palmer-worms, are protected by barbed hairs that are intensely irritating. They, too, can afford to be conspicuous. But a sweet and edible caterpillar, if conspicuous, is eaten, and thus by the elimination of the conspicuous the numerous dull green or brown larvæ have survived.

A walk through the Bird Gallery in the National collection will afford examples of protective resemblance among birds. Look, for example, at the Kentish plover with its eggs and young—faithfully reproduced in our frontispiece—and the way in which the creature is thus protected in early stages of its life will be evident. The stone-curlew, the ptarmigan, and other birds illustrate the same fact, which is also seen with equal clearness in many mammals, the hare being a familiar example.

Many oceanic organisms are protected through general resemblance. Some, like certain medusæ, are transparent. The pellucid or transparent sole of the Pacific (*Achirus pellucidus*), a little fish about three inches long, is so transparent that sand and seaweed can be seen distinctly through its tissues. The salpa is transparent save for the intestine and digestive gland, which are brown, and look like shreds of seaweed. Other forms, like the physalia, are cærulean blue. The exposed parts of flat-fish are brown and sandy coloured or speckled like the sea-bottom; and in some the sand-grains seem to adhere to the skin. So, too, with other fish. "Looking *down* on the dark back of a fish," says Mr. A. R. Wallace, "it is almost invisible, while to an enemy looking *up* from below, the light under surface would be equally invisible against the light of clouds and sky." Even some of the most brilliant and gaudiest fish, such as the coral-fish (*Chætodon*, *Platyglossus*, and others), are brightly coloured in accordance with the beautiful tints of the coral-reefs which form their habitat; the bright-green tints of some tropical forest birds being of like import. No conception of the range of protective resemblance can be formed when the creatures are seen or figured isolated from their surroundings. The zebra is a sufficiently conspicuous animal in a menagerie or a museum; and yet Mr. Galton assures us that, in the bright starlight of an African night, you may hear one breathing close by you, and be positively unable to see the animal. A black animal would be visible; a white animal would be visible; but the zebra's black and white so blend in the dusk as to render him inconspicuous.

To cite but one more example, this time from the invertebrates. Professor Herdman found in a rock-pool on the west coast of Scotland " a peculiarly coloured specimen of the common sea-slug (*Doris tuberculata*). It was lying on a mass of volcanic rock of a dull-green colour, partially covered with rounded spreading patches of a purplish pink nullipore, and having numerous whitish yellow *Spirorbis* shells scattered over it—the general effect being a mottled surface of dull green and pink peppered over with little cream-coloured spots. The upper surface of the Doris was of precisely the same colours arranged in the same way. . . . We picked up the Doris, and remarked the brightness and the unusual character of its markings, and then replaced it upon the rock, when it once more became inconspicuous." *

Then, too, there are some animals with variable protective resemblance—the resemblance changing with a changing environment. This is especially seen in some Northern forms, like the Arctic hare and fox, which change their colour according to the season of the year, being brown in summer, white and snowy in winter. The chamæleon varies in colour according to the hue of its surroundings through the expansion and contraction of certain pigment-cells; while frogs and cuttle-fish have similar but less striking powers. Mr. E. B. Poulton's † striking and beautiful experiments show that the colours of caterpillars

---

* Proceedings Liverpool Biological Society, 1889.

† Since this chapter was written, Mr. Poulton has published his interesting and valuable work on "The Colours of Animals," from which I have contrived to insert one or two additional examples.

and chrysalids reared from the same brood will vary according to the colour of their surroundings.

If this process of protective resemblance be carried far, the general resemblance in hue may pass into special resemblance to particular objects. The stick-insect and

Fig. 18.—Caterpillar of a moth (*Ennomos tiliaria*) on an oak-spray. (From an exhibit in the British Natural History Museum.)

the leaf-insect are familiar illustrations, though no one who has not seen them in nature can realize the extent of the resemblance. Most of us have, at any rate, seen the stick-caterpillars, or loopers (Fig. 18), though, perhaps, few have noticed how wonderful is the protective resemblance to a twig when the larva is still and motionless, for the very reason that the resemblance is so marked that the organism at that time escapes, not only casual observation, but even careful search. Fig. 19 gives a representation of a locust with special protective resemblance to a leaf—not a perfect leaf, but a leaf with fungoid blotches. This insect and the stick-caterpillar may be seen in the insect exhibits on the basement at South Kensington, having been figured from them by the kind permission of Professor Flower.

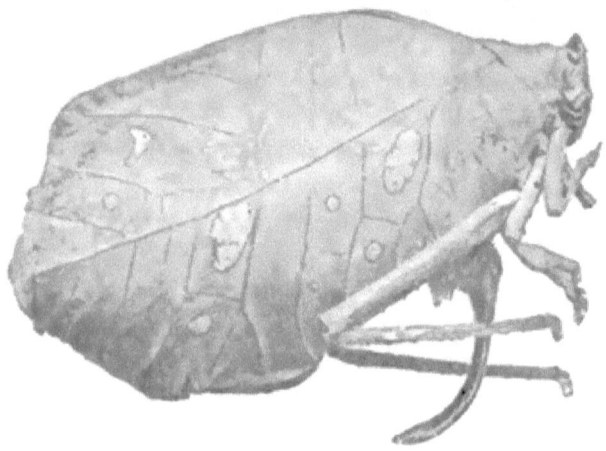

Fig. 19.—A locust (*Cycloptera speculata*) which closely resembles a leaf. (From an exhibit in the British Natural History Museum.)

Perhaps one of the most striking instances of special protective resemblance is that of the Malayan leaf-butterfly (*Kallima paralecta*). So completely, when the wings are closed, does this insect resemble a leaf that it requires a sharp eye to distinguish it. These butterflies have, moreover, the habit of alighting very suddenly. As a recent observer (Mr. S. B. T. Skertchly) remarks, they "fly rapidly along, as if late for an appointment, suddenly pitch, close their wings, and become leaves. It is generally done so rapidly that the insect seems to vanish."[*] Instances might

---

[*] *Ann. and Mag. Nat. Hist.*, September, 1889, p. 209, quoted by Poulton, "Colours of Animals," p. 55.

be multiplied indefinitely. Mr. Guppy thus describes a species of crab in the Solomon Islands: "The light purple colour of its carapace corresponds with the hue of the coral at the base of the branches, where it lives; whilst the light red colour of the big claws, as they are held up in their usual attitude, similarly imitates the colour of the branches. To make the guise more complete, both carapace and claws possess rude hexagonal markings which correspond exactly in size and appearance with the polyp-cells of the coral." *

When the special protective resemblance is not to an inanimate object, but to another organism, it is termed mimicry. It arises in the following way:—

Many forms, especially among the invertebrates, escape elimination by enemies through the development of offensive weapons (stings of wasps and bees), a bitter taste (the Heliconidæ among butterflies), or a hard external covering (the weevils among beetles). The animals which prey upon these forms learn to avoid these dangerous, nasty, or indigestible creatures; and the avoidance is often instinctive. It thus becomes an advantage to other forms, not thus protected, to resemble the animals that have these characteristics. Such resemblance is termed mimicry, concerning which it must be remembered that the mimicry is unconscious, and is reached by the elimination of those forms which do not possess this resemblance. Thus the *Leptalis*, a perfectly sweet insect, closely resembles the *Methona*, a butterfly producing an ill-smelling yellow fluid. The quite harmless *Clytus arietis*, a beetle, resembles, not only in general appearance, but in its fussy walk, a wasp. The soft-skinned *Doliops*, a longicorn, resembles the strongly encased *Pachyrhyncus orbifex*, a weevil. The not uncommon fly *Eristalis tenax* (Fig. 20), is not unlike a bee, and buzzes in an unpleasantly suggestive manner.†

* *Nature*, vol. xxxv. p. 77.

† Many other instances might be added. The hornet clear-wing moth (*Sphecia apiformis*) mimics the hornet or wasp; the narrow-bordered bee-hawk moth (*Sesia bombyliformis*) mimics a bumble-bee. These insects may be seen in the lepidoptera drawers in the Natural History Museum. But perhaps the most wonderful instance of insect-mimicry is that observed

Mimicry is not confined to the invertebrates. A harmless snake, the eiger-eter of Dutch colonists at the Cape, subsists mainly or entirely on eggs. The mouth is almost or quite toothless; but in the throat hard-tipped spines project into the gullet from the vertebræ of the column in this region. Here the egg is broken, and there is no fear of losing the contents. Now, there is one species of this snake that closely resembles the berg-adder. The head has naturally the elongated form characteristic of the harmless snakes. But when irritated, this egg-eater flattens it out till it has the usual viperine shape of the "club" on a playing-card. It coils as if for a spring, erects its head with every appearance of anger, hisses, and darts forward as if to strike its fangs into its foe, in every way imitating an enraged berg-adder. The snake is, however, quite harmless and inoffensive.*

Here we have mimicry both in form and habit. Another case of imperfect but no doubt effectual mimicry is given by Mr. W. Larden, in some notes from South America.† Speaking of the rhea, or South American ostrich, he says, "One day I came across an old cock in a nest that it had made in the dry weeds and grass. Its wings and feathers were loosely arranged, and looked not unlike a heap of dried grass; at any rate, the bird did not attract my attention until I was close on him. The long neck was stretched out close along the ground, the crest feathers were flattened, and an appalling hiss greeted my approach. It was a pardonable mistake if for a moment I thought I had come across a huge snake, and sprang back hastily under this impression."

Protective resemblance and mimicry have been con-

by Mr. W. L. Sclater, and given by Mr. E. B. Poulton, in his "Colours of Animals" (p. 252), where a (probably) homopterous insect mimics a leaf-cutting ant, *together with its leafy burden*—a membranous expansion in the mimic closely resembling the piece of leaf carried by the particular kind of ant he resembles.

* The late Mr. H. W. Oakley first drew my attention to this snake. Since then Mr. Hammond Tooke has described the facts in *Nature*, vol. xxxiv. p. 547.

† *Nature*, vol. xlii. p. 115.

sidered at some length because, on the hypothesis of natural selection, they admirably illustrate the results which may be reached through long-continued elimination by enemies.

Sufficient has now been said to show that this form of elimination is an important factor. We are not at present considering the question how variations arise, or why they should take any particular direction. But granting the fact that variations may and do occur in all parts of the organism, it is clear that, in a group of organisms surrounded by enemies, those individuals which varied in the direction of swiftness, cunning, inconspicuousness,* or resemblance to protected forms, would, other things being equal, stand a better chance of escaping elimination.

Elimination by competition is, as Darwin well points out, keenest between members of the same group and among individuals of the same species, or between different groups or different species which have, so to speak, similar aims in life. While enemies of various kinds are preying upon weaker animals, and thus causing elimination among them, they are also competing one with another for the prey. While the slower and stupider organisms are succumbing to their captors, and thus leaving more active and cunning animals in possession of the field, the slower and stupider captors, failing to catch their cunning and active prey, are being eliminated by competition. While protective resemblance aids the prey to escape elimination by enemies, a correlative resemblance, called by Mr. Poulton aggressive resemblance, in the captors aids them in stealing upon their prey, and so gives advantage in competition. Thus the hunting spider closely resembles the flies upon which he pounces, even rubbing his head with his fore legs after their innocent fashion.

---

* Since the above was written and sent to press, there has been added, at the Natural History Museum, in the basement hall, a case illustrating the adaptation of external colouring to the conditions of life. All the animals, birds, etc., there grouped were collected in the Egyptian desert, whence also the rocks, stones, and sand on which they are placed were brought. Though somewhat crowded, they exemplify protective resemblance very well.

As in the case of protective resemblance, so, too, in its aggressive correlative, the resemblance may be general or special, or may reach the climax of mimicry. And since the same organism is not only a would-be captor, but sometimes an unwilling prey, the same resemblance may serve to protect it from its enemies and to enable it to steal upon its prey. The mantis, for example, gains doubly by its resemblance to the vegetation among which it lives. Certain spiders, described by Mr. H. O. Forbes, in Java, closely resemble birds'-droppings. This may serve to protect them from elimination by birds; but it also enables them to capture without difficulty unwary butterflies, which are often attracted by such excreta. A parasitic fly (*Volucella bombylans*) closely resembles (Fig. 20) a bumble-bee (*Bombus muscorum*), and is thus enabled to enter the nest of the bee without molestation. Its larvæ feed upon the larvæ of the bee. The cuckoo bee *Psithyrus rupestris*, an idle queen, who collects no pollen, and has no pollen-baskets, steals into the nest of the bumble-bee *Bombus lapidarius*, and lays her eggs there. The resemblance between the two is very great, and it not only enables the mother bee to enter unmolested, but the young bees, when they are hatched, to escape. Another bee (*Nomada solidaginis*), which plays the cuckoo on *Halictus cylindricus*, does not resemble this bee, but is wasp-like, and thus escapes molestation, not because it escapes notice, but because it looks more dangerous than it really is.*

Many are the arts by which, in keen competition, organisms steal a march upon their congeners—not, be it remembered, through any conscious adaptation, but through natural selection by elimination. Mr. Poulton describes an Asiatic lizard (*Phrynocephalus mystaceus*) in which the "general surface resembles the sand on which it is found, while the fold of their skin at each angle of the mouth is of a red colour, and is produced into a flower-like shape

---

* I have to thank Mr. H. A. Francis for drawing my attention to this, and showing me the insects in his cabinet.

exactly resembling a little red flower which grows in the sand. Insects, attracted by what they believe to be flowers, approach the mouth of the lizard, and are, of course, captured."* The fishing frog, or angler-fish, is possessed of filaments which allure small fry, who think them worms, into the neighbourhood of the great mouth in which they are speedily engulfed; and certain deep-sea

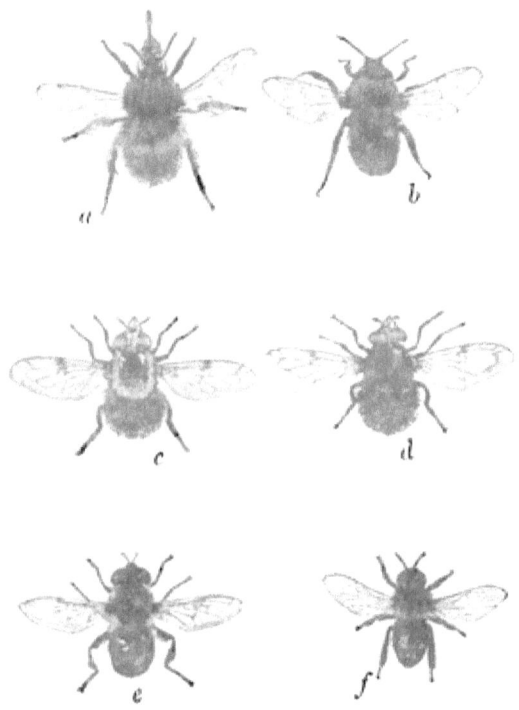

Fig. 20.—Mimicry of bees by flies.

a, b, *Bombus muscorum*; *c, d*, *Volucella bombylans*; *e*, *Eristalis tenax*; *f*, *Apis mellifica*. The underwings of the hive bee (*f*) were invisible in the photograph from which the figure was drawn. (From an exhibit in the British Natural History Museum.)

forms discovered during the *Challenger* expedition have the lure illumined by phosphorescent light.

We need say no more in illustration of the resemblances which have enabled certain organisms to escape elimination by competition. Once more, be it understood

* "Colours of Animals," p. 73.

that we are not at present considering *how* any of these resemblances have been brought about; we are merely indicating that, given certain resemblances, advantageous either for captor or prey, those organisms which possess them not will have to suffer elimination—elimination by enemies, or else elimination by competition.

The interaction between these two kinds of elimination is of great importance. Hunters and hunted are both, so to speak, playing the game of life to the best of their ability. Those who fail on either side are weeded out; and elimination is carried so far that those who are only as good as their ancestors are placed at a disadvantage as compared with their improving congeners. The standard of efficiency is thus improving on each side; and every improvement on the one side entails a corresponding advance on the other. Nor is there only thus a competition for subsistence, and arising thereout a gradual sharpening of all the bodily and mental powers which could aid in seeking or obtaining food; there is also in some cases a competition for mates, reaching occasionally the climax of elimination by battle. There is, indeed, competition for everything which can be an object of appetence to the brute intelligence; and, owing to the geometrical tendency in multiplication—the law of increase—the competition is keen and unceasing.

Such, then, in brief, are the three main modes of elimination: elimination by physical and climatic conditions; elimination by enemies; elimination by competition. Observe that it is a differentiating process. Unlike the indiscriminate destruction before alluded to, the incidence of which is on all alike, good, bad, and indifferent, it separates the well-adapted from the ill-adapted, dooming the latter to death, and allowing the former to survive and procreate their kind. The destruction is not indiscriminate, but differential.

Let us now turn to cases of selection, properly so called, where Nature is in some way working at the other end of the scale; where her method is not the elimination of the unfit, but the selection of the fit. Such a case may be found on Darwin's principles in brightly coloured flowers and fruits. "Flowers," he says, "rank amongst the most beautiful productions of nature; but they have been rendered conspicuous in contrast with the green leaves, and, in consequence, at the same time beautiful, so that they may be easily observed by insects. I have come to this conclusion from finding it an invariable rule that, when a flower is fertilized by the wind, it never has a gaily coloured corolla. Several plants habitually produce two kinds of flowers—one kind open and coloured, so as to attract insects; the other closed, not coloured, destitute of nectar, and never visited by insects. Hence we may conclude that, if insects had not been developed on the face of the earth, our plants would not have been decked with beautiful flowers, but would have produced only such poor flowers as we see on our fir, oak, nut, and ash trees, on grasses, spinach, docks, and nettles, which are all fertilized through the agency of the wind. A similar line of argument holds good with fruits; that a ripe strawberry or cherry is as pleasing to the eye as to the palate; that the gaily coloured fruit of the spindle-wood tree, and the scarlet berries of the holly, are beautiful objects,—will be admitted by every one. But this beauty serves merely as a guide to birds and beasts, in order that the fruit may be devoured and manured seeds disseminated: I infer that this is the case from having as yet found no exception to the rule that seeds are always thus disseminated when embedded within a fruit of any kind (that is, within a fleshy or pulpy envelope), if it be coloured of any brilliant tint, or rendered conspicuous by being white or black." *

Here we have a case of the converse of elimination—a case of genuine selection under nature. But even here the process of elimination also comes into play, for the visitations of flowers by insects involve cross-fertilization. The flowers of two distinct individuals of the same species of plants in this manner fertilize each other; and the act of

* "Origin of Species," p. 161.

crossing, as Darwin firmly believed, though it is doubted by some observers nowadays, gives rise to vigorous seedlings, which consequently would have the best chance of flourishing and surviving—would best resist elimination by competition. So that we here have the double process at work; the fairest flowers being selected by insects, and those plants which failed to produce such flowers being eliminated as the relatively unfit.

If we turn to the phenomena of what Darwin termed sexual selection, we find both selection and elimination brought into play. By the law of battle, the weaker and less courageous males are eliminated so far as the continuation of their kind is concerned. By the individual choice of the females (on Darwin's view, by no means universally accepted), the finer, bolder, handsomer, and more tuneful wooers are selected.

Let us again hear the voice of Darwin himself. "Most male birds," he says, "are highly pugnacious during the breeding season, and some possess weapons especially adapted for fighting with their rivals. But the most pugnacious and the best-armed males rarely or never depend for success solely on their power to drive away or kill their rivals, but have special means for charming the female. With some it is the power of song, or of emitting strange cries, or of producing instrumental music; and the males in consequence differ from the females in their vocal organs or in the structure of certain feathers. From the curiously diversified means for producing various sounds, we gain a high idea of the importance of this means of courtship. Many birds endeavour to charm the females by love-dances or antics, performed on the ground or in the air, and sometimes at prepared places. But ornaments of many kinds, the most brilliant tints, combs and wattles, beautiful plumes, elongated feathers, top-knots, and so forth, are by far the commonest means. In some cases, mere novelty appears to have acted as a charm. The ornaments of the males must be highly important to them, for they have been acquired in not a few cases at the cost of increased danger from enemies, and even at some loss of power in fighting with their rivals.* . . . What, then, are we to conclude from these facts and considerations? Does the male parade his charms with so much pomp and rivalry for no purpose? Are we not justified in believing that the female exerts a choice, and that she receives the addresses of the male who pleases her most?" †

Here again, then, we have the combined action of elimination and selection. And now we may note that selection involves intelligence—involves the play of appetence and choice. Hence it is that, when we come to consider the evolution of human-folk, the principle of elimination is so profoundly modified by the principle of selection. Not only are the weaker eliminated by the inexorable pressure of competition, but we select the more fortunate individuals and heap upon them our favours. This enables us also to soften the rigour of the blinder law; to let the full stress of competitive elimination fall upon the worthless, the idle, the profligate, and the vicious; but to lighten its incidence on the deserving but unfortunate.

Both selection and elimination occurring under nature, but elimination having by far the wider scope, we may now inquire what will be their effect as regards the three modes of variation—advantageous, disadvantageous, and neutral. It must be remembered that these modes are relative and dependent upon circumstances, so that variations, neutral under certain conditions, may become relatively disadvantageous under other conditions. Selection clearly leads to the preservation of advantageous variations alone, and these variations are advantageous in so far as they meet the taste of the selecting organism. For selection depends upon individual choice; and uniformity of selection is entirely dependent upon uniformity in the standard of taste. If, as Darwin contends, the splendid plumage and tuneful notes of male birds are the result of a selection of mates by the hens, there must be a remarkable uniformity

* "Descent of Man," summary of chap. xvi. pt. ii.   † Ibid. chap. xiv.

of taste among the hens of each particular species, since there is a uniformity of coloration among the cock-birds. It may be said that in all their mental endowments there is greater uniformity among animals than among men; and it is true that individuation has not been carried so far in them as in human-folk. Still, careful observers of animals see in them many signs of individual character; and this uniformity in the standard of taste in each species of birds seems to many naturalists a real difficulty in the way of the acceptance of sexual selection. We shall, however, return to this point. For the present it is clear that selection chooses out advantageous variations, that the advantage is determined by the taste of the selector, and that uniform selection implies uniformity of taste.

Turning to elimination, it is clear that it begins by weeding out, first the more disadvantageous, then the less disadvantageous variations. It leaves both the advantageous and the neutral in possession of the field. I imagine that many, perhaps most, of the variations tabulated by Mr. Wallace and other observers belong to the neutral category. Their fluctuating character seems to indicate that this is so. In any case, they are variations which have so far escaped elimination. And I think they are of great and insufficiently recognized importance. They permit, through interbreeding, of endless experiments in the combination of variations, some of which cannot fail to give favourable results.

It is just possible that it may be asked—If in natural elimination there is nothing more than the weeding out of the unfit and the suppression of disadvantageous variations, where is the possibility of advance? The standard may thus be maintained, but where is the possibility of progress? Such an objection would, however, imply forgetfulness of the fact that all the favourable variations remain to leaven the residual lump. Given a mean, with plus and minus variations: if in any generations the minus variations are got rid of, the mixture of the mean with the plus variations will give a new mean nearer the plus or advantageous end of the scale than the old mean. By how much the favourable variations tend to raise the mean standard, by so much will the race tend to advance. But in this process I see no reason why the neutral variations should be eliminated, except in so far as, in the keen struggle for existence, they become relatively unfavourable.

It is clear, however, that the intercrossing and interbreeding which occurs between average individuals on the one hand, and those possessing favourable variations on the other, while it tends gradually to raise the mean standard, tends also at the same time to reduce the advantageous variations towards the mean. It must tend to check advance by leaps and bounds, and to justify the adage, *Natura nil facit per saltum*. At the same time, it will probably have a greater tendency to reduce to a mean level neutral variations indefinite in direction than advantageous variations definite in direction. Still, it is a most important factor, and one not to be neglected. It tends to uniformity in the species, and checks individualism. It may act as a salutary brake on what we may figuratively term hasty and ill-advised attempts at progress. And at the same time, it favours repeated new experiments in the combination of variations, occasionally, we may suppose, with happy results.

But it does more than this. It tends to check, and, if the offspring always possessed the blended character of both parents, would be absolutely fatal to, divergence of character within the interbreeding members of a species. And yet no fact is more striking than this divergence of character. It is seen in the diversified products of human selection; for example, among pigeons. It is seen in the freedom of nature. Mr. Wallace gives many examples. "Among our native species," he says, "we see it well marked in the different species of titmice, pipits, and chats. The great titmouse, by its larger size and stronger bill, is adapted to feed on larger insects, and is even said sometimes to kill small and weak birds. The smaller and weaker coal-titmouse has adopted a more vegetarian diet, eating seeds as well as insects, and feeding on the ground as well as among trees. The delicate little blue titmouse, with its very small bill, feeds on the minutest insects and grubs, which it extracts from crevices of bark and from the buds of fruit trees. The marsh-titmouse, again, has received its name from the low and marshy localities it frequents; while the crested titmouse is a Northern bird, frequenting especially pine forests, on the seeds of which trees it partially feeds. Then, again, our three common pipits—the tree-pipit, the meadow-pipit, and the rock-pipit, or sea-lark—have each occupied a distinct place in nature, to which they have become specially adapted, as indicated by the different form and size of the hind toe and claw in each species. So the stone-chat, the whin-chat, and the wheat-ear are all slightly divergent forms of one type, with modifications in the shape of the wing, feet, and bill adapting them to slightly different modes of life."* There is scarcely a genus that does not afford examples of divergent species. The question then naturally occurs—How have these divergent forms escaped the swamping effects of intercrossing?

That perfectly free intercrossing, between any or all of the individuals of a given group of animals, is, so long as the characters of the parents are blended in the offspring, fatal to divergence of character, is undeniable. Through the elimination of less favourable variations, the swiftness, strength, and cunning of a race may be gradually improved. But no form of elimination can possibly differentiate the group into swift, strong, and cunning varieties, distinct from each other, so long as all three varieties freely interbreed, and the characters of the parents blend in the offspring. Elimination may and does give rise to progress in any given group as a group; it does not and cannot give rise to differentiation and divergence, so long as interbreeding with consequent interblending of characters be freely

* "Darwinism," p. 108.

permitted. Whence it inevitably follows, as a matter of simple logic, that where divergence has occurred, intercrossing and interblending must in some way have been lessened or prevented.

Thus a new factor is introduced, that of *isolation*, or *segregation*. And there is no questioning the fact that it is of great importance.* Its importance can, indeed, only be denied by denying the swamping effects of intercrossing, and such denial implies the tacit assumption that interbreeding and interblending are held in check by some form of segregation. The isolation explicitly denied is implicitly assumed.

There are several ways in which isolation, or segregation, may be effected. Isolation by geographical barriers is the most obvious. A stretch of water, a mountain ridge, a strip of desert land, may completely, or to a large extent, prevent any intercrossing between members of a species on either side of the barrier. The animals which inhabit the several islands of the Galapagos Archipelago are closely allied, but each island has its particular species or well-marked varieties. Intercrossing between the several varieties on the different islands is prevented, and divergence is thus rendered possible and proceeds unchecked. It is said that in the Zuyder Zee a new variety of herrings, the fry of which are very small compared with open-sea herrings, is being developed. And the salmon introduced into Tasmania seem to be developing a fresh variety with spots on the dorsal fin and a tinge of yellow on the adipose fin. In the wooded valleys of the Sandwich Islands there are allied but distinct species of land-shells. The valleys that are nearest each other furnish the most nearly related forms, and the degree of divergence is roughly measured by the number of miles by which they are separated. Here there is little or no intercrossing between

---

* Its importance in artificial selection was emphasized by Darwin: "The prevention of free crossing, and the intentional matching of individual animals, are the corner-stones of the breeder's art" ("Animals and Plants under Domestication," ii. 62).

the slow-moving molluscs in adjoining valleys; none at all between those at any distance apart.

But even if there are no well-marked physical barriers, the members of a species on a continent or large island tend to fall into local groups, between which, unless the animal be of a widely ranging habit, there will be little intercrossing. Hence local varieties are apt to occur, and varieties show the first beginnings of that divergence which, if carried further and more deeply ingrained, results in the differentiation of species. Geographically, therefore, we may have either complete isolation or local segregation, and in both cases the possibility of divergence.

Another mode of segregation arises also out of geographical conditions. If variations of habits occur (and structure is closely correlated with habit) such that certain individuals take to the mountains, others to the plains or valleys; or that certain individuals take to the forests, others to the open country; the probabilities are that the forest forms will interbreed frequently with each other, but seldom with those in the open, and so with the other varieties. The conditions of forest life or mountain life being thus similar throughout a large area, and life being through elimination slowly but surely adapted to its environment, there might thus arise two distinct varieties scattered throughout the length and breadth of the area, the one inhabiting the mountains, the other the forests. In illustration of this mode of segregation, we may take the case of two species of rats which have recently been found by Mr. C. M. Woodford on one of the Solomon Islands. These two quite distinct species are regarded by Mr. Oldfield Thomas as slightly modified descendants of one parent species, the modifications resulting from the fact that of this original species some individuals have adopted a terrestrial, others an arboreal life, and their respective descendants have been modified accordingly. Thus *Mus rex* lives in trees, has broad foot-pads, and a long rasp-like, probably semi-prehensile, tail; while *Mus imperator* lives on the ground, has smaller pads, and a short, smooth tail. The segregation of these two species has probably been effected by the difference of their mode of life, and each has been adapted to its special environment through the elimination of those individuals which were not in harmony with the condition of their life. It is probable that this mode of segregation has been an important one. And it is clear that in many cases competition would be a co-operating factor in this process, weaker organisms being forced into otherwise uncongenial habitats through the stress of competitive elimination, the weaker forms not perishing, but being eliminated from more favoured areas.

Protective coloration may also be a means of segregation. A species of insects having no protective resemblance might vary in two directions—in the direction of green tints, assimilating their hue to that of vegetation; and in the direction of sandy or dull earthy colours, assimilating them to the colour of the soil. In the one variety elimination would weed out all but the green forms, and these would be left to intercross. In the other variety, green forms would be eliminated, dull-brown forms being left to interbreed. Stragglers from one group into the other would stand a chance of elimination before interbreeding was effected.*

In the case of birds whose freedom of flight gives them a wide range, sometimes almost a world-wide range, it would seem at first sight that their facilities for interbreeding and intercrossing are so great that divergence is well-nigh impossible. And yet the examples of divergence I cited from Mr. Wallace were taken from birds, and it is well known that divergence is particularly well shown in this class. But when the habits of birds are studied attentively, it is found that, wide as is their range, their breeding area is often markedly restricted. The sanderling and knot

---

* From the absence of interblending in some cases (to be considered shortly), both brown *and* green forms may be produced; and under certain circumstances, even a power of becoming either brown *or* green in the presence of appropriate stimuli.

range freely during the winter throughout the Northern hemisphere; but their breeding area is restricted to the north polar region. The interbreeding within this area keeps the species one and homogeneous, notwithstanding its wide range, and, at the same time, prevents intercrossing with allied species with different breeding-grounds.

Another most important mode of segregation among animals arises out of habitual or instinctive preferences. Where varieties are formed there is a tendency for like to breed with like. In the Falkland Islands the differently coloured herds of cattle, all descended from the same stock, keep separate, and interbreed with each other, but not with individuals outside their own colour-caste. If two flocks of merino sheep and heath sheep be mixed together, they do not interbreed. In the Forest of Dean and in the New Forest, the dark and pale coloured herds of fallow deer have never been known to intermingle.* Here we have a case of selective *segregation through preferential mating*, and may find therein the basis of sexual selection in its higher ranges as advocated by Darwin.

The question of sexual selection will, however, be briefly considered in the chapter on "Organic Evolution." At present what we have to notice is that, through preferential mating, segregation is effected. The forms that interbreed have a distinguishing colour. From this it is but a step to the possession, not merely of a distinguishing colour, but of distinguishing colour-markings. Hence, through preferential mating, may arise those special markings which so frequently distinguish allied species. They not only enable *us* to recognize species as distinct, but enable the species which possess them to recognize the members of their own kind. Mr. Wallace calls these diacritical marks *recognition-marks*, and gives many illustrative examples.† They are especially noticeable in gregarious animals and in birds which congregate in flocks or which

* Wallace, "Darwinism," p. 172, where other examples are cited.
† Ibid. pp. 217, *et seq.*

migrate together. Mr. Wallace considers that they "have in all probability been acquired in the process of differentiation for the purpose of checking the intercrossing of allied forms;" for "one of the first needs of a new species would be to keep separate from its nearest allies, and this could be more readily done by some easily seen external mark of difference." This language seems, however, to savour of teleology (that pitfall of the evolutionist). The cart is placed before the horse. The recognition-marks were, I believe, not produced to prevent intercrossing, but intercrossing has been prevented because of preferential mating between individuals possessing special recognition-marks. To miss this point is to miss an important segregation-factor. Undoubtedly, other tendencies co-operate in maintaining the standard of the recognition-marks. Stragglers who failed in the matter of recognition would get separated from their fellows, and stand a greater chance of elimination by enemies; young who failed in this respect would be in like condemnation. Still, I cannot doubt that the foundations of recognition-marks were laid in preferential mating, and that in this we have an important factor in segregation.

We may here note, in passing, as also arising out of preference, how the selection of flowers by insects may lead to segregation; for insects seem often to have habitual or instinctive colour-preferences. Flowers of similar colour would be thus cross-fertilized, but would not intercross with those of different colour, whence colour-varieties might arise. It is important to note that in these cases there is a psychological factor in evolution.

We have so far assumed that intercrossing of parents and interblending of their characters in the offspring always go together. This, we must now notice, is not always the fact. If a blue-eyed Saxon marry a dark-eyed Italian, the children will have blue eyes *or* dark eyes, not eyes of an intermediate tint. The characters do not interblend. The *ancon*, or otter-sheep, a breed with a long body and short, bandy legs, appeared in Massachusetts as a chance sport in a single lamb. The offspring of this ram were either ancons or ordinary sheep. The ancon characters did not blend. Hence for a time a definite breed was maintained. We may call this mode of isolation *isolation by exclusive inheritance*.

A further mode of isolation or segregation, for which Mr. Romanes * claims a foremost, indeed, the foremost, place, is *physiological isolation* as due to differential fertility. One among the many variations to which organisms are subject is a variation in fertility, which may reach the climax of absolute sterility. But it is clear that a sterile variation carries with it its own death-warrant, since the sterile individual leaves no descendants to inherit its peculiarity. Relative infertility, too, unless it chances to be correlated with some unusual excellence, would be no advantage, would be transmitted to few descendants, and would tend to be extinguished. The same is not true, however, of differential fertility. "It is by no means rare," said Darwin, † "to find certain males and females which will not breed together, though both are known to be perfectly fertile with other males and females." Mr. Romanes assumes, as a starting-point, the converse of this, namely, that certain males and females will breed together, though they are infertile with all other members of the species.

Suppose, then, a variety to arise which is perfectly fertile within the limits of the varietal form, but imperfectly fertile or infertile with the parent species. Such a variety would have to run the risks of those ill effects which, as Darwin showed,‡ are attendant upon close interbreeding. But Mr. Wallace points out § that these ill effects may not be so marked under nature as they are under domestication. Suppose, then, that it escapes these ill effects. In this case, Mr. Romanes urges, it would neither be swamped by intercrossing nor die out on

---

* *Journal of the Linnæan Society*, vol. xix. No. 115: "Zoology."
† "Animals and Plants under Domestication," p. 145.
‡ Ibid. chap. xvii.   § "Darwinism," p. 326.

account of sterility. But although it could not be swamped by intercrossing, still, if it arose sporadically, here a case, there a case, and so on, the chances would be enormously against the perpetuation of the variety, unless some co-operating mode of segregation aided in bringing together the varying individuals. If, for example, there were a segregation of these variants in a particular habitat—all the variants meeting in some definite locality for breeding purposes; or if there were a further segregation through mutual preferences; or if, again, there were a further segregation in time; the variety might obtain a firm footing. But without these co-operating factors it is clear that if one male and one female in a hundred individuals varied in this particular way, the chances would be at least forty-nine to one against their happening to mate together.

It is interesting to note that almost the only particular example given by Mr. Romanes in illustration of his theory is one that involves the co-operation of one of these further segregation-factors. Suppose, he says, the variation in the reproductive system is such that the season of flowering or of pairing becomes either advanced or retarded. This particular variation being inherited, the variety breeding, let us say, in May, the parent species in July, there would arise two races, each perfectly fertile within its own limits, but incapable of crossing with the other. Thus is constituted "a barrier to intercrossing quite as effectual as a thousand miles of ocean." Yes! a time-barrier instead of a space-barrier. The illustration is faulty, inasmuch as it introduces a mode of segregation other than that in question. I think it very improbable that differential fertility alone, without the co-operation of other segregation-factors, would give rise to separate varieties capable of maintaining themselves as distinct species.

That distinct species are generally mutually infertile, or more frequently still, that their male offspring are sterile, is, however, an undoubted fact. But there are, exceptions. Fertile hybrids between the sheep and the goat seem to be well authenticated. Of rats Darwin says that "in some parts of London, especially near the docks, where fresh rats are frequently imported, an endless variety of intermediate forms may be found between the brown, black, and snake rat, which are all three usually ranked as distinct species."* Fertile hybrids have been produced between the green-tinted Japanese and the long-tailed Chinese pheasants. Mr. Thomas Moore, of Fareham, in Hants, has been particularly successful in producing a hybrid breed between the golden pheasant (*Thaumalia picta*), whose habitat is Southern and South-eastern China, and the Amherst pheasant (*Thaumalia amherstiæ*), which is found in the mountains of Yunnan and Thibet. In answer to my inquiries, Mr. Moore kindly informs me that he "has bred the half-bred gold and Amherst pheasant, crossed them again with gold, and re-crossed them with half-bred Amherst, and kept on crossing until only a strain of the gold pheasant remained. The result is that the birds so produced are far handsomer than either breed, since the feathers composing their tiplets as well as those under the chin are of so beautiful a colour that they beggar description. They all breed most freely, and are much more vigorous than the pure gold or Amherst, and their tails reach a length of over three feet. They are also exceedingly prolific. Out of a batch of forty-two eggs, forty chickens were hatched out, of which thirty-seven were reared to perfection."

Still, though there are exceptions, the general infertility of allied species when crossed is a fact in strong contrast with the marked fertility of varieties under domestication; concerning which, however, it should be noted that our domesticated animals have been selected to a very large extent on account of the freedom with which they breed in confinement, and that domestication has probably a tendency to increase fertility. The question, therefore, arises—Is the infertility between species, and the general

* "Animals and Plants under Domestication," vol. ii. p. 65. For Darwin's general conclusions on hybridism, see vol. ii. p. 162 of the same work.

sterility of their male offspring, a secondary effect of their segregation? or is their segregation the direct effect of their differential fertility? The former is the general opinion; the latter is held by Mr. Romanes. He contends that sterility is the primary distinction of species, other specific characters being secondary, and regards it as a pure assumption to say that the secondary differences between species have been historically prior to the primary difference. I do not propose to discuss this question. While it seems to me in the highest degree improbable that differential fertility, apart from other co-operating factors, has been or could be a practical mode of segregation, it has probably been a not unimportant factor in association with other modes of segregation or isolation. Suppose, for example, two divergent local varieties were to arise in adjacent areas, and were subsequently (by stress of competition or by geographical changes) driven together into a single area: we are justified in believing, from the analogy of the Falkland Island cattle, the Forest of Dean deer, and other similar observed habits, that preferential breeding, kind with kind, would tend to keep them apart. But, setting this on one side, let us say they interbreed. If, then, their unions are fertile, the isolation will be annulled by intercrossing—the two varieties will form one mean or average variety. But if the unions be infertile, the isolation will be preserved, and the two varieties will continue separate. Suppose now, and the supposition is by no means an improbable one, that this has taken place again and again in the evolution of species: then it is clear that those varietal forms which had continued to be fertile together would be swamped by intercrossing; while those varietal forms which had become infertile would remain isolated. Hence, in the long run, isolated forms occupying a common area would be infertile. Or suppose, once more, that, instead of the unions between the two varietal forms being infertile, they are fertile, but give rise to sterile (mule) or degenerate offspring, as is said to be the case in the unions of Japanese and Ainos: then it is clear that the sterile or degenerate offspring of such unions would be eliminated, and intercrossing, even though it occurred, would be inoperative while breeding within the limits of the variety continued unchecked.

Sufficient has now been said concerning the modes of isolation and segregation, geographical, preferential, and physiological. We must now consider their effects. Where the isolated varieties are under different conditions of life, there will be, through the elimination of the ill-adapted in each case, differential adaption to these different conditions. But suppose the conditions are similar: can there be divergence in this case? The supposition is a highly hypothetical one, because it postulates that all the conditions, climatal, environmental, and competitive, are alike, which would seldom, if ever, be likely to occur. Let us, however, make the supposition. Let us suppose that an island is divided into two equal halves by the submersion of a stretch of lowland running across it. Then the only possible causes of divergence would lie in the organisms themselves * thus divided into two equal groups. We have seen that variations may be advantageous, disadvantageous, or neutral. The neutral form a fluctuating, unfixed, indefinite body. But they afford the material with which nature may make, through intercrossing, endless experiments in new combinations, some of which may be profitable. Such profitable variations would escape elimination, and, if not bred out by intercrossing, would be preserved. In any case, the variety would tend to advance through elimination as previously indicated. But in the two equal groups we are supposing to have become geographically isolated, the chances are many to one against the same successful experiments in combination occurring in each of the two groups. Hence it follows that the progress or

---

* "In every case there are two factors, namely, the nature of the organism and the nature of the conditions. The former seems to be much the more important; for nearly similar variations sometimes arise under, as far as we can judge dissimilar conditions; and, on the other hand, dissimilar variations arise under conditions which appear to be nearly uniform" ("Origin of Species," p. 6).

advance in the two groups, though analogous, would not be identical, and divergence would thus be possible under practically similar conditions of life.

In his observations on the terrestrial molluscs of the Sandwich Islands, Mr. Gulick notes that different forms are found in districts which present essentially the same environment, and that there is no greater divergence when the climatic conditions are dissimilar than there is when those conditions are similar. As before noticed, the degree of divergence is, roughly speaking, directly as the distance the varietal forms are apart. Again, Darwin notes that the climate and environment in the several islands of the Galapagos group are much the same, though each island has a somewhat divergent fauna and flora. These facts lend countenance to the view that divergence can and does occur under similar conditions of life, if there be isolation. They seem, also, so far as they go, to negative the view that the species is moulded directly by the external conditions. For, if this factor were powerful, it would override the effects of experimental combination of characters when the conditions were similar, and would give rise to well-marked varietal forms when the conditions were diverse.

If we admit preferential breeding as a segregation-factor (and arising out of it sexual selection, in a modified form, as a determining one in the evolution of the plumage of male birds), it is evident that the standard of recognition-marks can only be maintained by a uniformity of preference or taste. Still, the uniformity is not likely to be absolute. In this matter, as in others, variations will occur, and after the lapse of a thousand generations, in which elimination has been steadily at work, it is hardly probable that the recognition standard would remain absolutely unchanged. For, though there may not be any direct elimination in this particular respect, there might well be colour-eliminations in other (*e.g.* protective) respects, and the mental nature would not remain quite unchanged. Moreover, we know that secondary sexual characters are remarkably subject to variation, as may be well seen in the case of ruffs (*Machetes pugnax*) in the British Natural History Museum. In the case of our two islands with isolated faunas, therefore, if they formed separate breeding-areas for birds, the chances would be many to one against the change in the standard of recognition-marks being identical in each area. Hence might arise those minute but definite specific distinctions which are so noteworthy in this class of the animal kingdom. Instance the Old and New World species of teal, the Eastern and Western species of curlew and whimbrel, and other cases numerous.[*] This, in fact, is probably in many cases the true explanation of the occurrence of representative species, slight specific variations of the same form as it is traced across a continent or through an archipelago of islands.

The question has been raised, and of late a good deal discussed, whether specific characters, those traits by which species are distinguishable, are always of use to the species which possess them. Here it is essential to define what is meant by utility. Characters may be of use in enabling the possessor to resist elimination; or, like the colours of flowers, they may be of use in attracting insects, and thus furthering selection; or, like recognition-marks, they may be of use in effecting segregation. This last form of utility is apt to be overlooked or lost sight of. In speaking of humming-birds, the Duke of Argyll says that "a crest of topaz is no better in the struggle for existence than a crest of sapphire. A frill ending in spangles of the emerald is no better in the battle of life than a frill ending in spangles of the ruby." But if these characters be recognition-marks, they may be of use in segregation. They are a factor in isolation. But it may be further asked—What is the use of the segregation? Wherein lies the utility of the divergence into two forms? This question, however, involves a complete change of view-point. The question before us is whether specific characters are of use to the

---

[*] See "Evolution without Natural Selection," by Charles Dixon. This author's facts are valuable; his theories are ill digested.

species which possesses them. To this question it is sufficient to answer that they are useful in effecting or preserving segregation, without which the species, *as a distinct species*, would cease to exist. We are not at present concerned with the question whether divergence in itself is useful or advantageous. If it be pressed, we must reply that, although divergence is undoubtedly of immense advantage to life in general, enabling, as Darwin said, its varying and divergent forms to become adapted to many and highly diversified places in the economy of nature, still in many individual cases it is neither possible nor in any respect necessary to our conception of evolution to assign any grounds of utility or advantage for the divergence itself.

In any case, we are dealing at present with the utility of specific characters to the species which possess them; and under the head of utility we are including usefulness in effecting or maintaining segregation. Now, we have already seen that variations may be either advantageous (useful), or neutral (useless), or disadvantageous (worse than useless). The latter class we may here disregard; elimination will more or less speedily dispose of them. With regard to neutral (useless) variations, we must also note that they may be correlated with variations of the other two classes. If correlated with disadvantageous variations, they will be eliminated along with them; if correlated with advantageous variations, they will escape elimination (or will be selected) together with them. There remain neutral, or useless, variations, not correlated with either of the other two classes. Are these in any cases distinctive of species?

It is characteristic of specific distinctions that they are relatively constant. Elimination, selection, or preferential breeding gives them relative fixity. On the other hand, it is characteristic of neutral variations that they are inconstant. There is nothing to give them fixity. It is, of course, conceivable that all the migrants to a new area were possessed of a useless neutral character, which those in the mother area did not possess; **or that such a useless** character was in them preponderant, and by intercrossing formed a less fluctuating, useless character than their progenitors exhibited. Still, the extensive occurrence of such neutral, or useless, characteristics would be in the highest degree improbable. Our ignorance often prevents us from saying in what particular way a character is useful. We must neither, on the one hand, demand proof that this, that, or the other specific character is useful, nor, on the other hand, demand negative evidence (obviously impossible to produce) that it is without utilitarian significance; but we may fairly request those who believe in the wide occurrence of useless specific characters to tell us by what means these useless characters have acquired their relative constancy and fixity. A suggestion on this head will be found in the next chapter.

We must now pass on to consider briefly a most important factor in the struggle for existence. Hitherto we have regarded this struggle as uniform in intensity; we must now regard it as variable, with alternations of good times and hard times, and indicate the causes to which such variations are due.

With variations of climate, such as we know to occur from year to year, or from decade to decade, there are variations in the productiveness of the soil; and when we remember how closely interwoven are the web and woof of life, we shall see that the increased or diminished productiveness of any area will affect for good or ill all the life which that area supports. The introduction of new forms of life into an area, or their preponderance at certain periods owing to climatic or other conditions particularly favourable to them as opposed to other forms, may alter the whole balance of life in the district. We are often unable to assign any reason for the sudden increase or diminution of the numbers of a species; we can only presume that it is the result of some favourable or unfavourable change of conditions. Thus Mr. Alexander Becker[*] has recently drawn attention to the fact that whereas for

[*] *Nature*, vol. xlii. p. 136.

several years various species of grasshoppers appeared in great numbers in South-east Russia, there came then one year of sudden death for most of them. They were sitting motionless on the grasses and dying. He gives similar cases of butterflies for a while numerous, and then rare, and states that a squirrel common near Sarepta suddenly disappeared in the course of one summer, probably, he adds, succumbing to some contagious disease. Such is the nice balance of life, that the partial disappearance of a single form may produce remarkable and little-expected effects. Darwin amusingly showed how the clover crops might be beneficially affected by the introduction of a family of old maids into a parish. The clover is fertilized by humble-bees, the bees are preyed upon by mice; the relations between cats and mice, and between old maids and cats, are well known and familiar: more old maids, more cats; more cats, less mice; less mice, more humble-bees; more humble-bees, better fertilization. A little thing may modify the balance of life, and increase or diminish the struggle for existence, and the rigour of the process of elimination.

But when we take a more extended view of the matter, and include secular changes of climate, the possible range of variation in the struggle for existence is seen to be enormously increased. It is well known to those who have followed the progress of geology, that in early Kainozoic times a mild climate extended to within the Arctic circle, while during the glacial epoch much of the north temperate zone was fast locked in ice, and the climate of the northern hemisphere was profoundly modified. The animals in the north temperate zone were driven southwards.* Not only was there much elimination from the severe climatic conditions, but the migrants were driven southwards into areas already well stocked with life, and the

---

* We may here note, in passing, the fact that the changes of life-forms in a succession of beds points in nine cases out of ten rather to substitution through migration than to transmutation. Still, there are notable cases of transmutation, as in the fresh-water *Planorbes* of Steinhem, in Wittenberg (described, after Hilgendorf, by O. Schmidt, "The Doctrine of Descent," p. 96).

I

competition for means of subsistence in these areas must have been rendered extremely severe. Elimination was at a maximum. Then followed the withdrawal of glacial conditions. The increasing geniality of the climate allowed an expansion of life within a given area, and the withdrawal of snow and ice further and further north set free new areas into which this expanding life could migrate and find subsistence. The hard times of the glacial period were succeeded by good times of returning warmth and an expanding area; and if, as some geologists believe, there was an inter-glacial period (or more than one such period) in the midst of the Great Ice Age, then hard times and good times alternated during the glacial epoch.

Expansion and contraction of life-areas have also been effected again and again in the course of geological history by elevations and subsidences of the land. At the beginning of Mesozoic times much of Europe was dry land. In Triassic and Rhætic times there were lakes in England and in Germany, and a warm Mediterranean Sea to the south. Subsidence of the European area brought with it a lessened land-area and an increased sea-area: bad times and increased competition for land animals; good times and a widening life-area for marine forms of life. This continued, with minor variations, till its culmination in the Cretaceous period. Then came the converse process: the land-areas increased, the sea was driven back. A good time had come for terrestrial life; the marine inhabitants of estuaries and inland seas felt the pressure of increased competition in a lessening area. And so there emerged the continental Europe of the beginning of the Kainozoic era. And it is scarcely necessary to remind those who are in any degree conversant with geology that during tertiary times there have been alternate expansions and contractions of life-areas, marine and terrestrial, the former bringing good times, **the latter hard times and a** heightened struggle for existence.

Now, what would be the result of this alternation of good times and hard times? During good times varieties, which would be otherwise unable to hold their own, might arise and have time to establish themselves. In an expanding area migration would take place, local segregation in the colonial areas would be rendered possible, differential elimination in the different migration-areas would produce divergence. There would be diminished elimination of neutral variations, thus affording opportunities for experimental combinations. In general, good times would favour variation and divergence.

Intermediate between good times and hard times would come, in logical order, the times in which there is neither an expansion nor a contraction of the life-area. One may suppose that these are times of relatively little change. There is neither the divergence rendered possible by the expansion of life-area, nor the heightened elimination enforced by the contraction of life-area.* Elimination is steadily in progress, for the law of increase must still hold good. Divergence is still taking place, for the law of variation still obtains. But neither is at its maximum. These are the good old-fashioned times of slow and steady conservative progress. They are, perhaps, well exemplified by the fauna of the Carboniferous period, and it is not at all improbable that we are ourselves living in such a quiet, conservative period.

On the other hand, hard times would mean increased elimination. During the exhibitions at South Kensington there were good times for rats. But when the show was over, there followed times that were cruelly hard. The keenest competition for the scanty food arose, and the poor animals were forced to prey upon each other. "Their cravings for food," we read in *Nature*, "culminated in a fierce onslaught on one another, which was evidenced by the piteous cries of those being devoured. The method of seizing their victims was to suddenly make a raid upon

---

* I would ask historians whether there have not been, in English history, good times of free and beneficial divergence exemplified in diverse intellectual activity, hard times of rigorous elimination, and intermediate times of placid, somewhat humdrum conservatism.

one weaker or smaller than themselves, and, after overpowering it by numbers, to tear it in pieces." Elimination by competition, passing in this way into elimination by battle, would, during hard times, be increased. None but the best organized and best adapted could hope to escape. There would be no room for neutral variations, which, in the keenness of the struggle, would be relatively disadvantageous. Slightly divergent varieties, before kept apart through local segregation, would be brought into competition. The weakest would in some cases be eliminated. In other cases, the best-adapted individuals of each variety might survive. If their experiments in intercrossing, should such occur, gave rise to fertile offspring, more vigorous and better adapted than either parent-race, these would survive, and the parent-forms would be eliminated. But if such experiments in intercrossing gave rise to infertile, weakly offspring, these would be eliminated. Thus sterility between species would become fixed. Wherever, during the preceding good times, divergence had taken place in two different directions of adaptation, and some intermediate forms, fairly good in both directions, had been able to escape elimination, the chances are that these intermediates would be in hard times eliminated, and the divergent forms left in possession of the field. Wherever, during good times, a species had acquired or retained a habit of flexibility, that habit would stand it in good stead in the midst of the changes wrought by hard times; but when it had, on the other hand, acquired rigidity (like the proverbially "inflexible goose"), it would be at a disadvantage in the stress of a heightened elimination.

The alternation of good times and hard times may be illustrated by an example taken from human life. The introduction of ostrich-farming in South Africa brought good times to farmers. Whereupon there followed divergence in two directions. Some devoted increased profits to improvements upon their farms, to irrigation works which could not before be afforded, and so forth. For others increased income meant increased expenditure and an easier, if not more luxurious, mode of life. Then came hard times. Others, in Africa and elsewhere, learnt the secret of ostrich-farming. Competition brought down profits, and elimination set in—of which variety need hardly be stated.

I believe that the alternation of good times and hard times, during secular changes of climate and alternate expansions and contractions of life-areas through geological upheavals and depression of the land, has been a factor of the very greatest importance in the evolution of varied and divergent forms of life, and in the elimination of intermediate forms between adaptive variations. It now only remains in this chapter to say a few words concerning convergence, adaptation, and progress.

Convergence, which is the converse of divergence, is brought about through the adaptation of different forms of life to similar conditions of existence. The somewhat similar form of the body and fin-like limbs of fishes, of ancient reptiles (the ichthyosaurus and its allies), of whales, seals, and manatees, is a case in point. Both birds, bats, and pterodactyls have keeled breastbones for the attachment of the large muscles for flight. A whole series of analogous adaptations, as the result of analogous modes of life, are found in the placental mammals of Europe and Asia, on the one hand, and the marsupial forms of Australia on the other hand. The flying squirrel answers to the flying phalanger, the fox to the vulpine phalanger, the bear to the koala, the badger to the pouched badger, the rabbit to the bandicoot, the wolverine to the Tasmanian devil, the weasel to the pouched weasel, the rats and mice to the kangaroo rats and mice, and so on. A familiar example of convergence is to be seen in our swallows and martins, on the one hand, and the swifts on the other. Notwithstanding their superficial similarity in external form and habits, they are now generally regarded as belonging to distinct orders of birds.

These are examples of convergence.* Animals of

---

* Two more technical examples may be noticed in a note. (1) Professor Haeckel has recently (*Challenger* Reports, vol. xxviii.) shown that the

diverse descent and ancestry have, through similarity of surrounding conditions or of habits of life, become, in certain respects, assimilated. But some zoologists go further than this. They maintain that the same genus or species may, through adaptation to similar circumstances, be derived from dissimilar ancestors. Some palæontologists, for example, believe that the horse has been independently evolved along parallel lines in Europe and in America. Professor Cope considers that in the one continent *Protohippus*, and in the other *Hipparion*, was the immediate ancestor of *Equus*. The probabilities are, however, so strongly against such a view, that it cannot be accepted until substantiated by stronger evidence than is yet forthcoming.

A special and particular form of convergence, at any rate in certain obvious, if superficial, characters, has already been noticed in our brief consideration of mimicry. In the first place, among a number of closely related species of inedible butterflies, the tendency to divergence is checked, so far as external markings and coloration are concerned, that all may continue to profit by the resemblance, and that the numbers tasted by young birds in gaining their experience (for the avoidance seems to be at most incompletely instinctive) may be divided amongst all the species, thus lessening the loss to each. Secondly, there may be a convergence of certain genera of distantly related inedible groups (*e.g.* among the Heliconidæ and the Danaidæ), which gain by being apparently one species, since the loss from young birds is shared between them. And lastly, there is the true mimicry of quite distinct families of butterflies, not themselves inedible, but sheltering themselves under

---

Siphonophora include two groups, closely resembling each other, but of different ancestry: (*a*) The Disconanthæ, traceable to trachomedusoid ancestors; (*b*) the Siphonanthæ, traceable to anthomedusoid ancestors like Sarsia. (2) M. Paul Pelseneer has been led to the conclusion that the pteropod molluscs also include two groups resembling each other, but of different ancestry: (*a*) The Thecosomes, traceable to tornatellid ancestors; (*b*) the Gymnosomes, traceable to aphysiid ancestors. In each case, the ancestral sea-slug has been modified for a free-swimming life.

the guise and sharing the bad reputation of the mimicked forms. Such forms of convergence are in special adaptation to a very special environment.

We must remember that in all cases adaptation is a matter of life and environment. And these, we may now note, may be related in one or more of three ways. In the first place, there is the adaptation of life to an unchanging environment; for example, the adaptation of all forms of life to the fixed and unchanging properties of inorganic matter. If we liken life to a statue and the environment to a mould in which it is cast, we have in this case a rigid mould and a plastic statue. Secondly, the adaptation may be mutual, as, for example, when the structures of insects and flowers are fitted each to the other, or when the speed of hunters and hunted is steadily increased through the elimination of the slow in either group. Here the mould and statue are both somewhat plastic, and yield to each other's influence. Thirdly, the environment may be moulded to life. This, again, is only relative, since life never wholly loses its plasticity. The bird that builds a nest, the beaver that constructs a dam, the insect that gives rise to a gall,—these, so far, mould the environment to the needs of their existence. Man in especial has the power, through his developed intelligence, of manufacturing his own environment. Here the statue is relatively rigid, and the mould plastic.

Progress may be defined as continuous adaptation. In modern phrase, this is called evolution. The continuity makes the difference between evolution and revolution. Both are natural. Both occur in the organic, the social, and the intellectual sphere. Evolution is the orderly progress of the organism or group of organisms, by which it becomes more and more in harmony with surrounding conditions. If the conditions become more and more complex, the organism will progress in complexity; but if the conditions be more and more simple, progress (if such it may still be called) will be towards simplicity of structure, unnecessary complexity being eliminated, or, in any case, disappearing. Hence, in parasites and some forms of life which live under simple conditions, we have the phenomena of degeneration, or a passage from a more complex to a more simple condition.

Revolution in organic life is the destruction of one organism or group of organisms, and the replacement in its stead of a wholly different organism or group of organisms. During hard times there may be much revolution, or replacement of one set of organic forms by another set of organic forms. It was by revolution that the dominant reptiles of the Mesozoic epoch were replaced by the dominant mammals of Kainozoic times. It was by revolution that pterodactyls were supplanted by birds. Revolution has exterminated many a group in geological ages. On the other hand, it was by evolution that the little-specialized Eocene ungulates gave rise to the horse, the camel, and the deer; by divergent evolution that the bears and dogs were derived from common ancestors. Palæontology testifies both to evolution and revolution.* That history does the same, I need not stay to exemplify. The same laws also apply to systems of thought. Darwinism has revolutionized our conceptions of nature. Darwin placed upon a satisfactory basis a new order of interpretation of the organic world. By it other interpretations have been supplanted. And now this new conception is undergoing evolution, not without some divergence.

In this chapter we have seen how evolution is possible under natural conditions. If the law of increase be true, if more are born than can survive to procreate their kind, natural selection is a logical necessity. We must not blame our forefathers for not seeing this. Until geology had extended our conception of time, no such conclusions could be drawn. If organisms have existed but six or seven thousand years, and if in the last thousand years little or no change in organic life has occurred, the supposition that they could have originated by any such

---

* For evidence in copious abundance, see Nicholson's "Manual of Palæontology," new edition, vol. i.: "Vertebrata," by R. Lydekker.

process as natural selection is manifestly absurd. Lyell was the necessary precursor of Darwin. Given, then, increase and elimination throughout geological time, natural selection is a logical necessity. No one who adequately grasps the facts can now deny it. It is an unquestionable factor in organic evolution. Whether it is the sole factor, is quite another matter, and one we will consider in the chapter on " Organic Evolution."

CHAPTER V.

HEREDITY AND THE ORIGIN OF VARIATIONS.

The law of heredity, I have said above, may be regarded as that of persistence exemplified in a series of organic generations. Variation results—it is clear that it must result—from some kind of differentiating influence. Such statements as these, however, though they are true enough, do not help us much in understanding either heredity or variation.

Let us first notice that normal cases of reproduction exemplify both phenomena—heredity with variation; hereditary similarity to the parents in all essential respects, individual variations in minor points. This is seen in man. Brothers and sisters may present family resemblances among each other and to their parents, but each has individual traits of feature and of character. Only in particular cases of so-called "identical twins" are the variations so slight as not to be readily perceptible by even a casual observer.

Now, when we seek an explanation of these well-known facts, we may be tempted to find it in the supposition that the character of the parents does not remain constant, that the character influences the offspring, and that therefore the children born at successive periods will differ from each other, while twins born in the same hour will naturally resemble each other. As Darwin himself says,* "The greater dissimilarity of the successive children of the same family in comparison with twins, which often resemble each other in external appearance, mental disposition, and

---

* "Animals and Plants under Domestication," vol ii. p. 239.

constitution, in so extraordinary a manner, apparently proves that the state of the parents at the exact period of conception, or the nature of the subsequent embryonic development, has a direct and powerful influence on the character of the offspring." But a little consideration will show that, though this might, in the absence of a better explanation, account for variation in character, it could not account for variation in form and feature, unless we regard these as in some way determined by the character. Moreover, as we shall see presently, it is open to question whether acquired modifications of structure or character in the parent can in any way influence the offspring. Again, in the litter of puppies born of the same bitch by the same dog there are individual variations, often as well marked as those in successive births.

The facts, then, to be accounted for are—first, the close hereditary resemblance in all essential points of offspring to parent; and, secondly, the individual differences in minor points among the offspring produced simultaneously or successively by the same parents. These are the facts as they occur in the higher animals. It will be well to lead up to our consideration of them by a preliminary survey of the facts as they are exemplified by some of the lower organisms.

In the simpler protozoa, where fission occurs, and where the organism is composed of a single cell, where also there is a single nucleus which apparently undergoes division into two equal and similar parts, it is easy to understand that the two organisms thus resulting from the halving of a single organism partake completely of its nature. If the fission of an amœba is such as to divide it into two similar parts, there is no reason why these two similar parts should not be in all respects alike, and should not, by the assimilation of new material, acquire the size and all the characteristics of the parent form. In the higher and more differentiated protozoa, the case is not quite so simple; for the two halves are not each like the whole parent, but have to be remodelled into a similar organism. But if we suppose, as we seem to have every right to suppose, that it is the nucleus that controls the formative processes in the cell, there is not much difficulty in understanding how, when the nucleus divides into two similar portions, each directs, so to speak, the similar refashioning of its own separated protoplasmic territory.

From the protozoa we may pass to such a comparatively simple metazoon as the hydra. Here the organism is composed, not of a single cell, but of a number of cells. These cells are, moreover, not all alike, but have undergone differentiation with physiological division of labour. There is an inner layer of large nutritive cells, and an outer layer of protective cells, some of which are conical with fine processes proceeding from the point of the cone; others are smaller, and fill in the interstices between the apices of the cones, while others have developed into thread-cells, each with a fine stinging filament. Between the two layers there is a thin supporting lamella. The essential point we have here to notice is that there are two distinct layers with cells of different form and function.

Now, it has again and again been experimentally proved that if a hydra be divided into a number of fragments, each will grow up into a complete and perfect hydra. All that is essential is that, in the separated fragment, there shall be samples of the cells of both layers. Under these conditions, the separated cells of the outer layer regenerate a complete external wall, and the separated cells of the inner layer similarly regenerate a complete internal lining. From these facts, it would appear that such a small adequately sampled fragment has the power, when isolated, of assimilating nutriment and growing by the multiplication of the constituent cells, and that the growth takes such lines that the original form of the hydra is reproduced.

Here we may note, by way of analogy, what takes place in the case of inorganic crystals. If a fragment of an alum crystal be suspended in a strong solution of alum, the crystal will be recompleted by the growth of new parts along the broken edges. We say that this is effected under the influence of molecular polarity. Similarly, we may say that the fragment of the hydra rebuilds the complete form under the influence of an hereditary morphological tendency residing in the nuclei of the several cells. The case, though still comparatively simple, is more complex than that of the higher protozoa. There the divided nucleus in two separated cells directs each of these in hereditary lines of morphological growth. Here not only do the cells and their nuclei divide, but they are animated by a common morphological principle, and in their multiplication *combine* to form an organism possessing the ancestral symmetry. If, however, we call this an hereditary morphological tendency or a principle of symmetry; or, with the older physiologists, a *nisus formativus;* or, with Darwin, "the co-ordinating power of the organization" (all of these expressions being somewhat unsatisfactory);—we must remember that these terms merely imply a play of molecular forces analogous to that which causes the broken crystal of alum to become recompleted in suitable solution. The inherent molecular processes in the nuclei * in the one case enable the cells to regenerate the hydra; the inherent molecular stresses in the crystalline fragment in the other case lead to the reproduction of the complete crystal. In either case there is no true explanation, but merely a restatement of the facts under a convenient name or phrase.

The power of regeneration of lost parts, which is thus seen in the hydra, is also seen, in a less degree so far as amount is concerned, but in a higher degree so far as complexity goes, in animals far above the hydra in the scale of life. The lobster that has lost a claw, the snail whose tentacle has been removed, the newt which has been docked of a portion of its tail or a limb, are able more or less completely to regenerate these lost parts. And the regeneration may involve complex structures. With the tentacle of the snail the eye may be removed, and this, not once only, but

* Or in certain "physiological units" (Herbert Spencer), or "plastidules" (Haeckel), which may be regarded as organic molecules exhibiting their special properties under vital conditions.

a dozen times. After such mutilation, no part of the eye remains, though the stump of its nerve is, of course, left; still the perfect organ is reconstructed again and again, as often as the tentacle is removed. The cells at the cut end of the nerve-stump divide and multiply, as do also those of the surrounding tissues, and the growing nerve terminates in an optic cup, as it did previously under the influences of normal development before the mutilation. Here we have phenomena analogous to, and in some respects more complex than, those which are seen in the regenerative process in hydra. It is well known, however, that, in the case of higher animals, in birds and mammals, this power of regenerating lost parts does not exist. When a bone is broken, osseous union of the broken pieces may indeed take place; and in flesh-wounds, the gash is filled in and heals over, not without permanent signs of its existence, as may often be seen in the faces of German students. But beyond this there is normally no regeneration. The soldier who has lost an arm in battle cannot return home and in quiet seclusion reproduce a new limb. That which seems to be among lower animals a well-established law of organic growth does not here obtain. This is probably due to the fact that the higher histological differentiation of the tissues in the more highly developed forms of life is a bar to regeneration. In their devotion to special and minute details of physiological work, the cells have, so to speak, forgotten their more generalized reproductive faculties. In any case, however the fact is to be explained, the higher organisms have in many cases almost completely lost the power of regenerating lost parts. But this loss of the regenerative power in the more highly differentiated animals does not alter or invalidate the law of organic growth we are considering. The law may be thus stated: *Whenever, after mutilation, free growth of the mutilated surface occurs, that growth is directed in such lines as to reproduce the lost part and restore the symmetrical integrity of the organism.* This is a matter of heredity. And we may regard the hereditary reconstructive power as residing either (1) in those cells at or adjoining the mutilated surface which are concerned in the regrowth of the lost part; or (2) in the general mass of cells of the mutilated organism.

There are difficulties in either view. Professor Sollas, supporting the former, says,* "This power [in the snail] of growing afresh so complex and specialized an organ as an eye is certainly, at first sight, not a little astonishing, but it appears to be capable of a very simple explanation. The cells terminating the cut stump of the tentacle are the ancestors of those which are removed; a fresh series of descendants are derived from them, similarly related to the ancestral cells as their predecessors which they replace; the first generation of descendants become in turn ancestors to a second generation, similarly related to them as were the second tier of extirpated cells; and this process of descent being repeated, the completed organ will at length be rebuilt." This explanation is, however, misleading in its simplicity. The cells terminating the cut stump are not the direct ancestors of those which are removed, except in the same sense as gorillas are ancestors of men. They are rather collateral descendants of common ancestors. I think that Professor Sollas would probably agree that, though the lens and "retina" are of epiblastic (outer layer) origin, their relationship with the epiblastic cells at the cut stump is a somewhat distant one. In the reproduction of the lens the cell-heredity is not direct, but markedly indirect. And it is somewhat difficult to understand by what means the ordinary epiblastic cells of the cut stump, which have had no part in the special and peculiar work of lens-production, should be enabled to produce cell-offspring, some of which, and those in a special relation to other deeper-lying cells, possess this peculiar power.

On the other hand, if we turn to the view that the reproduction is effected, not by the cells of the cut surface alone, but by the general mass of cells in the mutilated

* *Nature*, vol. xxxix. p. 486.

organism, we have to face the difficulty of understanding how the influence of cells other than those partaking in the regrowth can be brought to bear on these so as to direct their lines of development. If we say that the organism is pervaded by a principle of symmetry such that both growth and regrowth, whenever they take place, are constrained to follow the lines of ancestral symmetry, we are really doing little more than restating the facts without affording any real organic explanation. That which we want to know is in what organic way this symmetrical growth is effected—how the hereditary tendency is transmitted through the nuclear network which is concerned in cell-division. I do not think that we are at present in a position to give a satisfactory answer to this question.

Let us now return to the hydra, the artificial fission of which has suggested these considerations. Multiplication in this way is probably abnormal. Under suitable conditions, however, if well fed, the hydra normally multiplies by budding. At some spot, generally not far from the "foot," or base of attachment, a little swelling occurs, and the growth of the cells in this region takes such lines that a new hydra is formed. This is at first in direct connection with the parent stem, the two having a common internal cavity; but eventually it separates and lives a free existence as a distinct organism (see Fig. 9, p. 45).

Now, here we may notice, as an implication from these facts, that the size of the organism is limited. When the normal limits of size are reached, any further assimilation of nutriment ministers, not to the further growth of the organism, but to the formation of a new outgrowth, or bud. What determines that the outgrowth, or bud, should originate in this or that group of cells, we do not know. But, like the isolated fragment in the hydra subdivided by fission, the little group in which budding commences contains a fair sample of the various kinds of cells which constitute the hydra. And here, too, we see that their growth and development follow definite lines of hereditary symmetry.

But there is a third method of multiplication in hydra: this is the sexual mode of reproduction, and occurs generally in the autumn. On the body-wall of certain individuals, near the tentacles, conical swellings appear. Within these swellings are great numbers of minute sperms, with small oval heads and active, thread-like tails. They appear to originate from the interstitial cells of the outer layer (see p. 124). Nearer the foot, or base of attachment, and generally, but not quite always, in separate individuals, there are other larger swellings, different in appearance, of which there is generally only one in the same individual at the same time. Each contains a single ovum, or egg-cell, surrounded by a capsule. It, too, and the cells which surround it would appear to be developed from the interstitial cells. It grows rapidly at the expense of the surrounding tissue, but when mature, it bursts through the enveloping capsule, and is freely exposed. A sperm-cell, which seems, in some cases at least, to be produced by the same individual, now unites with it; the egg-cell then begins to undergo division, becomes detached, falls to the bottom, and develops into a young hydra.

Here, then, we have that sexual mode of reproduction which occurs in all the higher animals. It is, however, in some respects peculiar in hydra. In the first place, the ovum is nearly always in other animals (but occasionally not in hydra) fertilized by the sperm from a separate and distinct individual. In the second place, the germinal cells are generally produced, not from the outer layer, but from the middle layer, which appears between the two primitive layers. In some allies of hydra, however, they take their origin in the inner layer; and it has been suggested that, even in hydra, the true germinal cells may migrate from the inner to the outer layer. But of this there does not seem to be at present sufficient evidence. In any case, however, the essential fact to bear in mind is that a new individual is produced by the union of a single cell produced by one organism and of another cell produced in most cases (but not always in the hydra) from a different individual. In the higher forms of animal life, the organisms are either female (egg-producing) or male (sperm-producing). But there are many hermaphrodite forms which produce both eggs and sperms, as in the common snail and earthworm. Even in these cases, however, there are generally special arrangements by which it is ensured that the sperm from one individual should fertilize the ovum produced by another individual.

What, we must next inquire, is the relation in the higher forms of life—for we may now leave the special consideration of hydra—of the ovum or sperm to the organism which produces it? This is but one mode of putting a very old question—Does the hen produce the egg, or does the egg produce the hen? Of course, in a sense, both are true; for the hen produces an egg which, if duly fertilized, will develop into a new hen. But the question has of late been asked in a new sense; and many eminent naturalists reply, without hesitation—The egg produces the hen, but under no circumstances does the hen produce the egg. What, then, it may be asked, does produce the egg? To this it is replied—The egg was produced by a previous egg. At first sight, this may seem a mere quibble; for it may be said that, of course, if an egg produces a hen which contains other eggs, these eggs may be said to be produced by the first. But it is really more than a quibble, and we must do our best clearly to grasp what is meant.

We have seen that, in development, the fertilized egg-cell undergoes division into two cells, each of which again divides into two, and so on, again and again, until from one there arises a multitude of cells. Nor is this all. The multitude are organized into a whole. The constituent cells have different forms and structures, and perform diverse functions. Some are skeletal, such as bone and connective tissue; some are protective, such as those which give rise to feathers or scales; some form nerves or nerve-centres; some, muscles; some give rise to glandular tissue; and lastly, some form the essential elements in reproduction. If, now, we express the development of tissues and the sequence of organisms in the following scheme, the continuity of the reproductive cells will be apparent:—

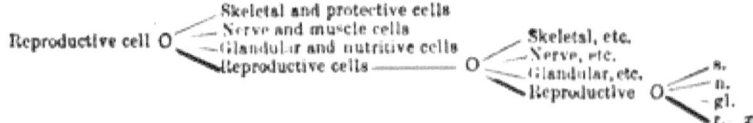

It is clear that there is a continuity of reproductive cells, which does not obtain with regard to nerve, gland, or skeleton. If, then, we class together as body-cells those tissue-elements which constitute what we ordinarily call the body, *i.e.* the head, trunk, limbs—all, in fact, except the reproductive cells, our scheme becomes—

Reproductive cell O ⎯⎯ Body / Reproductive cells O ⎯⎯ Body / Reproductive cells O ⎯⎯ b. / r.

From this, again, it is clear that the body does not produce the egg, or reproductive cell, but that the reproductive cell does produce the body. Of course, it should be noted that we are here using the term "body" as distinguished from, and not as including, the reproductive cells. But this is convenient, in that it emphasizes the fact that the muscular, nervous, skeletal, and glandular cells take (on this view) no part whatever in producing those reproductive cells which are concerned in the continuance of the species.

Such, in brief, is the view that the egg produces the hen. We will return to it presently when we have glanced at the alternative view that the hen produces the egg.

On this view, the reproductive elements are not merely cells, the result of normal cell-division, which have been set aside for the continuance of the species. They are, so to speak, the concentrated extract of the body, and contain minute or infinitesimal elements derived from all the different tissues of the organism which produces them. Darwin* suggested that all the cells of the various tissues produce minute particles called gemmules, which circulate

---

* Darwin, "Animals and Plants under Domestication," 2nd edit., vol. ii. chap. xxvii., from which the following description and quotations are taken.

freely throughout the body, but eventually find a home in the reproductive cells. Just as the organism produces an ovum from which an organism like itself develops, so do the cells of the organism produce gemmules, which find their way to the ovum and become the germs of similar cells in the developing embryo. "The child, strictly speaking," says Darwin, "does not grow into a man, but includes germs which slowly and successively become developed and form the man." "Each animal may be compared with a bed of soil full of seeds, some of which soon germinate, some lie dormant for a period, whilst others perish." Or, to vary the analogy, "an organic being is a microcosm—a little universe formed of a host of self-propagating organisms." This is Darwin's provisional hypothesis of pangenesis, which has recently been accepted in a modified form by Professor W. K. Brooks in America, to some extent by De Vries on the Continent, by Professor Herdman of Liverpool, and by other biologists. The ovum on this view is to be regarded as a composite germ containing the germs of the cellular constituents of the future organism. The scheme representing this view will stand thus—

Reproductive cell O ―― Skeletal and protective cells ―― >O― sk. and pr. ―― >O― s.
           ―― Nerve and muscle cells ――   ― n. and m. ――   ― n.
           ―― Glandular and nutritive cells ――   ― gl. and nu. ―― ― gl.

It is clear that, on this hypothesis, we may frame an apparently simple and, on first sight, satisfactory theory of heredity. Since all the body-cells produce gemmules, which collect in or give rise to the reproductive cells, and since each gemmule is the germ of a similar cell, what can be more natural than that the ovum, thus composed of representative cell-germs, should develop into an organism resembling the parent organism? Modifications of structure acquired during the life of the organism would thus be transmitted from parent to offspring; for the modified cells of the parent would give rise to modified gemmules, which would thus hand on the modification. The inheritance of ancestral traits from grandparent or great-grandparent might be accounted for by supposing that some of the gemmules remained latent to develop in the second or third generation. The regeneration of lost parts receives also a ready explanation. If a part be removed by amputation, regrowth is possible because there are disseminated throughout the body gemmules derived from each part and from every organ. A stock of nascent cells or of partially developed gemmules may even be retained for this special purpose, either locally or throughout the body, ready to combine with the gemmules derived from the cells which come next in due succession. Similarly, in budding, the buds may contain nascent cells or gemmules in a somewhat advanced stage of development, thus obviating the necessity of going through all the early stages in the genesis of tissues. The gemmules derived from each part being, moreover, thoroughly dispersed through the system, a little fragment of such an organism as hydra may contain sufficient to rebuild the complete organism; or, if it contains an insufficient number, we may assume that the gemmules, in their undeveloped state, are capable of multiplying indefinitely by self-division. Finally, variations might arise from the superabundance of certain gemmules and the deficiency of others, and from the varying potency of the gemmules contained in the sperm and ovum. Where the maternal and paternal gemmules are of equal potency, the cell resulting from their union will be a true mean between them; where one or other is prepotent, the resulting cell will tend in a corresponding direction. And since the parental cells are subject to modification, transmitted through the gemmules to the reproductive elements, it is clear that there is abundant room and opportunity for varietal combinations.

It is claimed, as one of the chief advantages of some form of pangenetic hypothesis, that it, and it alone, enables us to explain the inheritance of characters or modifications of structure acquired by use (or lost by disuse) during the life of the organism, or imprinted on the tissues by environmental stresses. The evidence for the transmission of such acquired characters we shall have to consider hereafter. We may here notice, however, that at first sight the hypothesis seems to prove too little or too much. For while modifications of tissues, the effects of use and disuse, are said to be inherited, the total removal of tissues by amputation, even if repeated generation after generation, as in the docking of the tails of dogs and horses, formerly so common, does not have the effect of producing offspring similarly modified. Professor Weismann has recently amputated the tails of white mice so soon as they were born, for a number of generations, but there is no curtailment of this organ in the mice born of parents who had not only themselves suffered in this way, but whose parents, grandparents, and great-grandparents were all rendered tailless. The pangenetic answer to this objection is that gemmules multiply and are transmitted during long series of generations. We have only to suppose that the gemmules of distantly ancestral tails have been passing through the mutilated mice in a dormant condition, awaiting an opportunity to develop, and the constant reappearance of tails is seen to be no real anomaly. In this case the gemmules of the parental and grandparental tail are simply absent. But if the muscles of the parental tail were modified through unwonted use, the modified cells would give rise to modified gemmules, which would unite in generation with ancestral gemmules, and to a greater or less degree modify them. The difference is between the mere absence of gemmules and the presence of modified gemmules. And the fact that it takes some generations for the effects of use or disuse to become marked is (pangenetically) due to the fact that it takes some time for the modified gemmules to accumulate and be transmitted in sufficient numbers to affect seriously the numerous ancestral gemmules.

The direction in which Professor W. K. Brooks has recently sought to modify Darwin's pangenetic hypothesis may here be briefly indicated. He holds that it is under unwonted and abnormal conditions that the cells are stimulated to produce gemmules, and that the sperm is the special centre of their accumulation. Hence it is the paternal influence which makes for variation, the maternal tendency being conservative. The reproductive cell is not merely or chiefly a microcosm of gemmules. It is a cell produced by ordinary cell-division from other reproductive cells. The ovum remains comparatively unaffected by changes in the body; but it receives from the sperm, with which it unites, gemmules from such tissues in the male as were undergoing special modification. The hen does not produce the egg; but the cock does produce the sperm; and the union of the two hits the happy mean between the conservatism of the one view and the progressive possibilities of the other.

Mr. Francis Galton, in 1876,* suggested a modification of Darwin's hypothesis, which included, as does that of Professor Brooks, the idea of germinal continuity which had been suggested by Professor (now Sir Richard) Owen, in 1849. He calls the collection of gemmules in the fertilized ovum the "stirp." Of the gemmules which constitute the stirp only a certain number, and they the most dominant, develop into the body-cells of the embryo. The rest are retained unaltered to form the germinal cells and keep up a continuous tradition. Mr. Galton's place in the history of theories of heredity can scarcely be placed too high. Only one further modification of pangenesis can here be mentioned, namely, that proposed in 1883 by Professor Herdman, of Liverpool. He suggested "that the body of the individual is formed, not by the development of gemmules alone and independently into cells, but by the gemmules in the cells causing, by their affinities and repulsions, these cells so to divide as to give rise to new cells, tissues, and organs."

Such are Darwin's provisional hypothesis of pangenesis, and some more recent modifications thereof. Bold and ingenious as was Darwin's speculation, supported as it at

* For an excellent account of the genesis and growth of the modern views of heredity, see Mr. J. Arthur Thomson's paper on "The History and Theory of Heredity:" Proceedings of the Royal Society of Edinburgh, 1889.

first sight seems to be by organic analogies, it finds to-day but few adherents. With all our increased modern microscopical appliances, no one has ever seen the production of gemmules. Although it appears sufficiently logical to say that, just as a large organism produces a small ovum, so does each small cell produce an exceedingly minute gemmule; when closely investigated, the analogy is not altogether satisfactory. It is denied, as we have seen, by many biologists that the organism does produce the ovum. Multiplication is normally by definite, visible cell-division. Nuclear fission can be followed in all its phases. But where is the nuclear fission in the formation of gemmules? It is true that the conjugation of monads is followed by the pouring forth of a fluid which must be crowded with germs from which new monads arise, and that these germs are so minute as to be invisible, even under high powers of the microscope. It might be suggested, then, that in every tissue some typical cell or cells might thus break up into a multitude of invisible gemmules. But there is at present no evidence that they do so. And even if this were the case, it would not bear out Darwin's view, that every cell is constantly throwing off numerous gemmules. It is known, however, or at least generally believed, that there is a constant replacement of tissues during the life of the organism. It is said, for example, that in the course of seven years the whole cellular substance of the human body is entirely renewed. The fact is, I think, open to question. Granting it, however, it might be suggested that the effete cells, ere they vanish, give rise to minute gemmules, which find their way to the ova. But it must be remembered that the new tissue-cells in the supposed successional renewal of the organs are the descendants of the old tissue-cells; that these are, therefore, already reproducing their kind directly; and that the formation of gemmules would thus be a special superadded provision for a future generation. Still, there is no reason why cells should not have this double mode of reproduction, if any definite evidence of its existence could be brought forward.

Without such definite evidence, we may well hesitate before we accept it even provisionally.

The existence of gemmules, then, is unproven, and their supposed mode of origin not in altogether satisfactory accordance with organic analogies. Furthermore, the whole machinery of the scheme of heredity is complicated and hyper-hypothetical. It is difficult to read Darwin's account of reversion, the inheritance of functionally acquired characters, and the non-inheritance of mutilation, or to follow his skilful manipulation of the invisible army of gemmules, without being tempted to exclaim—What cannot be explained, if this be explanation? and to ask whether an honest confession of ignorance, of which we are all so terribly afraid, be not, after all, a more satisfactory position.

That the hen produces the egg, that "gemmules are collected from all parts of the system to constitute the sexual elements, and that their development in the next generation forms a new being," is further rendered improbable by direct observation upon the mode of origin of the germinal cells, ova, or sperms.

It will be remembered that the view that the egg produces the hen, while the hen does not produce the egg, suggested the question—What, then, does produce the egg? to which the answer was—The egg is the product of a previous egg. On this view, then, the germinal cells, ova, or sperms are the direct and unmodified descendants of an ovum and sperm which have entered into fertile union. Now, in certain cases, notably in the fly *Chironomus*, studied by Professor Balbiani, but also in a less degree in some other invertebrate forms, it is possible to trace the continuity of the germinal cells with the fertilized ovum from which they are derived. In *Chironomus*, for example, "at a very early stage in the embryo, the future reproductive cells are distinguishable and separable from the body-forming cells. The latter develop in manifold variety, into skin and nerve, muscle and blood, gut and gland; they differentiate, and lose almost all protoplasmic likeness to the mother ovum. But the reproductive cells are set apart; they take no share in the differentiation, but remain virtually unchanged, and continue unaltered the protoplasmic tradition of the original ovum."* In such a case, then, observation flatly negatives the view that the germinal cells are "constituted" by gemmules collected from the body-cells, though, of course (on a modified pangenetic hypothesis), they might be the recipients of such gemmules.

It is only in a minority of cases, however, that the direct continuity of germinal cells *as such* is actually demonstrable. In the higher vertebrates, for instance, the future reproductive cells can first be recognized only after differentiation of some of the body-cells and the tissues they constitute is relatively advanced. While in cases of alternation of generations, "an entire asexual generation, or more than one, may intervene between one ovum and another." In all such cases the continuity of the chain of recognizably germinal cells cannot be actually demonstrated.

The impracticability of actually demonstrating a continuity of germinal cells in the majority of cases has induced Professor Weismann to abandon the view that there is a continuity of germinal cells, and to substitute for it the view that there is a continuity of germ-plasm (*keimplasma*). "A continuity of germ-*cells*," he says,† "does not now take place, except in very rare instances; but this fact does not prevent us from adopting a theory of the continuity of the germ-*plasm*, in favour of which much weighty evidence can be brought forward." It might, however, be suggested that, although a continuity of germ-cells cannot be *demonstrated*, such continuity may, nevertheless, obtain, the future germinal cells remaining undifferentiated, while the cells around them are undergoing differentiation. The comparatively slight differentiation of the body-cells in hydroids renders such a view by no means improbable. But Professor Weismann does not regard such an idea as admissible, at all events, in certain

---

* Geddes and Thomson, "The Evolution of Sex," p. 92.
† Weismann, "Essays on Heredity," English translation, p. 173.

cases. "It is quite impossible," he says,* "to maintain that the germ-cells of hydroids, or of the higher plants, exist from the time of embryonic development, as undifferentiated cells, which cannot be distinguished from others, and which are only differentiated at a later period." The number of daughter-cells in a colony of hydroid zoophytes is so great that "all the cells of the embryo must for a long time act as body-cells, and nothing else." Moreover, actual observation (*e.g.* in *Coryne*) convinces Dr. Weismann that ordinary body-cells are converted into reproductive cells. After describing the parts of the body-wall in which a sexual bud arises as in no way different from surrounding areas, he says, "Rapid growth, then, takes place at a single spot, and some of the young cells thus produced *are transformed into germ-cells* which did not previously exist as separate cells."†

This transformation of body-cells or their daughter-cells into germ-cells seems therefore, if it be admitted, to negative the continuity of germ-cells as such. But this fact, says Weismann, does not prevent us from adopting a theory of the continuity of germ-plasm. "As a result of my investigations on hydroids," he says,‡ "I concluded that the germ-plasm is present in a very finely divided and therefore invisible state in certain body-cells, from the very beginning of embryonic development, and that it is then transmitted, through innumerable cell-generations, to those remote individuals of the colony in which the sexual products are formed."

This germ-plasm resides in the nucleus of the cell; and it would seem that by a little skilful manipulation it

---

* Weismann, "Essays on Heredity," p. 205.

† A few pages earlier (p. 200) in the same essay, Professor Weismann says, "A sudden transformation of the nucleo-plasm of a somatic cell into that of a germ-cell would be almost as incredible as the transformation of a mammal into an amœba." This at first sight does not seem quite consistent with the subsequent sentence which I have quoted in the text; for here, at any rate, the daughters of "mammals" are said to be converted into "amœbæ." But this is no doubt because the amœbæ (germ-plasms) are *contained in* the mammals (body-cells). (See the quotations that follow in the text.)

‡ Weismann, "Essays on Heredity," p. 207.

can be made to account for anything that has ever been observed or is ever likely to be observed. It is one of those convenient invisibles that will do anything you desire. The regrowth of a limb shows that the cells contained some of the original germ-plasm. A little sampled fragment of hydra has it in abundance. It lurks in the body-wall of the budding polype. It is ever ready at call. It conveniently accounts for atavism, or reversion; for * "the germ-plasm of very remote ancestors can occasionally make itself felt. Even a very minute trace of a specific germ-plasm possesses the definite tendency to build up a certain organism, and will develop this tendency as soon as the nutrition is, for some reason, favoured above that of the other kinds of germ-plasm present in the nucleus."

In place, then, of the direct continuity of germ-cells as distinct from body-cells, we have here the direct continuity of germ-plasm as opposed to body-plasm. The germ-plasm can give rise to body-plasm to any extent; the body-plasm can never give rise to germ-plasm. If it seems to do so, this is only because the nuclei of the body-cells contain some germ-plasm in an invisible form. The body-plasm dies; but the life of the germ-plasm is, under appropriate conditions, indefinitely continuous.

So far as heredity is concerned, it matters not whether there be a continuity of germ-cells or of germ-plasma. In either case, the essential feature is that body-cells as such cannot give rise to the germ—that the hen cannot produce the egg. On either view, characters acquired by the body cannot be transmitted to the offspring through the ova or sperms. The annexed diagram illustrates how, on the view that the hen produces the egg, dints hammered into the body by the environment will be handed on; while, on the view that the hen does not produce the egg, the dints of the environment are not transmitted to the offspring. On the hypothesis of continuity, heredity is due to the fact that two similar things under similar conditions will give similar products. The ovum from which the mother is

* Weismann, "Essays on Heredity," p. 179.

developed, and the ovum from which the daughter is developed, are simply two fragments separated at different times from the same continuous germ-plasm.* Both develop under similar circumstances, and their products cannot, therefore, fail to be similar. How variation is possible under these conditions we shall have to consider presently.

Now, although I value highly Professor Weismann's

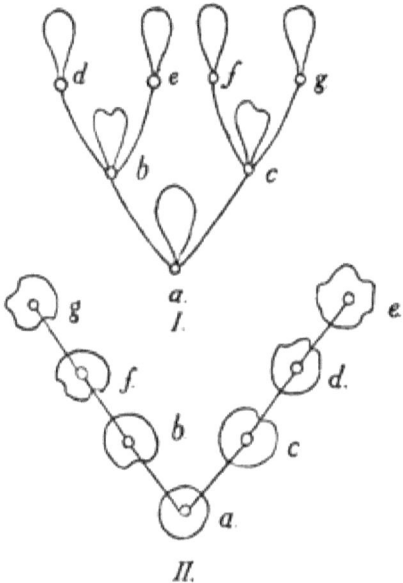

Fig. 21.—Egg and hen.

I. "The egg produces the hen." *II.* "The hen produces the egg." In *I.* the dints produced by the environment are not inherited; in *II.* they are. The letters indicate successive individuals. The small round circles indicate the eggs.

luminous researches, and read with interest his ingenious speculations, I cannot but regard his doctrine of the continuity of germ-plasm as a distinctly retrograde step. His germ-plasm is an unknowable, invisible, hypothetical entity. Material though it be, it is of no more practical value than a mysterious and mythical germinal principle. By a little skilful manipulation, it may be made to account for any-

* It will, of course, be understood that a minute fragment of germ-plasm is capable of almost unlimited growth by assimilation of nutritive material, its properties remaining unchanged during such growth.

thing and everything. The fundamental assumption that whereas germ-plasm can give rise to body-plasm to any extent, body-plasm can under no circumstances give rise to germ-plasm, introduces an unnecessary mystery. Biological science should set its face against such mysteries. The fiction of two protoplasms, distinct and yet commingled, is, in my opinion, little calculated to advance our knowledge and comprehension of organic processes. For myself, I prefer to take my stand on protoplasmic unity and cellular continuity.

The hypothesis of cellular continuity is one that the researches of embryologists tend more and more to justify. The fertilized ovum divides and subdivides, and, by a continuance of such processes of subdivision, gives rise to all the cells of which the adult organism is composed. It is true that in some cases, as in that of peripatus, as interpreted by Mr. Adam Sedgwick, the cells of the embryo run together or remain continuous as a diffused protoplasmic mass with several or many nuclei. But this seemingly occurs only in early stages as a step towards the separation of distinct cells. And even if the process should be proved of far wider occurrence, it would not disprove the essential doctrine of cellular continuity. The nucleus is the essence of the cell. And the doctrine of cellular continuity emphasizes the fact that the nuclei of all the cells of the body are derived by a process of divisional growth from the first segmentation-nucleus which results from the union of the nuclei of the ovum and the sperm. In this sense, then, however late the germinal cells appear as such, they are in direct continuity with the germinal cell from which they, in common with all the cells of the organism, derive their origin. In this sense there is a true continuity of germ-cells.

Now, it has again and again been pointed out that the simple cell of which an amœba is composed is able to perform, in simple fashion, the various protoplasmic functions. It absorbs and assimilates food; it is contractile and responds to stimulation; it respires and exhibits metabolic processes; it undergoes fission and is reproductive. The metazoa are cell-aggregates; and in them the cells exemplify a physiological division of labour. They differentiate, and give rise to muscle and nerve, gut and gland, blood and connective or skeletal tissue, ova and sperms. Are these germinal cells mysteriously different from all the other cells which have undergone differentiation? No. *They are the cells which have been differentiated and set apart for the special work of reproduction, as others have been differentiated and set apart for other protoplasmic functions.*

Cell-reproduction is, however, in the metazoa of two kinds. There is the direct reproduction of differentiated cells, by which muscle-cells, nerve-cells, or others reproduce their kind in the growth of tissues or organs; and there is the developmental reproduction, by which the germinal cells under appropriate conditions reproduce an organism similar to the parent. The former is in the direct line of descent from the simple reproduction of amœba. The latter is something peculiarly metazoan, and is, if one may be allowed the expression, specialized in its generality.

That the metazoa are derived from the protozoa is generally believed. How they were developed is to a large extent a matter of speculation. But, however originating, their evolution involved the production, from cells of one kind, of cells of two or more kinds, co-operating in the same organism. Whenever and however this occurred, the new phase of developmental reproduction must have had its origin. And if in cell-division there is any continuity of protoplasmic power, the faculty of producing diverse co-operating cells would be transmitted. On any view of the origin of the metazoa, this diverse or developmental reproduction is a new protoplasmic faculty; on any view, it must have been transmitted, for otherwise the metazoa would have ceased to exist. This new faculty of developmental reproduction, then, with the inception of the metazoa, takes its place among other protoplasmic faculties, and, with the progress of differentiation and the division of labour, will become the special business of certain cells. On this view, the specialization of the reproductive faculty and of germinal cells takes its place in line with other cell-differentiations with division of labour; and the difficulties of comprehending and following the process of differentiation in this matter are similar to those which attend physiological division of labour in general.

It is probable that, in the lower metazoa, in which differentiation has not become excessively stereotyped, the power of developmental reproduction is retained by a great number of cells, even while it is being specialized in certain cells. Hence the ability to produce lost parts and the reproduction of hydra by fission. But, on the other hand, the special differentiation of a tissue on particular lines has always a tendency to disqualify the cells from performing other protoplasmic faculties, and that of developmental reproduction among the number. I do not know of any definite, well-observed cases on record in the animal kingdom of ova or sperms being derived from cells which are highly differentiated in any other respect. In the vertebrata, the mesoblastic, or mid-layer, cells, from which the germinal epithelium arises, have certainly not been previously differentiated in any other line. And in the case of the hydroid zoophytes, quoted by Professor Weismann, the cells which give rise to the germinal products have never been so highly differentiated as to lose the protoplasmic faculty of developmental reproduction.

Some such view of developmental reproduction, based upon cellular continuity and the division of labour, seems to me more in accord with the general teachings of modern biology than a hypothetical and arbitrary distinction between a supposed germ-plasm and a supposed body-plasm.

To which category, then, does this hypothesis belong? Does it support the view that the hen produces the egg or that the egg produces the hen? Undoubtedly the latter. It is based on cellular continuity, and is summarized by the scheme on p. 131. It adequately accounts for hereditary continuity, for there is a continuity of the germinal cells, the bearers of heredity. But how, it may be asked, on this view, or on any continuity hypothesis, are the origin of variations and their transmission to be accounted for? To this question we have next to turn. But before doing so, it will be well to recapitulate and summarize the positions we have so far considered.

We saw at the outset that the facts we have to account for are those of heredity with variation. To lead up to the facts of sexual heredity, we considered fission, the regeneration of lost parts, and budding in the lower animals. We saw that, if a hydra be divided, each portion reproduces appropriately the absent parts. But we found it difficult to say whether this power resides, in such cases, in the cells along the plane of section or in the general mass of cells which constitute the regenerating portion.

Having led up to the sexual mode of reproduction, we inquired whether the egg produces the hen or the hen produces the egg. We saw that there is a marked difference between a *direct continuity* of reproductive cells, giving rise to body-cells as by-products, and an *indirect continuity* of reproductive cells, these cells giving rise to the hen, and then the hen to fresh reproductive cells, which, on this view, are to be regarded as concentrated essence of hen.

Darwin's hypothesis of pangenesis as exemplifying the latter view was considered at some length, and the modifications suggested by Professor Brooks, Mr. Galton, and Professor Herdman were indicated. The hypothesis, so far as it is regarded as a theory of the main facts of heredity, was rejected.

It was then pointed out that only in a few cases has a direct continuity of germinal cells *as such* been actually demonstrated. Whence Professor Weismann has been led to elaborate his doctrine of the continuity of germ-plasm. This germ-plasm can give rise to, but cannot originate from, body-plasm. It may lurk in body-cells, which may, by its subsequent development, be transformed into germ-cells. But any external influences which may affect these body-cells produce no change on the germ-plasm which they may contain. We regarded this hypothesis as a retrograde step, much as we admire the genius of its propounder, and considered that the fiction of two protoplasms, distinct and yet commingled, is little calculated to advance our comprehension of organic processes.

In the known and observed phenomena of cellular continuity and cell-differentiation, we found a sufficiently satisfactory hypothesis of heredity. The reproductive cells are the outcome of normal cell-division, and have been differentiated and set apart for the special work of developmental reproduction, as others have been differentiated and set apart for other protoplasmic functions. Such a view adequately accounts for hereditary continuity, for there is a continuity of the germinal cells, the bearers of heredity. But how, we repeat, on this view or any other hypothesis of direct continuity, are the origin of variations and their transmission to be accounted for?

Every individual organism reacts more or less markedly under the stress of environing conditions. The reaction may take the form of passive resistance, or it may be exemplified in the performance of specially directed motor-activities. The power to react in these ways is inborn; but the degree to which the power is exercised depends upon the conditions of existence, and during the life of the individual the power may be increased or diminished according to whether the conditions of life have led to its exercise or not. The effects of training and exercise on the performance of muscular feats and in the employment of mental faculties are too well known to need special exemplification. By manual labour the skin of the hand is thickened; and by long-continued handling of a rifle a bony growth caused by the weapon in drilling, the so-called *exercierknochen* of the Germans, is developed. Now, it is clear that if these acquired structures or faculties are transmitted from parent to offspring, we have here a most important source and origin of variations—a source from which spring variations just in the particular direction in which they are wanted. The question is—Are they transmitted? and if so, how?

Let us begin with the protozoa. Dr. Dallinger made some interesting experiments on monads. They extended over seven years, and were directed towards ascertaining whether these minute organisms could be gradually acclimatized to a temperature higher than that which is normal to them. Commencing at 60° Fahr., the first four months were occupied in raising the temperature 10° without altering the life-history. When the temperature of 73° was reached, an adverse influence appeared to be exerted on the vitality and productiveness of the organism. The temperature being left constant for two months, they regained their full vigour, and by gradual stages of increase 78° was reached in five months more. Again, a long pause was necessary, and during the period of adaptation a marked development of vacuoles, or internal watery spaces, was noticed, on the disappearance of which it was possible to raise the temperature higher. Thus by a series of advances, with periods of rest between, a temperature of 158° Fahr. was reached. It was estimated that the research extended over half a million generations. Here, then, these monads became gradually acclimatized to a temperature more than double that to which their ancestors had been accustomed to—a temperature which brought rapid death to their unmodified relatives.

Now, in such observations it is impossible to exclude elimination. It is probable that there were numbers of monads which were unable to accommodate themselves to the changed conditions, and were therefore eliminated. But in any case, the fact remains that the survivors had, in half a million generations, acquired a power of existing at a temperature to which no individual in its single life could become acclimatized. Here, then, we have the hereditary transmission of a faculty. But the organisms experimented on were protozoa. In them there is no distinction between germ-cell and body-cell. Multiplication is by fission. And if the cell which undergoes fission has been modified, the two separate cell-organisms which result from that fission will retain the special modification. In such cases the transmission of acquired characters is readily comprehensible. We have an hereditary summation of effects.

With the metazoa the case is different. In the higher forms the germinal cells are internal and sheltered from environing influences by the protecting body-wall. It is the body-cells that react to environmental stresses; it is muscle and nerve in which faculty is strengthened by use and exercise, or allowed to dwindle through neglect. The germ-cells are shielded from external influences. They lead a sheltered and protected life within the body-cavity. It is no part of their business to take part in either passive resistance or responsive activity. During the individual life, then, the body may be modified, may acquire new tissue, may by exercise develop enhanced faculties. But can the body so modified affect the germ-cells which it carries within it?

Biologists are divided on this question. Some say that the body cannot affect the germ; others believe that it can and does do so.

It might seem an easy matter to settle one way or another. But, in truth, it is by no means so easy. Suppose that a man by strenuous exercise brings certain muscles to a high degree of strength or co-ordination. His son takes early to athletics, and perhaps excels his parent. Is this a case of transmitted fibre and faculty? It may be. But how came it that the father took to athletics, and was enabled to develop so lithe and powerful a frame? It must have been "in him," as we say. In other words, it must have been a product of the germ-cells from which he was developed. And since his son was developed, in part at least, from a germ-cell continuous with these, what more natural than that he too should have an inherent athletic habit? Every faculty that is developed in any individual is potential in the germ-stuff from which he springs; the tendency to develop any particular faculty is there too; and both faculty and tendency to exercise it are handed on by the continuity of germ-protoplasm or germ-cells. Logically, there is no escape from the argument if put as follows: The body and all its faculties (I use the term "faculties" in the broadest possible sense) are the product of the germ; the acquisition of new characters or the strengthening of old faculties by the body is therefore a germinal product; there is continuity of the germs of parent and child; hence the acquisition by the child of characters acquired by the parent is the result of germinal or cellular continuity. It is not the acquired character which influences the germ, but the germ which develops what appears to be an acquired character. Finally, if an acquired character, so called, is better developed in the child than in the parent, what is this but an example of variation? And if, in a series of generations, the acquired character continuously increases in strength, this must be due to the continued selection of favourable variations. It is clear that the organism that best uses its organs has, other things equal, the best chance of survival. It will therefore hand on to its offspring germinal matter with an inherent tendency to make vigorous use of its faculties.

Those who argue thus deny that the body-cells can in any way affect the germ-cells. To account for any continuous increase in faculty, they invoke variation and the selection of favourable varieties. What, then, we may now ask, is, on their view, the mode of origin of variations?

In sexual reproduction, with the union of ovum and sperm, we seem to have a fertile source of variation. The parents are not precisely alike, and their individual differences are, *ex hypothesi*, germinal products. In the union of ovum and sperm, therefore, we see the union of somewhat dissimilar germs. And in sexual reproduction we have a constantly varying series of experiments in germinal combinations, some of which, we may fairly suppose, will be successful in giving rise to new or favourable variations. This view, however, would seem to involve an hypothesis which may be true, but which, in any case, should be indicated. For it is clear that if new or favourable variations arise in this way, the germinal union cannot be a mere mixture, but an organic combination.

An analogy will serve to indicate the distinction implied in these phrases. It is well known that if oxygen and hydrogen be mixed together, at a temperature over 100° C., there will result a gaseous substance with characters intermediate between those of the two several gases which are thus commingled. But if they are made to combine, there will result a gas, water-vapour, with quite new properties and characters. In like manner, if, in sexual union, there is a mere mixture, a mere commingling of hereditary characters, it is quite impossible that new characters should result, or any intensification of existing characters be produced beyond the mean of those of ovum and sperm. If, for example, it be true, as breeders believe, that when an organ is strongly developed in both parents it is likely to be even more strongly developed in the offspring, and that weakly parts tend to become still weaker, this cannot be the result of germinal mixture. Let us suppose, for the sake of illustration, that a pair of organisms have each an available store of forty units of growth-force, and that these are distributed among five sets of organs, a to e, as in the first two columns. Then the offspring will show the organs as arranged in the third column.*

	Parents.		Offspring.
a	10	10	10
b	8	10	9
c	9	5	7
d	7	9	8
e	6	6	6
	40	40	40

* Latency is here neglected. Mr. Francis Galton has shown, statistically, that the offspring, among human folk, inherit ¼ from each parent, 1/16 from each grandparent, and the remaining ¼ from more remote ancestors. In domesticated animals, reversion to characters of distant ancestors sometimes occurs. This, however, does not invalidate the argument in the text, which

There is no increase in the set of organs a, which are strongly developed in both parents; and no decrease in the set of organs e, which are weakly developed in both parents. By sexual admixture alone there can be no increase or decrease beyond the mean of the two parental forms. If, then, the union of sperm and ovum be the source of new or more favourable variations other than or stronger than those of either parent, this must be due to the fact that the hereditary tendencies not merely commingle, but under favourable conditions combine, in some way different indeed from, but perhaps analogous to, that exemplified in chemical combination.

Such organic combination, as opposed to mere commixture, is altogether hypothetical, but it may be worth while to glance at some of its implications. If it be analogous to chemical combination, the products would be of a definite nature; in other words, the variations would be in definite directions. Selection and elimination would not have to deal with variations in any and all directions, but would have presented to them variations specially directed along certain lines determined by the laws of organic combination. As Professor Huxley has said, "It is quite conceivable that every species tends to produce varieties of a limited number and kind, and that the effect of natural selection is to favour the development of some of these, while it opposes the development of others along their predetermined line of modification." Mr. Gulick * and others have been led to believe in a tendency to divergent evolution residing in organic life-forms. Such a tendency might be due to special modes of organic combination giving rise to particular lines of divergence. Again, we have seen that some naturalists believe that specific characters are not always of utilitarian significance. But, as was before pointed out, on the hypothesis of all-

is that sexual admixture tends towards the mean of the race (ancestors included), and cannot be credited with new and unusually favourable variations. The prepotency of one parent is also here neglected.

* See his valuable paper on "Divergent Evolution," Lin. Soc. Zool., No. cxx.

round variation, there is nothing to give these non-useful specific characters fixity and stability, nothing to prevent their being swamped by intercrossing. If, however, on the hypothesis of combination, we have definite organic compounds, instead of, or as well as, mere hereditary mixtures; if, in other words, variations take definite lines determined by the laws of organic combination (as the nature and properties of chemical compounds are determined by the laws of chemical combination), then this difficulty disappears. There is no reason why a neutral divergence—one neither useful nor deleterious—should be selected or eliminated. And if its direction is predetermined, there is no reason why it should not persist, though, of course, it will not be kept at a high standard by elimination. It has again and again been pointed out as a difficulty in the path of natural selection that, in their first inception, certain characters or structures cannot yet be of sufficient utility to give the possessor much advantage in the struggle for existence. If, however, these be definite products of organic combination, this difficulty also disappears. So long as they are not harmful, they will not be eliminated, and by fortunate combinations will progress slowly until natural selection gets a hold on them and pushes them forward, developing to the full the inherent tendency. Finally, we must notice that, on this hypothesis, our conception of panmixia, or intercrossing, would have to be modified. As generally held, this doctrine is based upon hereditary mixture, not organic combination. It is a doctrine of means and averages. There is a good deal of evidence that intercrossing does not, at least in all cases, produce mean or average results. And according to the hypothesis of organic combination, it need not always do so. According to this hypothesis, then, divergent modifications might arise and be perpetuated without the necessity of isolation. Sterility might result from the fact that divergence had been carried so far that organic combination was no longer possible; reversion, due to intercrossing, from the fact that combinations long rendered impossible by the isolation of the necessary factors in distinct varieties, are again rendered possible when these varieties interbreed.

On this hypothesis of organic combination, to which we shall recur in the chapter on "Organic Evolution," the varied forms of animal life are the outcome of definite organic products with definite organic structure, analogous to the definite chemical compounds with definite crystalline and molecular structure; and the analogy between the regeneration of hydra and the reconstruction of a crystal is carried on a step further. I do not say that I am myself at present prepared to adopt the hypothesis, at least in this crude form; but it is, perhaps, worth a passing consideration. Its connection with Mr. Herbert Spencer's doctrine of physiological units is obvious. The analogy there is with crystallization; here it is with chemical combination.

We must now return to the point which gave rise to this digression, and repeat that mere hereditary commixture in the union of ovum and sperm cannot give rise to new characters or raise existing structures (1) where there is free intercrossing beyond the mean of the species, and (2) where there is rigorous elimination beyond the existing maximum of the species. Variations beyond this existing maximum must be due to some other cause.

Professor Weismann has suggested, as a cause of variation, the extrusion of the polar cells from the ovum. It has before been mentioned that, generally previous to fertilization, the ripe ovum buds off two minute polar bodies. The nucleus of the ovum divides, and one half is extruded in the first polar cell; the nucleus then (except in parthenogenetic * forms, where there is no union of ovum and sperm) again divides, and a second polar cell is extruded. In accordance with his special view of the absolute distinction between the body-plasm and the germ-plasm, the first polar cell is formed to carry off the body-plasm of the

---

* One parthenogenetic form—the drone—has been shown by Blochmann to extrude a second polar cell. This observation is in serious opposition to Dr. Weismann's theory.

ovum-nucleus. For the ovum, besides being a germ-bearer, is a specialized cell, and its special form is determined by the body-plasm it contains. This is got rid of in the first polar cell, and nothing but germ-plasm remains. Now, if nothing further took place, all the ova of this same individual containing similar germ-plasm would be identical, and similarly with all the sperms from the same parent. The union of these similar ova from one parent with similar sperms from another should therefore give rise to similar offspring. But the offspring are not all similar; they vary. Professor Weismann here makes use of the second polar cell.* "A reduction of the germ-plasm," he says, "is brought about by its formation, a reduction not only in quantity, but above all, in the complexity of its constitution. By means of the second nuclear division, the excessive accumulation of different kinds of hereditary tendencies or germ-plasms is prevented. With the nucleus of the second polar body, as many different kinds of plasm are removed from the egg as will be afterwards introduced by the sperm-nucleus." "If, therefore, every egg expels half the number of its ancestral germ-plasms during maturation, the germ-cells of the same mother cannot contain the same hereditary tendencies, unless we make the supposition that corresponding ancestral germ-plasms chance to be retained by all eggs—a supposition that cannot be sustained."

The two polar cells are therefore, on this view, of totally different character; and the nuclear division in each case of a special kind and *sui generis*. I do not think that the evidence afforded by observation lends much support to this view. But with that we are not here specially concerned. We have to consider how this reduction of the number of ancestral germ-plasms can further the kind of variation required. Now, it is difficult to see, and Professor Weismann does not explain, how the getting rid of certain ancestral tendencies can give rise to new characters or the enhancement of old characters. One can understand how this "reducing division," as Dr. Weismann calls it, can

* Weismann, "Essays on Heredity," pp. 355, 378.

reduce the level of now one and now another character. But how it can raise the level beyond that attained by either parent is not obvious. It is perhaps possible, though Professor Weismann does not, I think, suggest it, that, by a kind of compensation,* the reduction of certain characters may lead to the enhancement of others. Let us revert to the illustration on p. 150, where each individual has an available store of forty units of growth-force; and let us express by the minus sign the units lost in the parents by the extrusion of the polar cell and an analogous process which may occur in the genesis of the sperm. Then the units of growth-force which may thus be lost by a "reducing division" in b, c, and e may be, in the offspring, applied to the further growth of a; thus—

	Parents.		Offspring.
a	10	10	14
b	8 - 1	10 - 3	7
c	9 - 1	5 - 1	6
d	7	9	8
e	6 - 2	6	5

Here the reduction of the characters b, c, and e has led to the enhancement of a, which thus stands at a higher level than in either parent.

On such an hypothesis we may, perhaps, explain the fact to which breeders of stock testify—that the organ strongly developed in both parents (a) is yet more strongly developed in some of their offspring, and that weakly parts (e) tend to become still weaker. I know not whether this way of putting the matter would commend itself to Professor Weismann or his followers; but some such additional hypothesis of transference of growth-force from one set of organs to another set of organs seems necessary to complete his hypothesis.

Professor Weismann's view, then, assumes (1) that the cell-division which gives rise to the ova in the ovary is so absolutely equal and similar that all ova have precisely

---

* The law of compensation of growth or balancement was suggested at nearly the same time by Goethe and Geoffroy Saint-Hilaire. The application in the text has not, so far as I know, been before suggested.

the same characters; (2) that the first polar cell leaves the germinal matter unaffected, merely getting rid of formative body-plasm; (3) that the nuclear division giving rise to the second polar cell is unequal and dissimilar, effecting the differential reduction of ancestral germ-plasms. Concerning all of which one can only say that it may be so, but that there is not much evidence that it is so. And, without strong confirmatory evidence, it is questionable whether we are justified in assuming these three quite different modes of nuclear division.

There remains one more question for consideration, on the hypothesis that the germ-cells cannot in any special way be affected by the body-cells. In considering the union of ovum and sperm as a source of variation, we have taken for granted the existence of variations. We have been dealing with the mixture or combination of already existing variations. How were variations started in the first instance?

We have already seen that in the protozoa parent and offspring are still, in a certain sense, one and the same thing; the child is a part, and usually half, of the parent. If, therefore, the individuals of a unicellular species are acted upon by any of the various external influences, it is inevitable that hereditary individual differences will arise in them; and, as a matter of fact, it is indisputable that changes are thus produced in these organisms, and that the resulting characters are transmitted. Hereditary variability cannot, however, arise in the metazoa, in which the germ-plasm and the body-plasm are differentiated and kept distinct. It can only arise in the lowest unicellular organisms. But when once individual difference had been attained by these, it necessarily passed over into the higher organisms when they first appeared. Sexual reproduction coming into existence at the same time, the hereditary differences were increased and multiplied, and arranged in ever-changing combinations. Such is Professor Weismann's solution of the difficulty, told, for the most part, in his own words.

I do not know that Professor Weismann has anywhere distinctly stated what he conceives to be the relation of body-plasm and germ-plasm in the protozoa. Are the two as yet undifferentiated? This can hardly be so, seeing the fundamental distinction he draws between them. Is it the germ-plasm or the body-plasm that is influenced by external stresses? If the former, does it transfer its influence to the body-plasm during the life of the individual? If the latter, then the body-plasm must either directly influence the germ-plasm in unicellular organisms (it would seem that, according to Professor Weismann, it cannot do so in the metazoa), or the changed body-plasm, which shares in the fission of the protozoon, must participate in that so-called immortality which is often said to be the special prerogative of germinal matter.

These, however, are matters for Professor Weismann and his followers to settle. I regard the sharp distinction between body-plasm and germ-plasm as an interesting biological myth. For me, it is sufficient that the protoplasm of the protozoon is modified, and the modification handed on in fission. And it is clear that Professor Weismann is correct in saying that the commixture or combination of characters takes its origin among the protozoa. If the unicellular individuals are differently modified, however slightly, then, whenever conjugation occurs between two such individuals, there will be a commingling or combination of the different characters. The transmissible influence of the environment, however, ceases when the metazoon status is reached, and special cells are set apart for reproductive purposes—ceases, that is to say, in so far as the influence on the body is concerned. There may, of course, be still some direct[*] influence on the germinal cells themselves. Except for this further influence, the metazoon starts with the stock of variations

---

[*] Darwin spoke of changed conditions acting "directly on the organization or indirectly through the reproductive system." Now, since Professor Weismann has taught us to reconsider these questions, we speak of such conditions as acting directly on the germ or indirectly through the body. The germ is no longer subordinate to the body, but the body to the germ.

acquired by that particular group of protozoa—whatever it may be—from which it originated. All future variations in even the highest metazoa arise from these.

Now, it is obvious that no mere commingling and rearrangement of protozoan characters could conceivably give rise to the indefinitely more complex metazoan characters. But if there be a combination and recombination of these elements in ever-varying groups, the possibilities are no longer limited. Let us suppose that three simple protozoan characters were acquired. The mere commixture of these three could not give much scope for further variation. It would be like mixing carbon, oxygen, and hydrogen in varying proportions. But let them in some way combine, and you have, perhaps, such varied possibilities as are open to chemical combinations of oxygen, hydrogen, and carbon, whose name is legion, but whose character is determined by the laws of chemical combination.

Summing up now the origin of variations, apart from those which are merely individual, on the hypothesis that particular modifications of the body-cells cannot be transmitted to the germ-cells, we have—

1. In protozoa, the direct influence of the environment and the induced development of faculty.

2. In metazoa—

(*a*) Some direct and merely general influence of the environment on the germ, including under the term "environment" the nutrition, etc., furnished by the body.

(*b*) The combination and recombination of elementary protoplasmic faculties (specific molecular groupings) acquired by the protozoa.

(*c*) Influences on the germ, the nature of which is at present unknown.

We may now pass on to consider the position of those who give an affirmative answer to the question—Can the body affect the germ? Two things are here required. First, definite evidence of the fact that the body does so affect the germ; *i.e.* that acquired characters are inherited. Secondly, some answer to the question—How are the body-cells able to transmit their modifications to the germ-cells? We will take the latter first, assuming the former point to be admitted.

Let us clearly understand the question. An individual, in the course of its life, has some part of the epidermis, or skin, thickened by mechanical stresses, or some group of muscles strengthened by use, or the activity of certain brain-cells quickened by exercise: how are the special modifications of these cells, here, there, or elsewhere in the body, communicated to the germ, so that its products are similarly modified in the offspring? The following are some of the hypotheses which have been suggested:—

(a) Darwin's pangenesis.
(b) Haeckel's perigenesis; Spencer's physiological units.
(c) The conversion of germ-plasm into body-plasm, and its return to the condition of germ-plasm (Nägeli).
(d) The unity of the organism.

(a) Concerning pangenesis, nothing need be added to what has already been said. Although, as we have seen, it has been adopted with modifications by Professor Brooks; although Mr. Francis Galton, a thinker of rare ability and a pioneer in these matters, while contending for continuity, admitted a little dose of pangenesis; although De Vries has recently renewed the attempt to combine continuity and a modified pangenesis;—this hypothesis does not now meet with any wide acceptance.

(b) With the pamphlet in which Professor Haeckel brought forward his hypothesis termed the perigenesis of the plastidule, I cannot claim first-hand acquaintance. According to Professor Ray Lankester, who gave some account of it in *Nature*,* protoplasm is regarded by Haeckel as consisting of certain organic molecules called plastidules. These plastidules are possessed of special undulatory movements, or vibrations. They are liable to have their undulations affected by every external force, and, once

* July 15, 1876. Since reprinted in "The Advancement of Science," p. 273.

modified, the movement does not return to its pristine condition. By assimilation, they continually increase to a certain size and then divide, and thus perpetuate in the undulatory movement of successive generations the impressions or resultants due to the action of external agencies on the individual plastidules. On this view, then, the form and structure of the organism are due to the special mode of vibration of the constituent plastidules. This vibration is affected by external forces. The modified vibration is transmitted to the plastidules by the germ, which, therefore, produce a similarly modified organism. As Mr. J. A. Thomson says, "In metaphorical language, the molecules remember or persist in the rhythmic dance which they have learned."

Darwin's hypothesis was frankly and simply organic—the gemmules are little germs. This of Professor Haeckel tries to go deeper, and to explain organic phenomena in terms of molecular motion. Mr. Herbert Spencer long ago suggested that, just as molecules are built up, through polarity, into crystals, so physiological units are built up, under the laws of organic growth, into definite and special organic forms. Both views involve special units. With Mr. Herbert Spencer, their "polarity" is the main feature; with Professor Haeckel, their "undulatory movements." According to Mr. Spencer, "if the structure of an organism is modified by modified function, it will impress some corresponding modification on the structures and polarities of its units."* According to Professor Haeckel, the vibrations of the plastidules are permanently affected by external forces. In either case, an explanation is sought in terms of molecular science, or rather, perhaps, on molecular analogies. So far good. Such "explanation," if hypothetical, may be suggestive. It may well be that the possibilities of fruitful advance will be found on these lines.

But though, as general theories, these suggestions may be valuable, they do not help us much in the comprehension of our special point. To talk vaguely about "undula-

* Herbert Spencer, "Principles of Biology," vol. i p. 256.

tory movements" or "polarities" does not enable us to comprehend with any definiteness how this particular modification of these particular nerve-cells is so conveyed to the germ that it shall produce an organism with analogous nerve-cells modified in this particular way.

(c) The hypothesis that the germ-plasm may be converted into body-plasm, which, on its return again to the condition of germ-plasm, may retain some of the modifications it received as body-plasm, seems to be negatived, so far as most animals are concerned, by the facts of embryology and development. The distinction of germ-plasm and body-plasm I hold to be mythical. And there is no evidence that cells specially differentiated along certain lines can become undifferentiated again, and then contribute to the formation of ova or sperms. From the view-point of cell-differentiation, which seems to me the most tenable position, there does not seem any evidence for, or any probability of, the occurrence of any roundabout mode of development of the germinal cells which could enable them to pick up acquired characters *en route*.

(d) We come now to the contention that the organism, being one and continuous, if any member suffers, the germ suffers with it. The organs of the body are not isolated or insulated; the blood is a common medium; the nerves ramify everywhere; the various parts are mutually dependent: may we not, therefore, legitimately suppose that long-continued modification of structure or faculty would soak through the organism so completely as eventually to modify the germ? The possibility may fairly be admitted. But how is the influence of the body brought to bear on the germ? The common medium of the blood, protoplasmic continuity, the influence of the products of chemical or organic change,—these are well enough as vague suggestions. But how do they produce their effects? Once more, how is this increased power in that biceps muscle of the oarsman able to impress itself upon the sperms or the ova? No definite answer can be given.

We are obliged to confess, then, that no definite and

satisfactory answer can be given to the question—How can the body affect the germ so that this or that particular modification of body-cells may be transmitted to the offspring? We may make plausible guesses, or we may say—I know not how the transmission is effected; but there is the indubitable fact.

This leads us to the evidence of the fact.

It must be remembered that no one questions the modifiability of the individual. That the epidermis of the oarsman's hand is thickened and hardened; that muscles increase by exercise; that the capacity for thinking may be developed by steady application;—these facts nobody doubts. That well-fed fish grow to a larger size than their ill-fed brethren; that if the larger shin-bone (the tibia) of a dog be removed, the smaller shin-bone (the fibula) soon acquires a size equal to or greater than that of the normal tibia; that if the humerus, or arm-bone, be shifted through accident, a new or false joint will be formed, while the old cavity in which the head of the bone normally works, fills up and disappears; that canaries fed on cayenne pepper have the colour of the plumage deepened, and bullfinches fed on hemp-seed become black; that the common green Amazonian parrot, if fed with the fat of siluroid fishes, becomes beautifully variegated with red and yellow; that climate affects the hairiness of mammals;—these and many other reactions of the individual organism in response to environing conditions, will be admitted by every one.* That constitutional characters of germinal origin are inherited is also universally admitted. The difficulty is to produce convincing evidence that what is acquired is really inherited, and what is inherited has been really acquired.

Attempts have been made to furnish such evidence by showing that certain mutilations have been inherited. I question whether many of these cases will withstand rigid

---

* Mr. J. A. Thomson has published a most valuable "Synthetic Summary of the Influence of the Environment upon the Organism" (Proceedings Royal Physiological Society, Edinburgh: vol. ix. pt. 3, 1888). The case of the Amazonian parrots was communicated to Darwin by Mr. Wallace ("Animals and Plants under Domestication," vol. ii. p. 269).

criticism. Nor do I think that mutilations are likely to afford the right sort of evidence one way or the other. We must look to less abnormal influences. What we require is evidence in favour of or against the supposition that *modifications* of the body-cells are transmitted to the germ-cells. Now, these modifications must clearly be of such a nature as to be receivable by the cells without in any way destroying their integrity. The destruction or removal of cells is something very different from this. If it were proved that mutilations are inherited, this would not necessarily show that normal cell-modifications are transmissible. And if the evidence in favour of inherited mutilations breaks down, as I believe it does, this does not show that more normal modifications such as those with which we are familiar, as occurring in the course of individual life, are not capable of transmission. I repeat, we must not look to mutilations for evidence for or against the supposition that acquired characters are inherited. We must look to less abnormal influences.

These readily divide themselves into two classes. The first includes the direct effects on the organism of the environment—effects, for example, wrought by changes of climate, alteration of the medium in which the organism lives, and so forth. The second comprises the effects of use and disuse—the changes in the organism wrought by the exercise of function.

Taking the former first, we have the remarkable case of *Saturnia*, which was communicated to Darwin by Moritz Wagner. Mr. Mivart thus summarizes it: "A number of pupæ were brought, in 1870, to Switzerland from Texas of a species of *Saturnia*, widely different from European species. In May, 1871, the moths developed out of the cocoons (which had spent the winter in Switzerland), and resembled entirely the Texan species. Their young were fed on leaves of *Juglans regia* (the Texan form feeding on *Juglans nigra*), and they changed into moths so different, not only in colour, but also in form, from their parents, that they were reckoned by entomologists as a distinct species."[*] Professor Mivart also reminds us that English oysters transported to the Mediterranean are recorded by M. Costa to have become rapidly like the true Mediterranean oyster, altering their manner of growth, and forming prominent diverging rays; that setters bred at Delhi from carefully paired parents had young with nostrils more contracted, noses more pointed, size inferior, and limbs more slender than well-bred setters ought to have; and that cats at Mombas, on the coast of Africa, have short, stiff hair instead of fur, while a cat from Algoa Bay, when left only eight weeks at Mombas, underwent a complete metamorphosis — having parted with its sandy-coloured fur. Very remarkable is the case of the brine-shrimp *Artemia*, as observed and described by Schmankewitsch. One species of this crustacean, *Artemia salina*, lives in brackish water, while *A. milhausenii* inhabits water which is much salter. They have always been regarded as distinct species, differing in the form of the tail-lobes and the character of the spines they bear. And yet, by gradually altering the saltness of the water, either of them was transformed into the other in the course of a few generations. So long as the altered conditions remained the same, the change of form was maintained.

Many naturalists believe that climate has a direct and determining effect on colour, and contend or imply that it is hereditary. Mr. J. A. Allen correlates a decrease in the intensity of colour with a decrease in the humidity of the climate. Mr. Charles Dixon, in his "Evolution without Natural Selection," says, "The marsh-tit (*Parus palustris*) and its various forms supply us with similar facts [illustrative of the effects of climate on the colours of birds]. In warm, pluvial regions we find the brown intensified; in dry, sandy districts it is lighter; whilst in Arctic regions it is of variable degrees of paleness, until, in the rigorous climate of Kamschatka, it is almost white." Mr. Dixon does not think that these changes are the result of natural selection. "Depend upon it," he says, with some

[*] St. George Mivart, "On Truth," p. 378.

assurance,* in considering a different case, "it is the white of the ptarmigan (modified by climatic influence) that has sent the bird to the snowy wastes and bare mountain-tops, and rigorously keeps it there; not the bird that has assumed, by a long process of natural selection, a white dress to conceal itself in such localities." Professor Eimer† contends that in the Nile valley the perfectly gradual transition in the colour of the inhabitants from brownish-yellow to black in passing from the Delta to the Soudan is particularly conclusive for the direct influence of climate, for the reason that various races of originally various colours dwell there.

Mr. A. R. Wallace says‡ of the island of Celebes "that it gives to a large number of species and varieties (of Papilionidæ) which inhabit it, (1) an increase of size, and (2) a peculiar modification in the form of the wings, which stamp upon the most dissimilar insects a mark distinctive of their common birthplace." But this similarity may largely, or at least in part, be due to mimicry. Most interesting and valuable are the results of Mr. E. B. Poulton's experiments on caterpillars and chrysalids.§ They show that there is a definite colour-relation between the caterpillar (e.g. the eyed hawk-moth, *Smerinthus ocellatus*) and its food-plant, adjustable within the limits of a single life; that the predominant colour of the food-plant is itself the stimulus which calls up a corresponding larval colour; that there is also a direct colour-relation between the chrysalids of the small tortoiseshell butterfly (*Vanessa urticæ*) and the surrounding objects, the pupæ being dark grey, light grey, or golden, according to the nature and

---

* *Op. cit.*, p. 47. I venture to say, "with some assurance," because Charles Darwin, who had also considered this matter, writes, "Who will pretend to decide how far the thick fur of Arctic animals, or their white colour, is due to the direct action of a severe climate, and how far to the preservation of the best-protected individuals during a long succession of generations?" ("Animals and Plants under Domestication," p. 415).

† "Organic Evolution," English translation, p. 88.

‡ "Contributions to Natural Selection," p. 197.

§ Since this was written, Mr. Poulton has described his results in an interesting volume on "The Colours of Animals" (*q.v.*).

colour of the surroundings; and that the larvæ of the emperor moth (*Saturnia carpini*) spin dark cocoons in dark surroundings, but white ones in lighter surroundings. These are but samples of the interesting results Mr. Poulton has obtained.

What shall we say of such cases? Some of them seem to indicate the very remarkable and interesting fact that changes of salinity of the medium, or changes of food, or the more general influence of a special climate, may modify organisms in *particular* and little-related ways. The larvæ of a Texan *Saturnia* fed on a new food-plant develop into imagos so modified as to appear new species. Changes of salinity of the water modify one species of *Artemia* into another. If these be adaptations, the nature of the adaptation is not obvious. If the new character produced in this way be of utilitarian value, where the utility comes in is not clear. The facts need further confirmation and extension, which may lead to very valuable results. Mr. Poulton's observations, on the other hand, give us evidence of direct adaptation to colour-surroundings. But the effects are, in the main, restricted to the individual. What is hereditary is the power to assume one of two or three tints, that one being determined by the surrounding colour. His experiments neither justify a denial nor involve an assertion of the transmissibility of environmental influence. Secondly, some of the cases above cited seem to show clearly that, under changed conditions of life, the changes which have been wrought in one generation may *reappear* in the next. But are they inherited? Is there sufficient evidence to show conclusively that the body-cells have been modified, and have handed on the modification to the germ? Can we exclude the direct action of the more or less saline water, or the products of the unwonted food on the germinal cells? Can we be sure that there is really a summation of results—that each generation is not affected *de novo* in a similar manner? No one questions that the individual is modifiable, and that such modification is most readily effected in the early and plastic stages of life. If each plastic embryo is moulded in turn by similar influence, how can we conclusivly prove hereditary summation? Take a case that has been quoted in support of hereditary modification. Greyhounds transported from England to the uplands of Mexico are unable to course, owing to the rarity of the atmosphere. Their pups are, however, able to run down the fleetest hares without difficulty. Now, this may be due to the fact that the dogs acquire a certain amount of accommodation to a rare atmosphere, and hand on their acquired power to their offspring, which carry it on towards perfection. But it may also be due to the fact that the pups, subject from the moment of birth to the conditions of a rarified atmosphere, are developed in accordance with these conditions.

Or take another case that has been brought forward. English dogs are known in hot climates, like that of India, to degenerate in a few generations. Let us suppose that these degenerate dogs are removed back to England, and that their pups, born in English air and in our temperate climate, are still degenerate: would not this, it may be asked, show that the influence of climate on the body is inherited? I do not think that such a case would be convincing. For the climate might well influence the germ through the body. The body being unhealthy and degenerate, the germ-cells must, one may suppose, suffer too. The degenerate pup born in England might well owe its degeneracy to effects wrought upon the germinal cells. In other words, such a case would indicate some *general* influence of the environment (including the environing body) on the germ. It does not convince us that *particular* modifications of body-cells as such are transmitted under normal and healthy conditions.

On the whole, it seems to me that the evidence we at present possess on this head is not convincing or conclusive in favour of the effects on the body alone being transmitted to offspring. If cases can be brought forward in which there can be no direct influence on the germ, in which elimination is practically excluded, and in which there is a *gradual and increasing* accommodation of successive generations of organisms to changed conditions *which remain constant*, then such transmission will be rendered probable. I do not know that there are observations of this kind of sufficient accuracy to warrant our accepting this conclusion *as definitely proved*.

Attention may here be drawn to a peculiar and remarkable mode of influence. If a pure-bred mare have foals by an ill-bred sire, they will be ill-bred. This we can readily understand. But if she subsequently have a foal by a perfectly well-bred sire, that foal, too, may in some cases be tainted by the blemish of the previous sire. So, too, with dogs. If a pure-bred bitch once produce a mongrel litter, no matter how carefully she be subsequently matched, she will have a tendency to give birth to pups with a mongrel taint. This subsequent influence of a previous sire is a puzzling fact. It may be that some of the male germ-nuclei are absorbed, and influence the germ-cells of the ovary. But this seems an improbable solution of the problem. It is more likely, perhaps, that in the close relation of mother and fœtus during gestation, each influences the other (how it is difficult to say). On this view the bitch retains the influence of the mongrel puppies —is herself, in fact, partially mongrelized—and therefore mongrelizes subsequent litters. It would not be safe, however, to base any far-reaching conclusions on so peculiar a case, the explanation of which is so difficult. At all events, it is impossible to exclude the possibility of direct action on the germ, though the *particular* nature of the results of such influence are noteworthy.

We may pass now to the evidence that has been adduced in favour of a cumulative effect in the exercise of function, or of the inheritance of the results of use or disuse. Here, again, it must be remembered that no one questions the effects of use and disuse in the individual. What we seek is convincing evidence that such effects are inherited.

Physiologically, the effects of use or disuse are, in the main, effects on the relative nutrition, and hence on the differential growth of organs. When an organ is well exercised, there is increased nutrition and increased growth of tissue, muscular, nervous, glandular, or other. When an organ is, so to speak, neglected, there is diminished blood-supply, diminished growth, and diminished functional power. The development of a complex activity would necessitate a complex adjustment of size and efficiency of parts, involving a nice balance of differential growth dependent on delicately regulated nutrition. What is the evidence that adjusted nutrition can be inherited?

With regard to man, there is some evidence which bears upon this subject. Mr. Arbuthnot Lane, in his valuable papers in the *Journal of Anatomy and Physiology*, has shown that certain occupations, such as shoemaking, coal-heaving, etc., produce recognizable effects upon the skeleton, the muscular system, and other parts of the organization. And he believes* that such effects are inherited, being very much more marked in the third generation than they were in the first. Sir William Turner informed Professor Herdman that, in his opinion, the peculiar habits of a tribe, such as tree-climbing among the Australians, or those natives of the interior of New Guinea whose houses are built in the upper branches of lofty trees, not only affect each generation individually, but have an intensified action through the influence of heredity.†

Mr. Francis Galton's results mainly deal with human faculty; and though faculty has undoubtedly an organic basis, I do not propose to consider the evidence afforded by instinct, intelligence, or intellectual faculties in this chapter. Mention should, however, be made of the interesting results of his study of twins. Twins are either of the same sex, in which case they are remarkably alike, or of different sexes, in which case they are apt to differ even more widely than is usual with brothers and sisters. The former are believed to be developed from one ovum

---

* See *Journal of Anatomy and Physiology*, vol. xxii. p. 215.
† See Professor Herdman's Inaugural Address, Liverpool Biological Society, 1888.

which has divided into two halves, each of which has given rise to a distinct individual; the latter from two different ova. Mr. Galton collected a large mass of statistics concerning twins of both classes. The result of this analysis seems to be that, in the case of "identical twins," the resemblances are not superficial, but extremely intimate; that they are not apt to be modified to any large extent by the circumstances of life; that where marked diversity sets in it is due to some form of illness; and, on the whole, that innate tendencies outmaster acquired modifications. "Nature is far stronger than nurture within the limited range that I have been careful to assign to the latter." On the other hand, speaking of dissimilar twins, Mr. Galton says, "I have not a single case in which my correspondents speak of originally dissimilar characters having become assimilated through identity of nurture." "The impression that all this evidence leaves on the mind is one of some wonder whether nurture can do anything at all, beyond giving instruction and professional training." "There is no escape from the conclusion that nature prevails enormously over nurture where the differences of nurture do not exceed what is commonly to be found among persons of the same rank of society and in the same country." *

Combining the results of Messrs. Lane and Galton, we may say that it requires persistent and long-continued influence to modify the individual, and change, even by a little, the structure inherited or given by nature; but that if this structure is thus modified, there may be a tendency for such modification to increase by hereditary summation of effects. We require, however, further and fuller observations to render the evidence of such hereditary summation to any extent convincing.

Turning now from the evidence afforded by man † to

---

* Francis Galton, "**Inquiries into Human Faculty,**" p. 216.

† That the epidermis **is thicker on the palms of** the hands and the soles of the feet in the infant **long before birth, may be** attributable to the inherited effects of use or pressure. It **can hardly be held that** the thickening of the skin in these parts is of elimination value.

that afforded by animals, we may consider first that presented by domesticated breeds. They might be expected to afford exceptionally good examples. Their modifiability and the readiness with which they interbreed are two of the determining causes of their selection for domestication. They have, moreover, been placed under new conditions of life, and they undoubtedly exhibit changes of structure, many of which Darwin* regarded as attributable to the effects of use and disuse. In domestic ducks, the relative weight and strength of the wing-bones have been diminished, while conversely the weight and strength of the leg-bones have been increased. The bones of the shoulder-girdle have been decreased in weight and "the prominence of the crest of the sternum, relatively to its length, is also much reduced in all the domestic breeds. These changes," says Darwin, "have evidently been caused by the lessened use of the wings." The shoulder-girdle and breast-bone of domestic fowls have been similarly reduced. After a careful consideration of numerous facts concerning the brains of rabbits, Darwin concluded that this "most important and complicated organ in the whole organization is subject to the law of decrease in size from disuse." And Sir J. Crichton Browne has recently shown that, in the wild duck, the brain is nearly twice as heavy in proportion to the body as it is in the comparatively imbecile domestic duck. In pigs, the nature of the food supplied during many generations has apparently affected the length of the intestines; for, according to Cuvier, their length to that of the body in the wild boar is as 9 to 1, in the common domestic boar as 13·5 to 1, and in the Siam breed as 16 to 1. With regard to horses, Darwin tells us that "veterinarians are unanimous that horses are affected with spavins, splints, ring-bones, etc., from being shod and from travelling on hard roads, and they are almost unanimous that a tendency to these malformations is transmitted."

These are samples of the effects of domestication. It has been suggested, however, that, quite apart from any

* The instances cited are from "Animals and Plants under Domestication."

diminution from disuse, the reduction of size in parts or organs may be the result of the absence or cessation of selection. If an organ be subject to selection, the mean size in adult creatures will be that of the selected individuals; but if selection ceases, it will be the mean of those born. Let us suppose that nine individuals are born, and that the size of some organ varies in these from 1, the most efficient, to 9, the least efficient. The birth-mean will therefore be, as shown on the left-hand side of the following table, at the level of number 5, four being more efficient, and four less efficient. But if, of these nine, six be eliminated, then the mean of the survivals will be as shown on the right-hand side of the table:—

$$\begin{array}{l}1\\2\text{—Survival-mean.}\\3\\\left.\begin{array}{l}4\\\text{Birth-mean—}5\\6\\7\\8\\9\end{array}\right\}\text{Eliminated individuals.}\end{array}$$

The result, then, of the cessation of selection will be to reduce the survival-mean to the birth-mean, and that without any necessary effect of disuse. But unless this be accompanied by a tendency to diminution due to economy of growth or some other cause, this cannot produce any well-marked or considerable amount of reduction. I very much question, for example, whether the cessation of selection, even with the co-operation of the principle of economy of growth, will adequately account for the reduction to nearly one-half its original proportion of the brain of the duck. The subject will be more fully discussed, however, in the next chapter.

There is probably but little tendency for disused parts to be reduced in size through artificial selection. An imbecile duck does not probably taste nicer than one with bigger brains. On the other hand, the increase of size in organs may presumably, in certain cases, be increased by selection. Pigs, for example, have been selected according to their fattening capacity. Those with longer intestines, and therefore increased absorbent surface, may well have an advantage in this respect. Hence, in selecting pigs for fattening, breeders may have been unconsciously selecting those with the longest intestines. Of course, on this view, the longer intestine must be there to be selected, and the increased length must be due to variation. But this may be all-round variation (cause unknown), not variation in one direction, the result of increased function.

Another point that has to be taken into consideration is the amount of *individual* increment or decrement, owing to individual use or disuse, apart from any possible summation of results.

Seeing, then, that it is difficult to estimate the amount of purely individual increment or decrement, and that it is difficult, if not impossible, to exclude the disturbing effects of cessation of selection with economy of growth on the one hand, reducing the size of organs, and artificial selection on the other hand, increasing the size or efficiency of parts, it is clear that such cases cannot afford convincing evidence that the observed variations are the directly inherited results of use and disuse. Indeed, I am not aware of any experiments or direct observations on animals which are individually conclusive in favour of the hereditary summation of functionally produced modifications.

It may, however, be said—Although no absolutely convincing experiments or observations are forthcoming (for, from the nature of the case, it is almost impossible logically to prove that this interpretation of the facts is alone possible), still there are cases which are much more readily explained on the hypothesis that the effects of use and disuse are inherited, than on any other hypothesis. But, so far as Professor Weismann and his followers are concerned, such an argument is wholly beside the question. They are ready to admit that inherited modifications of the body, if they could be proved, would render the explanation of many results of evolution much easier. It would, no doubt, they say, be easier to account for the shifting of the eye of a flat-fish from one side of the head to the other on the supposition that individual efforts were inherited, until, by an hereditary summation of effort, the eye at last came round. The question is—Are we justified in accepting the easier explanation if it be based on a mere assumption, at present unproved, the *modus operandi* of which is inexplicable?

Let us consider very briefly these two points—first, the "mere assumption;" secondly, "the inexplicable *modus operandi*." Is there any reason why we should not assume the inheritance of effects of use or disuse as a working hypothesis, if it is not in opposition to any known biological law, and if it does enable us to explain certain observed phenomena? I see no such reason. We do not know enough about the causes of variation to be rigidly bound by the law of parcimony. I am not aware of any biological law that would render the acceptance of this view as a provisional hypothesis unjustifiable.

But how, it is asked, can we accept it if its *modus operandi* is inexplicable? I question the validity of this argument. I fear our knowledge of organic nature is not at present so full and exact as to justify us in excluding an hypothesis because we are not able to give an adequate answer to the question—How are these effects produced? Of course, if it can be shown that no *modus operandi* is possible, there is an end of the matter. But who shall dare thus to limit the possibilities of organic nature? And, if possible, then that natural selection in which the neo-Darwinians place their sole trust would certainly develop so advantageous a mode of influence. It is clear that a species sensitive to every shock of the environment on the organism would be unstable, and hence at a disadvantage. But, on the other hand, the ability to answer by adaptation to long-continued and persistent environmental influence or to oft-repeated and consistent performance of function would be so distinct an advantage to the species which possessed it, that, if it lay within the possibilities of organic nature, natural selection, always, as we are told, on the look out for every possible advantage, would assuredly seize upon it and develop it.

Those who believe in the absolute sway of natural selection have not at present given any adequate answer to the question—How are particular variations (*e.g.* the twisted skull of flat-fish) produced? They say that constitutional variations, which are alone inheritable, are due to variations in the germs. When asked how these variations are produced, they are forced to reply—We cannot say. But when it is suggested that they may be in some unknown way transmitted to the germ from the body, they are up in arms, and exclaim—You have no right to believe that, or ask us to believe it, unless you can tell us plainly how the effect is produced. Unable themselves to give the *modus operandi* of the origin of particular variations, they demand the exact *modus operandi* from those who suggest that variations may arise through this mode of influence of the body on the germ.

We shall have to consider this question from a more general standpoint in the next chapter on "Organic Evolution." We may now very briefly summarize some of the results we have reached in this chapter.

The ova and sperms are specially differentiated cells which have, in the division of labour, retained and emphasized the function of developmental reproduction.

There is a continuity of such cells. The cells which become ova or sperms have never become differentiated into anything else.

Hereditary similarity is due to the fact that parents and offspring are derived eventually from the same germinal cells.

Variation in the existing world is partly due to sexual union. But if there be mere admixture, new characters cannot arise in this way, nor can old characters be strengthened beyond the existing maximum.

Some mode of organic combination (analogous to chemical combination) might afford an explanation of the occurrence of new variations and the increase of existing characters.

In the protozoa there may be a summation of the effects of the environment in succeeding generations.

There is no convincing evidence that in the metazoa special modifications of the body so influence the germ as to become hereditary.

But there is no reason why such influence should not be assumed as a provisional hypothesis.

CHAPTER VI.

ORGANIC EVOLUTION.

It is difficult to realize the wealth, the variety, the diversity, of "animal life." Even if we endeavour to pass in review all that we have seen in woodland and meadow, in pond or pool, in the air, on the earth, in the waters, in temperate or tropical regions; even when we try to remember the results of all anatomical and microscopic investigation displaying new wonders and new diversities hidden from ordinary and unaided vision; even when we call to mind the multifarious contents, recent and fossil, of all the natural history museums we have ever visited, and throw in such mental pictures as we have formed of all the diverse adaptations we have read about or heard described;—even so we cannot but be conscious that not one-tenth, not one-hundredth, part of the diversity and variety of animal life has passed before our mental vision even in sample. It is said that our greatest living poet once, when a young man, left his companions to gaze into the waters of a clear, still pool. "What an imagination God has!" he said, as he rejoined his friends. Fit observation for the poet, whose sensitive nature must be keenly alive to the varied endowments which Nature has lavishly showered upon her animate children.

Certain it is that words, mere words, can never present, though they may aid in recalling, an adequate picture of either the wealth or the beauty of animal life. Fortunately for those who visit London (and who nowadays does not?), we have, in our national collection in South Kensington, the means of getting some insight into the wealth of life.

And much is being done there to aid the imagination and to facilitate study for those who are not professed students. Many of the birds are now to be seen set in their natural surroundings, with their life-history illustrated. Our frontispiece is taken from one of these cases. And this admirable system will, no doubt, so far as space permits, be extended; and, perhaps, dramatic incidents may be introduced, like those (notably in the life of heron and hawk) which form so marked a feature in the little museum at Exeter. Anything which leads us to understand the life of animals, and to go forth and study it for ourselves, has an educational value.

In our National Museum, again, much is being wisely done to illustrate the diversity and variety of structure and the principles that underlie them. Observe, as you enter the central hall, the case containing stuffed specimens of ruffs (*Machetes pugnax*). Among the young autumn birds there is not much difference between males and females, the male being distinguished chiefly by its somewhat larger size. Nor do the old birds, male and female, differ much during the winter months. But in pairing-time, May and June, the females are somewhat richer in colour; while the males not only don the ruff to which the bird owes its popular name, but develop striking colour-tints. Among different individuals it will be seen that the colour-variation is tolerably wide; but the same individual keeps strictly, we are told, in successive seasons, to the same summer dress. Note, next, in a bay to the right, the great variety of form, ornamentation, and colouring among the molluscan shells there exhibited. Observe that the rich colours are often hidden during life by the dull epidermis. Half an hour's attentive study of these varied molluscan forms will give a better idea of the beauty and diversity of these life-products than pages of mere description.

Pass on, too, to note, in a further bay to the right, the extraordinary modifications of the antenna, or feeler, in insects. There is the long, whip-like form in the locust;

the clubbed whip in the ant-lion and the butterfly; the feathered form in certain moths and flies; the hooked form characteristic of the sphinx-moths; the many-leaf form in the lamellicorn beetles, like the cockchafer; and the feathered plate of other beetles. Equally wonderful are the diverse developments of the mouth-organs of insects, the spiral tube of the butterfly or moth, the strong jaws of the great beetles, the lancets of the gnat, the sucking-disc of the fly,—all of them special modifications of the same set of structures. Then, in the same bay, note some of the striking differences between the males and females of certain insects. In some there is an extraordinary difference in size (*e.g.* the locust *Xiphocera*, and the moth *Attacus*); in others, like the stag-beetle, it is the size of the jaws that distinguishes the males; in others, again, the most notable differences are in the length, development, or complexity of the antennæ, or feelers; in some beetles the males have great horns on the head or thorax; while in many butterflies it is in richness of colour that the difference chiefly lies—the brilliant green of the *Ornithoptera* there exhibited contrasting strongly with the sober brown of his larger mate.

The fact that the special characteristics of the male, which we have seen to be variable in the ruff, are also variable among insects, is well exemplified in the case of the stag-beetle, in some males of which the mandibles are far larger than in others. This is shown in Fig. 22, which is copied from the series displayed in the British Museum, by the kind permission of Professor Flower.

Crossing the hall to where the vertebrate structures are displayed, the development of hair, of feathers, of teeth, the modifications of the skull and of legs, wings, and fins are being exemplified. Note here and elsewhere the special adaptations of structure, of which we may select two examples. The first is that seen in the *Balistes*, or trigger-fish. The anterior dorsal fin is reduced to three spines, of which that which lies in front is a specially modified weapon of defence, while that which follows it is the so-

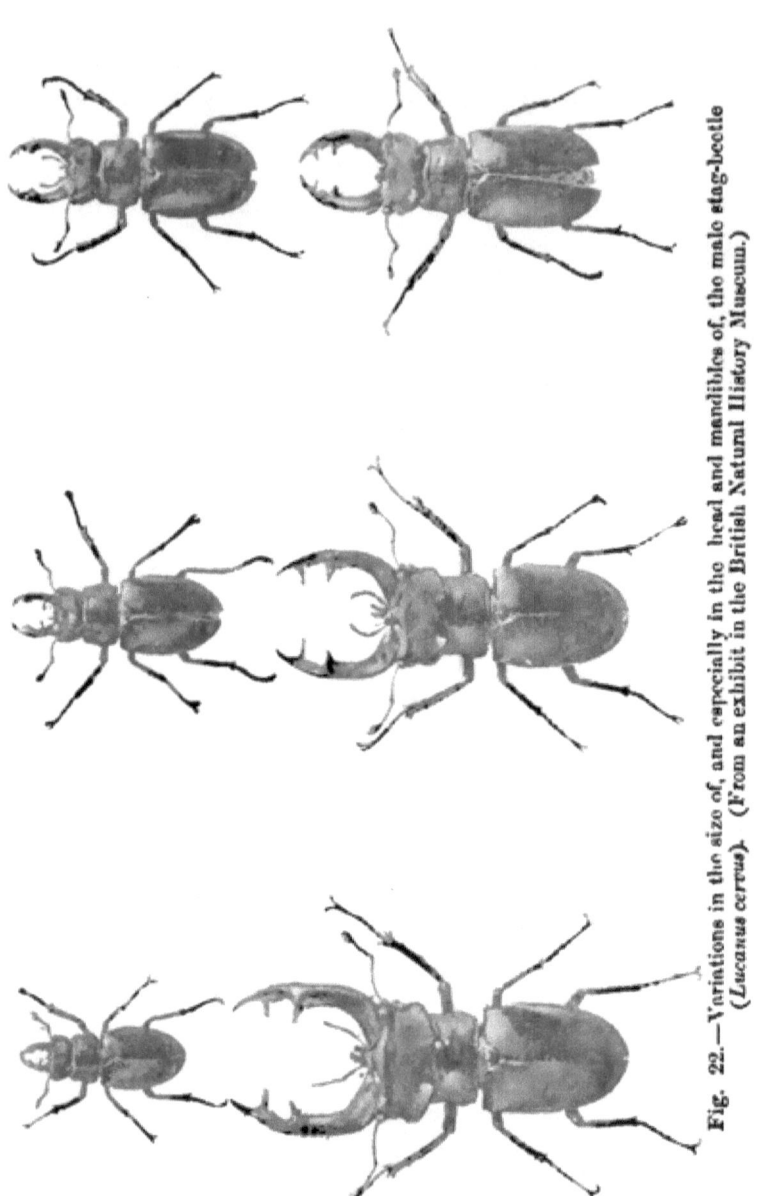

Fig. 22.—Variations in the size of, and especially in the head and mandibles of, the male stag-beetle (*Lucanus cervus*). (From an exhibit in the British Natural History Museum.)

called trigger. These two are so hinged to the underlying interspinous bones and so related to each other that, when once the defensive spine in front is erected, it cannot be forced down until the trigger is lowered. The second example of special adaptation is well displayed in specimens of the mud-tortoise *Trionyx*. Between the last vertebra of the neck and the first fixed vertebra of the dorsal series is a beautiful hinge-joint, enabling the neck to be bent back, S-fashion, when the creature withdraws its head within the carapace. These are only one or two particular instances of what any one who will visit the National Museum may see for himself admirably displayed and illustrated.

No one can, one would suppose, pass through the galleries in Cromwell Road and remain quite insensible to the beauties of animal life. Beauty of form and beauty of colour are conspicuously combined in many species of birds and insects. And much of this colour-beauty and splendid iridescence is known to be due to minute scales, to thin films of air or fluid, and to microscopically fine lines developed upon scales or feathers. But there is one phase of beauty which cannot be exhibited in the museum—the beauty that comes of life as opposed to death. For this we must go out into the free air of nature, where the animals not only have lived, but are still instinct with the glow of life, and where the silence of the museum galleries is replaced by the song of birds and the hum of insect-wings.

How have this wealth, this diversity, this beauty, this manifold activity, which we summarize under the term "animal life," been produced?

If we answer this question in a word—the word "evolution"[*]—we must remember that this word merely expresses our belief in a general fact; and we must not

---

[*] It is beyond the scope of this book to give the *evidences* of evolution. Such evidence from embryology, from distribution, and from palæontology, is now abundant. For palæontological evidence, see Nicholson's "Manual of Palæontology," 3rd edit., especially the second volume on "Vertebrates," by R. Lydekker.

forget that many questions remain behind, all centering round that little question, to which an adequate answer is so difficult to give, the question—How? Reduced to its simplest expression, the doctrine of evolution merely states that the animal world as it exists to-day is naturally developed out of the animal world as it existed yesterday, and will in turn develop into the animal world as it shall exist to-morrow. This is the central belief of the evolutionist. No matter what moment in the past history of life you select, the life at that moment was in the act of insensibly passing from the previous towards a future condition. Then at once arises the question—Does life remain the same yesterday, to-day, and to-morrow? A thousand indubitable facts at once make answer—No! Underlying the law of continuity there is a law of change. Life to-day is not what it was yesterday, nor will it be to-morrow the same as to-day. What, then, is the nature of this change? If it be replied that the change must be either for the better or the worse, we shall have to answer the further question—Better or worse in what respects?

Let us narrow our view from the contemplation of life as a whole to the more particular consideration of an organism as one of its constituent units. The individual life of that organism depends on (some would say consists in) its ceaseless adaptation to surrounding circumstances. The circumstances remaining the same, or only varying within constant limits, the adaptation may be *more* or *less* perfect. A change in the direction of more perfect adaptation will be a change for the better, a tendency to less perfect adaptation will be a change for the worse.

But the relation of an organism to its circumstances or environment is itself subject to change. The environment itself may alter, or the organism may be brought into relation with a new environment. We have to consider not only the changes in an organism in the direction of more or less perfect adaptation to its environment, but also changes in the environment. These changes are in the direction of increased simplicity or of increased complexity. So that we may say that the modification of life is in the direction of more or of less complete adaptation to simpler or to more complex conditions. Where the adaptation advances to more complex conditions, we speak of elaboration; where it retrogrades to less complex conditions, we speak of degeneration; but both fall under the head of evolution in its more general sense. Viewed as a whole, there can be little doubt that the general tendency of evolution is towards more complete adaptation to more diverse and complex environment. And this tendency is accompanied by a general increase of differentiation and of integration; of differentiation whereby the constituent elements of life, whether cells, tissues, organs, organisms, or groups of organisms, become progressively more specialized and more different from one another; of integration whereby these elements become progressively more interdependent one on the other. We may conveniently sum up the tendency towards more perfect adaptation to more complex circumstances in the word *progress*; the tendency to differentiation in the word *individuality*; and the tendency to integration in the word *association*.

Nobody now doubts the propositions thus briefly summarized, and it is therefore unnecessary to bring forward evidence in their favour.

We may pass, then, to the question—How? Evolution being continuity, associated with change, tending in certain directions, and accompanied by certain processes, how has it been effected? What are its methods?

Natural Selection.

Natural selection claims a foremost place. We have already devoted a chapter to its consideration. Animals vary; more are born than can survive to procreate their kind; hence a struggle for existence, in which the weaker and less adapted are eliminated, the stronger and better adapted surviving to continue the race.

It is scarcely possible to over-estimate what Darwin's labour and genius have done for the study of animal life. Through Darwin's informing spirit, biology has become a science. But now we must be on our guard. So long as natural selection was winning its way to acceptance, every application of the theory had to be made with caution, and was subjected to keen, if sometimes ignorant, criticism. Now there is, perhaps, some danger lest it should suffer the Nemesis of triumphant creeds, and be used blindly as a magic formula.

First, we should be careful not to use the phrase, "of advantage to the species," vaguely and indefinitely, but should in all cases endeavour clearly to indicate wherein lies the particular advantage, and how its possession enables the organism to escape elimination; next, we must remember that the advantage must be immediate and present, prospective advantage being, of course, inoperative; then we must endeavour to show that the advantage is really sufficient to decide the question of elimination or non-elimination; lastly, we must distinguish between indiscriminate and differential destruction, between mere numerical reduction by death or otherwise and selective elimination.

(1) In illustration of the first point, we may select a passage from the writings of even so great a biologist as Professor Weismann. As is well known, Professor Weismann believes that senility and death are no part of the natural heritage of animal life, but have been introduced among the metazoa on utilitarian grounds. In his earlier papers, he attributed the introduction of death, and the tissue-degeneration that precedes it, to the direct action of natural selection.* More lately, he attributes it to the cessation of selection.† Concerning this later view, we shall have somewhat to say presently; we may now consider the former as an example of too indefinite a use of such phrases as "of advantage to the species." "Worn-out individuals," says Professor Weismann, "are not only

---

* Weismann, "Essays on Heredity," p. 24.   † Ibid. p. 140.

valueless to the species, but they are even harmful, for they take the places of those which are sound. Hence, by the operation of natural selection, the life of our hypothetically immortal individual would be shortened by the amount which was useless to the species. It would be reduced to a length which would afford the most favourable conditions of existence of as large a number as possible of vigorous individuals at the same time." This may be so, but, as it stands, the *modus operandi* is not given, and is not obvious. We start with a hypothetically immortal metazoon. Barring accidents, it will go on existing indefinitely. But you cannot bar accidents for an indefinite time; hence, the longer the individual lives, the more defective and crippled it becomes. There is neither natural decay nor natural death here. The organism is gradually crippled through accident and injury. But the crippled individuals are harmful to the species, because they take the places of those which are sound. Therefore, says Professor Weismann, natural decay and death step in to take them off before they have time to become cripples. Now, the point I wish to notice is that there is no definite statement how or why natural decrepitude should thus be introduced. We must remember that it is not until a late stage in evolution that, through the association of its members, groups of organisms compete with other groups. In the earlier stages, when we must suppose decrepitude and death to arise on Professor Weismann's hypothesis, the law of the struggle for existence is—each for himself against all. The question, therefore, is—What advantage *to the individual* is there in natural decay and death to enable it, through the possession of these attributes, to escape elimination? Surely none as such. At the same time, it is quite conceivable that natural decay and death may be the penalty the individual has to pay for increased strength and vitality in the early stages of life. This, probably, was Professor Weismann's meaning. But, if so, it would surely have been better to state the matter in such a way as to lay the chief stress on the really important feature, and to say that, through natural selection, those individuals have survived which exhibited predominant strength and vitality for a shortened period, even at the expense of natural decay and death. The increased life-power, not the seeds of decay and death, was that which natural selection picked out for survival, or rather that which elimination allowed to survive.

In such ways—a short life with heightened activity being of advantage to some forms, a more prolonged existence at a lower level of vitality being essential to others—natural selection may have determined in some degree the relative longevity of different organisms. That it caused the introduction of senility as a preparation for death is a less tenable hypothesis.

And here we may note, in passing, that in using the phrase, "of advantage to the race or species," we must steadily bear in mind the fact that it is with *individuals* that the process of elimination deals. In the individual it is that every modification must make good its claim to existence and transmission. Where the principle of association for mutual benefit obtains, as in the case of social insects, it is still the individual that must resist elimination. Self-sacrifice, whether conscious or unconscious, must not be carried so far as to lead to the elimination of the self-sacrificing individual, for in this event it cannot but defeat its own ends. Within these limits, self-sacrifice is of advantage, as in the case of parental self-sacrifice, in that it enables certain other individuals to escape elimination. We should endeavour, then, not to use the phrase, "of advantage to the species," vaguely and indefinitely, but to indicate in what particular ways certain individuals are to be so advantaged as to escape the Nemesis of elimination.

(2) The second point that I mentioned above scarcely needs exemplification. That the advantage which enables an organism to escape elimination must be present and existent, not merely prospective, is obvious. Still, the mistake is sometimes made. I have heard it stated that feathers were evolved for the sake of flight. But clearly,

unless the wing sprang into existence already sufficiently developed for flight, this would be impossible. The same is true of the first stages of many structures which could not be of service for the purpose and use to which they were subsequently turned. Not impossibly, the earliest "wings" were for diving, and flight was, so to speak, an after-thought. Undoubtedly, structures which have been fostered under the wing of one form of advantage have been subsequently applied to new purposes, and fostered through new modes of adaptation. Teeth, for example, are probably modified scales, such as are found in the thornback skate. But the early development of these scales could have had no reference to their future application to purposes subservient to alimentation.

Again, such and such a structure is sometimes spoken of as a "prevision against emergencies." In his interesting and valuable work on "The Colours of Animals," for example, Mr. E. B. Poulton says, "Dimorphism [in the larvæ of butterflies and moths] is also valuable in another way: the widening range of a species may carry it into countries in which one of its forms may be especially well concealed, while in other countries the other form may be more protected. Thus a dimorphic form is more fully provided against emergencies than one with only a single form." And after giving, as an example, the fact that the convolvulus hawk-moth has a browner and a greener form of caterpillar, of which the browner is more prevalent under European conditions, and the greener under those which obtain in the Canary Islands, Mr. Poulton adds, "This result appears to have been brought about by the ordinary operation of natural selection, leading to the extermination of the less-protected variety." Now, I do not mean for one moment to imply that so careful and able a naturalist as Mr. Poulton believes that any character has been evolved through natural selection in prevision for future emergencies. But I do think that his statement is open to this criticism.

(3) It is sometimes said, in bold metaphor, that natural

selection is constantly on the watch to select any modification, however slight, which is of advantage to the species. And it is true that elimination is ceaselessly operative. But it is equally certain that the advantage must be of sufficient value to decide the question whether its possessor should be eliminated or should escape elimination. If it does not reach this value, Natural Selection, watch she never so carefully, can make no use of it. Elimination need not, however, be to the death; exclusion from any share in continuing the species is sufficient. To breed or not to breed, that is the question. Any advantage affecting this essential life-function will at once catch the eye of a vigilant natural selection. But it must be of sufficient magnitude for the machinery of natural selection to deal with. That machinery is the elimination of a certain proportion of the individuals which are born. Which shall be eliminated, and which shall survive, depends entirely on the way in which the individuals themselves come out in life's competitive examination. The manner in which that examination is conducted is often rude and coarse, too rough-and-ready to weigh minute and infinitesimal advantages.

What must be the value of a favourable or advantageous modification to decide the question of elimination, to make it an *available advantage*, must remain a matter of conjecture. It will vary with the nature and the pressure of the eliminative process. And perhaps it is scarcely too much to say that, at present, we have not observational grounds on which to base a reliable estimate in a single instance. We must not let our conviction of its truth and justice blind us to the fact that natural selection is a logical inference rather than a matter of direct observation. A hundred are born, and two survive; the ninety-eight are eliminated in the struggle for existence; we may therefore infer that the two escaped elimination in virtue of their possession of certain advantageous characters. There is no flaw in the logic that has thus convinced the world that natural selection is a factor in evolution. But by what percentage of elimination-marks the second of the two successful candidates beats the senior on the list of failures we do not know. We can only see that, on the hypothesis of natural selection, it must have been sufficiently appreciable to determine success or failure.

(4) And then, to come to our fourth point, we must remember that, apart from the differentiating process of elimination, there is much fortuitous destruction. A hundred are born, and but two survive. But of the ninety-eight which die, and fail to procreate, how many are eliminated, how many are fortuitously destroyed, we do not find it easy to say. And indiscriminate destruction gets rid of good, bad, and indifferent alike. It is a mistake to say that of the hundred born the two survivors are necessarily the very best of the lot. It is quite possible that indiscriminate destruction got rid of ninety of all sorts, and left only ten subject to the action of a true elimination. "In the majority of birds," says Professor Weismann, "the egg, as soon as it is laid, becomes exposed to the attacks of enemies; martens and weasels, cats and owls, buzzards and crows, are all on the look out for it. At a later period, the same enemies destroy numbers of the helpless young, and in winter many succumb in the struggle against cold and hunger, or to the numerous dangers which attend migration over land and sea—dangers which decimate the young birds." There is here, first, a certain amount of fortuitous destruction; secondly, some selection applied to the eggs; thirdly, a selection among the very young nestlings; and, fourthly, a selection among the young migratory birds. What may be the proportion of elimination to destruction at each stage it is difficult to say. Among the eggs and fry of fishes fortuitous destruction probably very far outbalances the truly differentiating process.

Panmixia and Disuse.

We may now pass on to consider shortly some of the phenomena of degeneration, and the dwindling or disappearance of structures which are no longer of use.

Many zoologists believe, or until lately have believed, that disuse is itself a factor in the process. Just as the well-exercised muscle is strengthened, so is the neglected muscle rendered weak and flabby. Until recently it was generally held that the effects of such use or disuse are inherited. But now Professor Weismann has taught us, if not to doubt ourselves, at least to admit that doubt is permissible. On the older view, the gradual dwindling of unused parts was readily comprehensible. But now, if Professor Weismann is right, we must seek another explanation of the facts; and, in any case, we may be led to recognize other factors (than that of disuse alone) in the process.

Professor Weismann regards panmixia, or free intercrossing, when the preserving influence of natural selection is suspended, as the efficient cause of a reduction or deterioration in the organ concerned. And Mr. Romanes had, in England, drawn attention to the fact that the "cessation of natural selection" would lead to some dwindling of the organ concerned, since it was no longer kept up to standard. In illustration of his panmixia, Professor Weismann says, "A goose or duck must possess strong powers of flight in the natural state, but such powers are no longer necessary for obtaining food when it is brought into the poultry-yard, so that a rigid selection of individuals with well-developed wings at once ceases among its descendants. Hence, in the course of generations, a deterioration of the organs of flight must necessarily ensue, and the other members and organs of the bird will be sensibly affected."* And, again, "As at each stage of retrogressive transformation individual fluctuations always occur, a continued decline from the original degree of development will inevitably, although very slowly, take place, until the last remnant finally disappears."† Now, I think it can be shown that panmixia, or the cessation of selection, alone cannot affect much reduction. It can only

---

* Weismann, "Essays on Heredity," p. 90.
† Ibid. p. 292. See also a discussion in *Nature*, in which Mr. Romanes and Professor Ray Lankester took part, beginning vol. xli. p. 437.

affect a reduction from the "survival-mean" to the "birth-mean." This was referred to in the chapter on "Heredity and the Origin of Variations," but may be again indicated. Suppose the number of births among wild ducks be represented by the number nine, of which six are eliminated through imperfections in the organs of flight. Let us place the nine in order of merit in this respect, as is done in the table on p. 172. The average wing-power of the nine will be found in No. 5, there being four ducks with superior wing-power (1—4), and four with inferior wing-power (6—9). The birth-mean will therefore be at the level of No. 5, as indicated to the left of the table. But if six ducks with the poorest wings be eliminated, only three survive. The average wing-power will now be found in No. 2, one duck being superior and one inferior to it in this respect. It is clear that this survival-mean is at a level of higher excellence than the birth-mean. Now, when the ducks are placed in a poultry-yard, selection in the matter of flight ceases, and, since all nine ducks survive, the survival-mean drops to the birth-mean. We may variously estimate this retrogression; but it cannot be a large percentage—I should suppose, in the case under consideration, one or two per cent. at most. But Professor Weismann says, "A *continued* decline from the original degree of development must inevitably take place." It is not evident why such decline should continue. If variations continue in the same proportion as before, the birth-mean will be preserved, since there are as many positive or favourable variations above the mean as there are negative or unfavourable variations below the mean. A continuous decline must result from a preponderance of negative over positive variations, and for this some other principle, such as atavism, or reversion to ancestral characters, must be called in. But in the case of so long-established and stable an organ as that of flight, fixed and rendered constant through so many generations, it is hardly probable that reversion would be an important factor. Mr. Galton has calculated that among human-folk the offspring inherits one-fourth from each parent, one-sixteenth from each grandparent, leaving one-fourth to be contributed by more remote ancestors. There is no doubt, however, that among domesticated animals reversion occurs to characters which have been lost for many generations. But we should probably have to go a very long way back in the ancestry of wild ducks for any marked diminution in wing-power. It must be remembered that, in the case of the artificial selection of domesticated animals, man has been working against and not with the stream of ancestral tendency. Reversion in their case is towards a standard which was long maintained and had become normal before man's interference. Reversion in domesticated ducks should therefore be towards the greater wing-power of their normal ancestry before domestication, not in the direction of lessened wing-power and diminished wing-structure. The whole question of reversion is full of interest, and needs further investigation.

In the dwindling of disused structures, Mr. Romanes has suggested "failure of heredity" as an efficient cause. I find it difficult, however, to distinguish this failure of heredity from the effects of disuse. To what other cause is the failure of heredity due? If natural selection has intervened to hasten this failure, this can only be because the failure is advantageous, since it permits the growth-force to be applied more advantageously elsewhere. And this involves a different principle. Even so it is difficult to exclude the possibility (to put it no stronger) that the diversion of growth-force from a less useful to a more useful organ is in part due to the use of the one and the disuse of the other. But of disuse Mr. Romanes says, "There is the gravest possible doubt lying against the supposition that any really inherited decrease is due to the inherited effects of disuse." We may fairly ask Mr. Romanes, therefore, to explain to what cause the failure of heredity is due. In any case, Professor Weismann and his school are not likely to accept this failure of heredity as an efficient factor in the process. Nor is Professor

Weismann likely to fall back upon any innate tendency to degeneration. Unless, therefore, some cause be shown why the negative variations should be prepotent over the positive variations, we must, I think, allow that unaided panmixia cannot affect any great amount of reduction.

In this connection we may notice Professor Weismann's newer view of the introduction of bodily mortality. He says, "The problem is very easily solved if we seek assistance from the principle of panmixia. As soon as natural selection ceases to operate upon any character, structural or functional, it begins to disappear. As soon, therefore, as the immortality of somatic [body-] cells became useless, they would begin to lose this attribute."[*] Even granting that panmixia could continuously reduce the size of ducks' wings, it is not easy to see how it could get rid of immortality. The essence of the idea of panmixia is that, when the natural selection which has raised an organ to a high functional level, and sustains it there, ceases or is suspended, the organ drops back from its high level. But on Professor Weismann's hypothesis, immortality has neither been produced nor is it sustained by natural selection. How, therefore, the cessation of selection can cause the disappearance of immortality—a character with which natural selection has had nothing whatever to do—Professor Weismann does not explain. He seems to be using "panmixia" in the same vague way that, in his previous explanation, he used "natural selection."

If panmixia alone cannot, to any very large extent, reduce an organ no longer sustained by natural selection, to what efficient cause are we to look? Mr. Romanes has drawn attention to the reversal of selection as distinguished from its mere cessation. When an organ is being improved or sustained by selection, elimination weeds out all those which have the organ in an ill-developed form. Under a reversal of selection, elimination will weed out all those which possess the organ well developed. In burrowing animals, the eyes may have been reduced in size, or even

---

[*] Weismann, "Essay on Heredity." p. 140.

buried beneath the skin, through a reversal of selection. The tuco-tuco (*Ctenomys*), a burrowing rodent of South America, is frequently blind. One which Darwin kept alive was in this condition, the immediate cause being inflammation of the nictitating membrane. "As frequent inflammation of the eyes," says Darwin, "must be injurious to any animal, and as eyes are certainly not necessary to animals having subterranean habits, a reduction in their size, with the adhesion of the eyelids and growth of fur over them, might in such cases be an advantage; and, if so, natural selection would aid the effect of disuse.* Granting that the inflammation of the eyes is a sufficient disadvantage to lead to elimination, such cases may be assigned to the effects of a reversal of selection.

Perhaps the best instances of the reversal of selection are to be found in the insects of wind-swept islands, in which, as we have already seen (p. 81), the power of flight has been gradually reduced or even done away with. Such instances are, however, exceptional. And one can hardly suppose that such reversal of selection can be very far-reaching in its effects, at least, through any direct disadvantage from the presence of the organ. One can hardly suppose that the presence of an eye in a cave-dwelling fish† could be of such direct disadvantage as to lead to the elimination of those members which still possess this structure.

But may it not be of indirect disadvantage? May not this structure be absorbing nutriment which would be more advantageously utilized elsewhere? This is Darwin's principle of economy. Granting its occurrence, is it effective? We may put the matter in this way: The crustacea which have been swept into a dark cave may be divided into three classes so far as fortuitous variations of eyes

---

* "Origin of Species," p. 110.

† With regard to blind cave-fish, Professor Ray Lankester has suggested that some selection has been effected. Those animals whose sight-sensitiveness enabled them to detect a glimmer of light would escape to the exterior, leaving those with congenitally weak sight to remain and procreate in the darkness of the cave.

and antennæ are concerned. First, those which preserve eyes and antennæ in the original absolute and relative proportion and value; secondly, those in which, while the eyes remain the same, the antennæ are longer and more sensitive; thirdly, those in which, while the antennæ are longer and more sensitive, the eyes are reduced in size and elaboration. According to the principle of economy, the third class have sufficient advantage over the first and second to enable them to survive and escape the elimination which removes those with fully developed eyes. It may be so. We cannot estimate the available advantage with sufficient accuracy to deny it. But we may fairly suppose that, in general, it is only where the useless organ in question is of relatively large size, and where nutriment is deficient, that economy of growth is an important factor.

We may here note the case of the hermit crab as one which exemplifies degeneration through the reversal of natural selection. This animal, as is well known, adopts an empty whelk-shell or other gasteropod shell as its own. The hinder part of the body which is thus thrust into the shell loses its protective armour, and is quite soft. Professor Weismann seems to regard this loss of the hardened cuticle as due entirely to panmixia. If what has been urged above has weight, this explanation cannot be correct. No amount of promiscuous interbreeding of crabs could reduce the cuticle to a level indefinitely below that of any of the interbreeding individuals. But it is clear that an armour-sheathed "tail" would be exceedingly ill adapted to thrusting into a whelk-shell. Hence there would, by natural selection, be an adaptation to new needs, involving not the higher development of cuticle, but the reverse. So far as the cuticle is concerned, it is a case of reversed selection. Whether this reversal alone will adequately account for the facts is another matter.

Mr. Herbert Spencer has made a number of observations and measurements of the jaws of pet dogs, which lead him to conclude that there has been a reduction in size and muscular power due to disuse. The creatures being fed on sops, have no need to use to any large extent the jaw-muscles. In this case, he argues, the principle of economy is not likely to be operative, since the pampered pet habitually overeats, and has therefore abundant nutriment and to spare to keep up the jaws. It is possible, however, that artificial selection has here been a factor. There may have been a competition among the old ladies who keep such pets to secure the dear little dog that never bites, while the nasty little wretch that does occasionally use his jaws for illegitimate purposes may have been speedily eliminated. Pet dogs are, moreover, a pampered, degenerate, and for the most part unhealthy race, often deteriorated by continued in-breeding, so that we must not build too much on Mr. Spencer's observations, interesting as they undoubtedly are.

There is one feature about the reduction of organs which must not be lost sight of. They are very apt to persist for a long time as remnants or vestiges. The pineal gland is the vestigial remnant of a structure connected with the primitive, median, or pineal eye. The whalebone whales and the duck-bill platypus have teeth which never cut the gum and are of no functional value. With regard to these, it may be asked—If disuse leads to the reduction of unused structures, how comes it that it has not altogether swept away these quite valueless structures? In considering this point, we must notice the unfortunate and misleading way in which disuse is spoken of as if it were a positive determinant, instead of the mere absence of free and full and healthy exercise. Few will question the fact that in the individual, if an organ is to be kept up to its full standard of perfection, it must be healthily and moderately exercised; and that, if not so exercised, it will not only cease to increase in size, but will tend to degenerate. The healthy, functionally valuable tissue passes into the condition of degenerate, comparatively useless tissue. Now, those who hold that the inheritance of functional modifications is still a tenable hypothesis, carry on into the history of the race that which they find to hold good in the history of the individual. They believe that, in the race, the continued functional activity of an organ is necessary for the maintenance of the integrity and perfection of its structure, and that, if not so exercised, the organ will inevitably tend to dwindle to embryonic proportions and to degenerate. The healthy, functionally valuable tissue passes at last into the condition of degenerate, comparatively useless tissue. The force of heredity will long lead to the production in the embryo of the structure which, in the ancestral days of healthy exercise, was to be of service to the organism. At this stage of life the conditions have not changed. The degeneration sets in at that period when the ancestral use is persistently denied. There is no reason why "disuse" should in all cases remove all remnants of a structure; but if the presence of the degenerate tissue is a source of danger to the organism which possesses it, that organism will be eliminated, and those (1) which possess it in an inert, harmless form, or (2) in which it is absent, will survive. Thus natural selection (which will fall under Mr. Romanes's reversed selection) will step in—will in some cases reduce the organ to a harmless and degenerate rudiment, and in others remove the last vestiges of the organ.

On the whole, even taking into consideration the effects of panmixia, of reversed selection, and of the principle of economy, the reduction of organs is difficult to explain, unless we call into play "disuse" as a co-operating factor.

Sexual Selection, or Preferential Mating.

It is well known that, in addition to and apart from the primary sexual differences in animals, there are certain secondary characters by which the males, or occasionally the females, are conspicuous. The antlers of stags, the tail of the peacock, the splendid plumes of the male bird of paradise, the horns or pouches of lizards, the brilliant frilled crest of the newt, the gay colours of male sticklebacks, the metallic hues of male butterflies, and the large horns or antennæ of other insects,—these and many other examples which will at once occur to the reader are illustrations of the fact.

As a contribution towards the explanation of this order of phenomena, Darwin brought forward his hypothesis of sexual selection, of which there are two modes. In the first place, the males struggle together for their mates; in this struggle the weakest are eliminated; those possessed of the most efficient weapons of offence and defence escape elimination. In the second place, the females are represented as exercising individual choice, and selecting (in the true sense of the word) those mates whose bright colours, clear voices, or general strength and vigour render them most pleasing and attractive. For this mode I shall employ the term "preferential mating." Combining these two in his summary, Darwin says, "It has been shown that the largest number of vigorous offspring will be reared from the pairing of the strongest and best-formed males, victorious in contests over other males, with the most vigorous and best-nourished females, which are the first to breed in the spring. If such females select the more attractive and, at the same time, vigorous males, they will rear a larger number of offspring than the retarded females, which must pair with the less vigorous and less attractive males. So it will be if the more vigorous males select the more attractive and, at the same time, healthy and vigorous females; and this will especially hold good if the male defends the female, and aids in providing food for the young. The advantage thus gained by the more vigorous pairs in rearing a larger number of offspring has apparently sufficed to render sexual selection efficient."*

With regard to the first of the two modes, little need be said. There can be no question that there are both elimination by battle and elimination by competition in the struggle for mates. It is well known that the emperor moth discovers his mate by his keen sense of smell residing

* Darwin, "Descent of Man," pt. ii. chap. viii.

probably in the large, branching antennæ. There can be little doubt that, if an individual is deficient in this sense, or misinterprets the direction in which the virgin female lies, he will be unsuccessful in the competition for mates; he will be eliminated from procreation. And it is a familiar observation of the poultry-yard that the law of battle soon determines which among the cock birds shall procreate their kind. The law of battle for mates is, indeed, an established fact among many animals, especially those which are polygamous, and the elimination of the unfit in this respect is a logical necessity.

It is when we come to the second of the two modes, that which involves selection proper, that we find differences of opinion among naturalists.

Darwin, as we have seen, suggested that those secondary sexual characters which can be of no value in aiding their possessor to escape elimination by combat result from the preferential choice of the female, the female herself remaining comparatively unaffected. But Mr. Wallace made an exceedingly valuable suggestion with regard to these comparatively dull colours of the female. He pointed out that conspicuousness (unless, as we have seen, accompanied by some protective character, such as a sting or a bitter taste) increased the risk of elimination by enemies. Now, the males, since they are generally the stronger, more active, and more pugnacious, could better afford to run this risk than their mates. They could to some extent take care of themselves. Moreover, when impregnation was once effected, the male's business in procreation was over. Not so the female; she had to bear the young or to lay the eggs, often to foster or nourish her offspring. Not only were her risks greater, but they extended over a far longer period of time. Hence, according to Mr. Wallace, the dull tints of the females, as compared with those of the males, are due to natural selection eliminating the conspicuous females in far greater proportion than the gaudy males.

There is clearly no reason why this view should not be combined with Darwin's; preferential mating being one factor, natural elimination being another factor; both being operative at the same time, and each contributing to that marked differentiation of male and female which we find to prevail in certain classes of the animal kingdom.

But Mr. Wallace will not accept this compromise. He rejects preferential mating altogether, or, in any case, denies that through its agency secondary sexual characters have been developed. He admits, of course, the striking and beautiful nature of some of these characters; he admits that the male in courtship takes elaborate pains to display all his finery before his would-be mate; he admits that the "female birds may be charmed or excited by the fine display of plumage by the males;" but he concludes that "there is no proof whatever that slight differences in that display have any effect in determining their choice of a partner."*

How, then, does Mr. Wallace himself suppose that these secondary sexual characters have arisen? His answer is that "ornament is the natural outcome and direct product of superabundant health and vigour," and is "due to the general laws of growth and development."† At which one rubs one's eyes and looks to the title-page to see that Mr. Wallace's name is really there, and not that of Professor Mivart or the Duke of Argyll. For, if the plumage of the argus pheasant and the bird of paradise is due to the general laws of growth and development, why not the whole animal? If Darwin's sexual selection is to be thus superseded, why not Messrs. Darwin and Wallace's natural selection?

Must we not confess that Mr. Wallace, for whose genius I have the profoundest admiration, has here allowed himself to confound together the question of origin and the question of guidance or direction? Natural selection by elimination and sexual selection through preferential

---

* "Darwinism," chap. x.

† "Darwinism," p. 295. Messrs. Geddes and Thomson, "The Evolution of Sex," p. 28, also contend that "combative energy and sexual beauty rise *pari passu* with male katabolism."

mating are, supposing them to be *veræ causæ*, guiding or selecting agencies. Given the variations, however caused, these agencies will deal with them, eliminating some, selecting others, with the ultimate result that those specially fitted for their place in nature will survive. Neither the one nor the other deals with the origin of variations. That is a wholly different matter, and constitutes the leading biological problem of our day. Mr. Wallace's suggestion is one which concerns the origin of variations, and as such is worthy of careful consideration. It does not touch the question of their guidance into certain channels or the maintenance of specific standards. Concerning this Mr. Wallace is silent or confesses ignorance. "Why, in allied species," he says, "the development of accessory plumes has taken different forms, we are unable to say, except that it may be due to that individual variability which has served as the starting-point for so much of what seems to us strange in form or fantastic in colour, both in the animal and vegetable world."* It is clear, however, that "individual variability" cannot be regarded as a *vera causa* of the maintenance of a specific standard—a standard maintained *in spite of* variability.

The only directive agency (apart from that of natural selection) to which Mr. Wallace can point is that suggested by Mr. Alfred Tylor, in an interesting, if somewhat fanciful, posthumous work on "Coloration in Animals and Plants," "namely, that diversified coloration follows the chief lines of structure, and changes at points, such as the joints, where function changes." But even if we admit that coloration-bands or spots originate at such points or along such lines—and the physiological rationale is not altogether obvious—even if we admit that in butterflies the spots and bands usually have reference to the form of the wing and the arrangement of the nervures, and that in highly coloured birds the crown of the head, the throat, the ear-coverts, and the eyes have usually distinct tints, still it can hardly be maintained that this affords us any

---

* "Darwinism," p. 293.

adequate explanation of the *specific* colour-tints of the humming-birds, or the pheasants, or the Papilionidæ among butterflies. If, as Mr. Wallace argues, the immense tufts of golden plumage in the bird of paradise owe their origin to the fact that they are attached just above the point where the arteries and nerves for the supply of the pectoral muscles leave the interior of the body, are there no other birds in which similar arteries and nerves are found in a similar position? Why have these no similar tufts? And why, in the birds of paradise themselves, does it require four years (for it takes so long for the feathers of the male to come to maturity) ere these nervous and arterial influences take effect upon the plumage? Finally, one would inquire how the colour is determined and held constant in each species. The difficulty of the Tylor-Wallace view, even as a matter of origin, is especially great in those numerous cases in which the colour is determined by delicate lines, thin plates, or thin films of air or fluid.*

Under natural selection, as we have seen, the development of colour is fostered under certain conditions. The colour is either protective, rendering the organism inconspicuous amid its normal surroundings, or it is of warning value, advertising the organism as inedible or dangerous, or, in the form of recognition-marks, it is of service in enabling the members of a species to recognize each other. Now, in the case of both warning colour and recognition-marks, their efficacy depends upon the perceptual powers of animals. Unless there be a rapidly acquired and close association of the quality we call nastiness with the quality we call gaudiness (though, for the animal, there is no such *isolation* of these qualities as is implied in our words †), such that the sight of the gaudy insect suggests that it will be unpleasant to eat, the gaudiness will be of no avail. And if there is any truth in the doctrine of mimicry, the association is particular. It is not merely that bright

---

* Mr. Poulton, who takes a similar line of argument in his "Colours of Animals," lays special stress upon the production of *white* (see p. 326).

† See Chapter VIII.

colours are suggestive of a nasty taste. The insect-eating birds associate nastiness especially with certain markings and coloration—"the tawny *Danais*, the barred *Heliconias*, the blue-black *Euplœas*, and the fibrous *Acræas*;" and this is proved by the fact that sweet insects mimicking these particular forms are thereby protected.

So, too, with recognition-marks. If the bird or the mammal have not sufficient perceptive powers to distinguish between the often not very different recognition-marks, of what service can they be?

Recognition-marks and mimicry seem, therefore, to show that in the former case many animals, and in the latter the insect-eating birds, mammals, lizards, and other animals concerned, have considerable powers of perception and association.

Among other associations are those which are at the base of what I have termed preferential mating. We must remember how deeply ingrained in the animal nature is the mating instinct. *We* may find it difficult to distinguish closely allied species. But the individuals of that species are led to mate together by an impelling instinct that is so well known as to elicit no surprise. Instinct though it be, however, the mating individuals must recognize each other in some way. The impulse that draws them together must act through perceptual agency. It is not surprising, therefore, to find, when we come to the higher animals, that, built upon this basis, there are well-marked mating preferences. And this, as we have before pointed out, following Wallace, is an efficient factor in segregation. Let us, however, hear Mr. Wallace himself in the matter.

There is, he says,* "a very powerful cause of isolation in the mental nature—the likes and dislikes—of animals; and to this is probably due the fact of the rarity of hybrids in a state of nature. The differently coloured herds of cattle in the Falkland Islands, each of which keeps separate, have been already mentioned. Similar facts occur, however, among our domestic animals, and are

* "Darwinism," p. 172.

well known to breeders. Professor Low, one of the greatest authorities on our domesticated animals, says, 'The female of the dog, when not under restraint, makes selection of her mate, the mastiff selecting the mastiff, the terrier the terrier, and so on.' And again, 'The merino sheep and the heath sheep of Scotland, if two flocks are mixed together, each will breed with its own variety.' Mr. Darwin has collected many facts illustrating this point.* One of the chief pigeon-fanciers in England informed him that, if free to choose, each breed would prefer pairing with its own kind. Among the wild horses in Paraguay those of the same colour and size associate together; while in Circassia there are three races of horses which have received special names, and which, when living a free life, almost always refuse to mingle and cross, and will even attack one another. In one of the Faröe Islands, not more than half a mile in diameter, the half-wild native black sheep do not readily mix with imported white sheep. In the Forest of Dean and in the New Forest the dark and pale coloured herds of fallow deer have never been known to mingle; and even the curious ancon sheep, of quite modern origin, have been observed to keep together, separating themselves from the rest of the flock when put into enclosures with other sheep. The same rule applies to birds, for Darwin was informed by the Rev. W. D. Fox that his flocks of white and Chinese geese kept distinct. This constant preference of animals for their like, even in the case of slightly different varieties of the same species, is evidently a fact of great importance in considering the origin of species by natural selection, since it shows us that, so soon as a slight differentiation of form or colour has been effected, isolation will at once arise by the selective association of the animals themselves."

Mr. Wallace thus allows, nay, he lays no little stress on, preferential mating, and his name is associated with the hypothesis of recognition-marks. But he denies that preferential mating, acting on recognition-marks, has had

* See "Animals and Plants under Domestication," vol. ii. p. 80.

any effect in furthering a differentiation of form or colour. He admits that so soon as a slight differentiation of form or colour has been effected, segregation will arise by the selective association of the animals themselves; but he does not admit that such selective association can carry the differentiation further.

Now, it is clear that mating preferences must be either fixed or variable. If fixed, how can differentiation occur in the same flock or herd? And how can selective association be a means of isolation? Or, granting that differentiation has occurred, if the mating preferences are then stereotyped, all further differentiation, so far as colour and form are concerned, will be rendered impossible; for divergent modifications, not meeting the stereotyped standard of taste, will for that reason fail to be perpetuated. We must admit, then, that these mating preferences are subject to variation. And now we come to the central question with regard to sexual selection by means of preferential mating. What guides the variation along special lines leading to heightened beauty? This, I take it, is the heart and centre of Mr. Wallace's criticism of Darwin's hypothesis. Sexual selection of preferential mating involves a standard of taste; that standard has advanced from what we consider a lower to what we consider a higher æsthetic level, not along one line, but along many lines. What has guided it along these lines?

Not as in any sense affording a direct answer to this question, but for illustrative purposes, we may here draw attention to what seems to be a somewhat parallel case, namely, the development of flowers through insect agency. In his "Origin of Species," Darwin contended that flowers had been rendered conspicuous and beautiful in order to attract insects, adding, "Hence we may conclude that, if insects had not been developed on the earth, our plants would not have been decked with beautiful flowers, but would have produced only such poor flowers as we see on our fir, oak, nut, and ash trees, on grasses, docks, and nettles, which are all fertilized through the agency of the wind." "The argument in favour of this view," says Mr. Wallace,* who quotes this passage, "is now much stronger than when Mr. Darwin wrote;" and he cites with approval the following passage from Mr. Grant Allen's "Colour-Sense:" "While man has only tilled a few level plains, a few great river-valleys, a few peninsular mountain slopes, leaving the vast mass of earth untouched by his hand, the insect has spread himself over every land in a thousand shapes, and has made the whole flowering creation subservient to his daily wants. His buttercup, his dandelion, and his meadowsweet grow thick in every English field. His thyme clothes the hillside; his heather purples the bleak grey moorland. High up among the Alpine heights his gentian spreads its lakes of blue; amid the snows of the Himalayas his rhododendrons gleam with crimson light. Even the wayside pond yields him the white crowfoot and the arrowhead, while the broad expanses of Brazilian streams are beautified by his gorgeous water-lilies. The insect has thus turned the whole surface of the earth into a boundless flower-garden, which supplies him from year to year with pollen or honey, and itself in turn gains perpetuation by the baits that it offers to his allurement." †

Mr. Grant Allen is perfectly correct in stating that the insect has produced all this beauty. It is the result of insect choice, a genuine case of selection as contrasted with elimination. And when we ask in this case, as we asked in the case of the beautiful colours and forms of animals, what has guided their evolution along lines which lead to such rare beauty, we are given by Mr. Wallace himself the answer, "The preferential choice of insects." If these insects have been able to produce through preferential selection all this wealth of floral beauty (not, indeed, for the sake of the beauty, but incidentally in the practical business of their life), there would seem to be no *a priori* reason why the same class and birds and mammals should not have been able to produce, through preferential selection, all the wealth of animal beauty.

* "Darwinism," p. 332. † "The Colour-Sense," by Grant Allen, p. 95.

It should be noted that the answer to the question is in each case a manifestly incomplete one. For if we say that these forms of beauty, floral and animal, have been selected through animal preferences, there still remains behind the question—How and why have the preferences taken these *æsthetic* lines? To which I do not see my way to a satisfactory answer, though some suggestions in the matter will be made in a future chapter.* At present all we can say is this—to be conspicuous was advantageous, since it furthered the mating of flowers and animals. To be diversely conspicuous was also advantageous. As Mr. Wallace says, "It is probably to assist the insects in keeping to one flower at a time, which is of vital importance to the perpetuation of the species, that the flowers which bloom intermingled at the same season are usually very distinct, both in form and colour." † But conspicuousness is not beauty. And the question still remains—From what source comes this tendency to beauty?

Leaving this question on one side, we may state the argument in favour of sexual selection in the following form: The generally admitted doctrine of mimicry involves the belief that birds and other insect-eating animals have delicate and particular perceptual powers. The generally received doctrine of the origin of flowers involves the belief that their diverse forms and markings result from the selective choice of insects. There are a number of colour and form peculiarities in animals that cannot be explained by natural selection through elimination. There is some evidence in favour of preferential mating or selective association. It is, therefore, permissible to hold, as a provisional hypothesis, that just as the diverse forms of flowers result from the preferential choice of insects, so do the diverse secondary sexual characters of animals result, in part at least, from the preferential choice of animals through selective mating.

If this be admitted, then the elaborate display of their finery by male birds, which Mr. Wallace does admit, may

* That on "The Emotions of Animals" (X.).   † "Darwinism," p. 318.

fairly be held to have a value which he does not admit. For if preferential mating is *à priori* probable, such display may be regarded as the outcome of this mode of selection. At the same time, it may be freely admitted that more observations are required. In a recent paper, " On Sexual Selection in Spiders of the Family *Attidæ*," * by George W. and Elizabeth G. Peckham, a full, not to say elaborate, description is given of the courtship, as they regard it, of spiders. The "love-dances" and the display of special adornments are described in detail. And the observers, as the result, be it remembered, of long and patient investigation and systematic study, come to the conclusion that female spiders exercise selective choice in their mates. And courtship must be a serious matter for spiders, for if they fail to please, they run a very serious risk of being eaten by the object of their attentions. Some years ago I watched, on the Cape Flats, near Capetown, the courtship of a large spider (I do not know the species). In this case the antics were strange, and, to me, amusing; but they seemed to have no effect on the female spider, who merely watched him. Once or twice she darted forward towards him, but he, not liking, perhaps, the gleam in her eyes, retreated hastily. Eventually she seemed to chase him off the field.

We must remember how difficult it is to obtain really satisfactory evidence of mating preferences in animals. In most cases we must watch the animals undisturbed, and very rarely can we have an opportunity of determining whether one particular female selects her mate out of her various suitors. We watch the courtship in this, that, or the other case. In some we see that it is successful; in others that it is unsuccessful. How can we be sure that in the one case it was through fully attaining, in the other through failing to reach, the standard of taste? And yet it is evidence of this sort that Mr. Wallace demands. After noting the rejection by the hen of male birds which had lost their ornamental plumage, he says, " Such cases do

* Natural History Society of Wisconsin, vol. i. (1889).

not support the idea that males with the tail-feathers a trifle longer, or the colours a trifle brighter, are generally preferred, and that those which are only a little inferior are as generally rejected,—and this is what is absolutely needed to establish the theory of the development of these plumes by means of the choice of the female." * If Mr. Wallace requires direct observational evidence of this kind, I do not suppose he is likely to get any large body of it. But one might fairly ask him what body of direct observational evidence he has of natural selection. The fact is that direct observational evidence is, from the nature of the processes involved, almost impossible to produce in either case. Natural selection is an explanation of organic phenomena reached by a process of logical inference and justified by its results. It is not claimed for the hypothesis of selective mating that it has a higher order of validity.

Use and Disuse.

As we have already seen, biologists are divided into two schools, one of which maintains that the effects of use and disuse † have been a potent factor in organic evolution; the other, that the effects of use and disuse are restricted to the individual. My own opinion is that we have not a sufficient body of carefully sifted evidence to enable us to dogmatize on the subject, one way or the other. But, the position of strict equilibrium being an exceedingly difficult and some would have us believe an undesirable attitude of mind, I may add that I lean to the view that use and disuse, if persistent and long-continued, take effect, not only on the individual, but also on the species.

It is scarcely necessary to give examples of the kind of change which, according to the Lamarckian school, are wrought by use and disuse. Any organ persistently used will have a tendency, on this view, to become in successive

* "Darwinism," p. 286.
† On the negative character of disuse, see p. 196.

generations more and more adapted to its functional work. To give but one example. It is well known that certain hoofed creatures are divisible into two groups—first, those which, like the horse, have in each limb one large and strong digit armed with a solid hoof; and, secondly, those which, like the ox, have in each limb two large digits, so that the hoof is cloven or split. It is also well known that the ancestral forms from which both horse-group and ox-group are derived were possessed of five digits to each limb. Professor Cope regards the differentiation of these two groups as the result of the different modes of use necessitated by different modes of life. "The mechanical effect," he says, "of walking in the mud is to spread the toes equally on opposite sides of the middle line. This would encourage the equal development of the digits on each side of the middle line, as in the cloven-footed types. In progression on hard ground the longest toe (the third) will receive the greatest amount of shock from contact with the earth."* Hence the solid-hoofed types. Here, then, the middle digit in the horse-group, or two digits in the ox-group, having the main burden to bear, increase through persistent use, while the other digits dwindle through disuse.†

On the other hand, one who holds the opposite view will

* Cope, "Origin of the Fittest," p. 374.
† It would appear, from certain passages of his "Darwinism," that Mr. A. R. Wallace (*e.g.* p. 139, note) holds or held similar views. "The genera *Ateles* and *Colobus*," he says, "are two of the most purely arboreal types of monkeys, and it is not difficult to conceive that the constant use of the elongated fingers for climbing from tree to tree, and catching on o branches while making great leaps, might require all the nervous energy and muscular growth to be directed to the fingers, the small thumb remaining useless." I should also have quoted Mr. Wallace's account of the twisting round of the eyes of flat-fishes—where he says that the constant repetition of the effort of twisting the eye towards the upper side of the head, when the bony structure is still soft and flexible, causes the eye gradually to move round the head till it comes to the upper side—had he not subsequently disclaimed this explanation (see *Nature*, vol. xl. p. 619). It is possible that Mr. Wallace, notwithstanding the words "constant use" in the passage I have quoted, merely intends to imply that the elongated fingers are of advantage in climbing, and are thus subject to natural selection, the thumb diminishing through economy of growth.

say—I do not believe that use and disuse have had anything whatever to do with the matter. Fortuitous variations in these digits have taken place. The conditions have determined which variations should be preserved. In the horse, variations in the direction of increase of functional value of the mid digit, and variations in the simultaneous decrease of the functional value of the lateral digits, have been of advantage, and have therefore survived the eliminating process of natural selection.

Now, since it is quite clear, in this and numberless similar cases, that we can explain the facts either way, it is obviously not worth while to spend much time or ingenuity in devising such explanations. They are not likely to convince any one worth convincing. What we need is (1) crucial cases which can only be explained one way or the other; or (2) direct observation or experiment leading to the establishment of one hypothesis or the other (or both).

1. Crucial cases are very difficult to find. We cannot exclude the element of use or disuse, for on both hypotheses it is essential. The difference is that one school says the organ is developed in the species *by* use; the other school says it is developed *for* use. What we must seek is, therefore, the necessary exclusion of natural selection; and that is not easy to prove, in any case, to a Darwinian. If it can be shown that there exist structures which are of use, but not of vital importance (that is to say, which have not what I called above the *available advantage* necessary to determine the question of elimination or not-elimination), then we are perhaps able to exclude the influence of natural selection. I think, if anywhere, such cases are to be found in faculties and instincts;[*] and as such they must be considered in a later chapter. I will, however, here cite one case in illustration of my meaning.

[*] I find, on re-reading one of his articles, that I have here unwittingly adopted one of Mr. Romanes's arguments (see *Nature*, vol. xxxvi. p. 406). The instance Mr. Romanes cites is the curious habit of dogs turning round before they lie down.

We have seen that certain insects are possessed of warning colours, which advertise their nastiness to the taste. Birds avoid these bright but unpleasant insects, and though there is some individual learning, there seems to be an instinctive avoidance of these unsavoury morsels. There is hesitation before tasting; and one or two trials are sufficient to establish the association of gaudiness and nastiness. Moreover, Mr. Poulton and others have shown that, under the stress of keen hunger, these gaudy insects may be eaten, and apparently leave no ill effects. Birds certainly instinctively avoid bees and wasps; and yet the sting of these insects can seldom be fatal. It is, therefore, improbable that nastiness or even the power of stinging can have been an eliminating agency. In the development of the instinctive avoidance, natural selection through elimination seems to be excluded, and the inheritance of individual experience is thus rendered probable. As before pointed out, it is not enough to say that a nasty taste or a sting in the gullet is disadvantageous; it must be shown that the disadvantage has an eliminating value. From my experiments (feeding frogs on nasty caterpillars, and causing bees to sting chickens), I doubt the eliminating value in this case. Hence elimination by natural selection seems, I repeat, to be excluded, and the inheritance of individual experience rendered probable.

Mr. Herbert Spencer has contended that, in certain modifications, natural selection is excluded on the grounds of the extreme complexity of the changes, and adduces the case of the Irish "elk" with its huge antlers, and the giraffe with its specially modified structure. He points out that in either case the conspicuous modification—the gigantic antlers or the long neck—involves a multitude of changes affecting many and sometimes distant parts of the body. Not only have the enormous antlers involved changes in the skull, the bones of the neck, the muscles, blood-vessels, and nerves of this region, but changes also in the fore limbs; while the long neck of the giraffe has brought with it a complete change of gait, the co-ordinated movements of the hind limbs sharing in the general modification. Mr. Spencer, therefore, argues that it is difficult to believe that these multitudinous co-ordinated modifications are the result of fortuitous variations seized upon by natural selection. For natural selection would have to wait for the fortunate coincidence of a great number of distinct parts, all happening to vary just in the particular way required. That natural selection should seize upon the favourable modification of a particular part is comprehensible enough; that two organs should coincidently vary in favourable directions we can understand; that half a dozen parts should, in a few individuals among the thousands born, by a happy coincidence, vary each independently in the right way is conceivable; but that the whole organization should be remodelled by fortunately coincident and fortuitously favourable variations is not readily comprehensible. It may be answered—Notwithstanding all this, we know that such happy coincidences have occurred, for there is the resulting giraffe. The question, however, is not whether these modifications have occurred or not, but whether they are due to fortuitous variation alone, or have been guided by functional use. The argument seems to me to have weight.*

Still, we should remember that among neuter ants—for example, in the Sauba ant of South America (*Oecodoma cephalotes*)—there are certain so-called soldiers with relatively enormous heads and mandibles. The possession of these parts so inordinately developed must necessitate many correlated changes. But these cannot be due to inherited use, since such soldiers are sterile.

Furthermore, according to Professor Weismann, natural selection is really working, not on the organism at large, but on the germ-plasm which produces it; and it is con-

---

* Mr. Darwin, while contending that the modifications need not all have been simultaneous, says, " Although natural selection would thus tend to give the male elk its present structure, yet it is probable that the inherited effects of use, and of the mutual action of part on part, have been equally or more important" ("Animals and Plants under Domestication," vol. ii. p. 328).

ceivable that the variation of one or more of the few cells in early embryonic life may introduce a great number of variations in the numerous derivative cells. In explanation of my meaning, I will quote a paragraph from a paper of Mr. E. B. Poulton's on "Theories of Heredity.[*] "It appears," he says, "that, in some animals, the great groups of cells are determined by the first division [of the ovum in the process of cleavage [†]]; in others, the right and left sides, or front and hind ends of the body; while the cells giving rise to the chief groups on each side would then be separated at some later division. This is not theory, but fact; for Roux has recently shown that, if one of the products of the first division of the egg of a frog be destroyed with a hot needle, development is not necessarily arrested, but, when it proceeds, leads to the formation of an embryo from which either the right or the left side is absent. When the first division takes place in another direction, either the hind or the front half was absent from the embryo which was afterwards produced. After the next division, when four cells were present, destruction of one produced an embryo in which one-fourth was absent." Now, it is conceivable that a single modification or variation of the primitive germ might give rise to many correlated modifications or variations of the numerous cells into which it develops; just as an apparently trivial incident in childhood or youth may modify the whole course of a man's subsequent life. It is difficult, indeed, to see how this could be effected; to understand what could be the nature of a modification of the germ which could lead simultaneously to many favourable variations of bones, muscles, blood-vessels, and nerves in different parts of the body. This, however, is a question of the origin of variations; and it is, at any rate, conceivable that, just as by the extirpation with a hot needle of one cell of the cleaved frog's ovum all the anterior part of the body should be absent in development, so by the appropriate modification of this one cell, or the germinal matter which produced it,

[*] *Midland Naturalist*, November, 1889.   [†] See *ante*, p. 52.

all the anterior part of the body should be appropriately modified.

These considerations, perhaps, somewhat weaken the force of Mr. Spencer's argument, which is not quite so strong now as it was when the "Principles of Biology" was published.

(2) We may pass now to the evidence afforded by direct observation and experiment. There is little enough of it. The best results are, perhaps, those which have been incidentally reached in the poultry-yard and on the farm in the breeding of domesticated animals. We have seen that, under these circumstances, certain parts or organs have very markedly diminished in size and efficiency; others have as markedly increased. Of the former, or decrease in size and efficiency, the imbecile ducks with greatly diminished brains have been already mentioned. Mr. Herbert Spencer draws attention[*] to the diminished efficiency in ear-muscles, giving rise to the drooping ears of many domesticated animals. "Cats in China, horses in parts of Russia, sheep in Italy and elsewhere, the guinea-pig formerly in Germany, goats and cattle in India, rabbits, pigs, and dogs in all long-civilized countries, have dependent ears."[†] Since many of these animals are habitually well fed, the principle of economy of growth seems excluded. Indeed, the ears are often unusually large; it is only their motor muscles that have dwindled either relatively or absolutely. If what has been urged above be valid, panmixia cannot have been operative; since panmixia *per se* only brings about regression to mediocrity. If the effects in these two cases, ducks' brains and dogs' ears, be not due to disuse, we know not at present to what they are due. In the correlative case of increase by use, we find it exceedingly difficult to exclude the disturbing effects of artificial selection. The large and distended udders of cows, the enhanced egg-laying powers of hens, the fleetness or strength of different breeds of horses,—all of these have been

---

[*] *Nature*, vol. xli. p. 511.
[†] "Animals and Plants under Domestication," vol. ii. p. 201.

subjects of long-continued, assiduous, and careful selection. One cannot be sure whether use has co-operated or not.

Sufficient has now, I think, been said to show the difficulty of deciding this question, the need of further observation and discussion, and the necessity for a receptive rather than a dogmatic attitude; and sufficient, also, to indicate my reasons for leaning to the view that use and disuse, long-continued and persistent, may be a factor in organic evolution.

The Nature of Variations.

The diversity of the variations which are possible, and which actually occur in animal life, is so great that it is not easy to sum up in a short space the nature of variations. Without attempting anything like an exhaustive classification, we may divide variations into three classes.

1. *Superficial variations* in colour, form, etc., not necessarily in any way correlated with

2. *Organic variations* in the size, complexity, and efficiency of the organs of the body;

3. *Reproductive and developmental variations.*

Any of these variations, if sufficient in amount and value to determine the question of elimination or not-elimination, selection or not-selection, may be seized upon by natural selection.

Our domesticated animals exemplify very fully the superficial variations which, through man's selection, have in many cases been segregated and to some extent stereotyped. It is unnecessary to do more than allude to the variations in form and coloration of dogs, cattle, fowls, and pigeons. These variations are not *necessarily* in any way correlated with any deeper organic variations. They are, however, in many cases so correlated. For example, the form of the pouter pigeon is correlated with the increased size of the crop, the length of the beak carries with it a modification of the tongue, the widely expanded tail of the fantail carries with it an increase in the size and number of the caudal vertebræ. And here we might take the whole series of secondary sexual characters. These and their like may be said to be direct correlations. But there are also correlations which are seemingly indirect, their connection being apparently remote. That in pigeons the size of the feet should vary with the size of the beak; that the length of the wing and tail feathers should be correlated; that the nakedness of the young should vary with the future colour of the plumage; that white dogs should be subject to distemper, and white fowls to the "gapes;" that white cats with blue eyes should be nearly always deaf;—in these cases the correlation is indirect. But from the existence of correlation, whether direct or indirect, it follows that variations seldom come singly. The organism is so completely a unity that the variation of one part, even in superficial matters, affects directly or indirectly other parts.

In the freedom of nature such superficial variations are not so obvious. But among the invertebrates they are not inconsiderable. The case of land-snails, already quoted, may again be cited. Taking variations in banding alone, Mr. Cockerell knows of 252 varieties of *Helix nemoralis* and 128 of *H. hortensis*. Still, among the wild relatives of our domestic breeds of animals and birds the superficial variations are decidedly less marked. And this is partly due to the fact that they are in a state of far more stable equilibrium than our domestic products, and partly to the constant elimination of all variants which are thereby placed at a serious or vital disadvantage. White rats, mice, or small birds, in temperate regions, would soon be seized upon by hawks and other enemies. If the eggs and young of the Kentish plover, shown in our frontispiece, were white or yellowish, like the eggs and young of our fowls, they would soon be snapped up. The varied protective resemblances, general and special, have been brought about by the superficial variations of organisms, and the elimination of those which, from non-variation or wrong variation, remained conspicuous. We need only further notice one thing here, namely, that, in the case of special resemblance to an inorganic object or to another organism, the variations of the several parts must be very closely, and sometimes completely, correlated. The correlations, however, need not, perhaps, have been simultaneous—the resemblance having been gradually perfected by the filling in of additional touches, first one here, then another there, and so on.

Concerning "organic variations," little need be said. It is clear that an organ or limb may vary in size, such variation carrying with it a correlative variation in power; or it may vary in complexity—the teeth of the horse tribe, for example, having increased in complexity, while their limbs have been rendered less complex; or it may vary in efficiency through the more perfect correlation and co-ordination of its parts.

The evidence of such variations from actual observation is far less in amount than that of superficial variations. And this is not to be wondered at, since in many cases it can only be obtained by careful anatomical investigation. Nevertheless, anatomists, both human and comparative, are agreed that such variations do occur. And no one can examine such a collection as that of the Royal College of Surgeons without acknowledging the fact.

Thirdly, "reproductive and developmental variations" are of very great importance. The following are among the more important modifications which may occur in the animal kingdom.

1. Variations in the mode of reproduction, sexual or asexual.

2. Variations in the mode of fertilization.

3. Variations in the number of fertilized ova produced.

4. Variations in the amount of food-yolk and in the way in which it is supplied.

5. Variations in the time occupied in development.

6. Variations in the time at which reproduction commences.

7. Variations in the duration and amount of parental protection and fosterage.

8. Variations in the period at which secondary sexual characters and the maximum efficiency of the several organs is reached.

It is impossible here to discuss these modes of variation *seriatim*. I shall therefore content myself with but a few remarks on the importance of protection and fosterage. It is not too much to say that, without fosterage and protection, the higher forms of evolution would be impossible. If you are to have a highly evolved form, you must allow time for its evolution from the egg; and that development may go on without let or hindrance, you must supply the organism with food and lighten the labour of self-defence. Most of the higher organisms are slow in coming to maturity, passing through stages when they are helpless and, if left to themselves, would inevitably fall a prey to enemies.

In those animals in which the system of fosterage and protection has not been developed a great number of fertilized ova are produced, only a few of which come to maturity. It might be suggested that this is surely an advantage, since the greater the number produced the greater the chances of favourable variations taking place. But it has before been pointed out that these great numbers are decimated, and more than decimated, not by elimination, but by indiscriminate destruction; embryos, good, bad, and indifferent, being alike gobbled up by those who had learnt the secret of fostering their young. The alternative has been between producing great numbers* of embryos which soon fend for themselves, and a few young who are adequately provided for during development. And the latter have proved the winners in life's race. If we compare two flat-fishes belonging to very different groups, the contrast here indicated will be readily seen. The skate is a member of the shark tribe, flattened sym-

---

* In the third chapter we saw that in such cases not only are there an enormous number of ova produced, but that (*e.g.* in aurelia and the liver-fluke) each ovum produces, through the intervention of asexual multiplication, many individuals.

metrically from above downwards. It lays, perhaps, eighty to a hundred eggs. Each of these is large, and has a rich supply of nutritive food-yolk. Each is also protected by a horny case with pointed corners—the so-called sea-purse of seaside visitors. These are committed by the skate to the deep, and are not further cared for. But the abundant supply of food-yolk gives the little skate which emerges a good start in life. On the other hand, the turbot, one of the bony fishes, flattened from side to side with an asymmetrical head, lays several millions of eggs, which float freely in the open sea. These are minute and glassy, and not more than one-thirtieth of an inch in diameter. When the fishes are hatched, they are not more than about one-fifth of an inch in length. The slender stock of food-yolk is soon used up, and henceforth the little turbot (at present more like a stump-nosed eel than a turbot) has to get its own living. Hundreds of thousands of them are eaten by other fishes.

Or, if we compare such different vertebrates as a frog, a sparrow, and a mouse, we find that the frog produces a considerable number of fertilized ova, though few in comparison with the turbot, each provided with a small store of food-yolk. The tiny tadpoles very soon have to obtain their own food and run all the risks of destruction. Few survive. The sparrow lays a few eggs; but each is supplied with a large store of food-yolk, sufficient to meet its developmental needs until, under the fostering influence of maternal warmth, it is hatched. Even on emerging from the eggs, the callow fledglings enjoy for a while parental protection and fosterage, and, when sent forth into the world, are very fairly equipped for life's struggle. The mouse produces minute eggs with little or no food-yolk; but they undergo development within the womb of the mother, and are supplied with nutrient fluids elaborated within the maternal organism. Even when born, they are cherished for a while and supplied with food-milk by the mother.

The higher stages of this process involve a mental

element, and are developed under the auspices of intelligence or instinct. But the lower stages, the supply of food-yolk and intra-uterine protection, are purely organic. A hen cannot by instinctive or intelligent forethought increase the amount of food-yolk stored up in the ovum, any more than the lily, which, by an analogous process, stores up in its bulb during one year material for the best part of next year's growth, can increase this store by a mental process.

It cannot therefore be questioned that variations in the amount of capital with which an embryo is provided in generation would very materially affect its chances of escaping elimination by physical circumstances, by enemies, and by competition.

Nor can it be questioned that variations in the time occupied in reaching maturity would, other things equal, not a little affect the chances of success of an organism in the competition of life. Hence we have the phenomena of what may be termed acceleration and retardation in development. These terms have, however, been used by American zoologists, notably Professors Hyatt and Cope, in a somewhat different and wider sense; for they include not merely time-changes, but also the loss of old characters or the acquisition of new characters. "It is evident," says Professor Cope, "that the animal which adds something to its structure which its parents did not possess has grown more than they; while that which does not attain to all the characteristics of its ancestors has grown less than they." "If the embryonic form be the parent, the advanced descendant is produced by an increased rate of growth, which phenomenon is called 'acceleration;' but if the embryonic type be the offspring, then its failure to attain the condition of the parent is due to the supervention of a slower rate of growth; to this phenomenon the term 'retardation' is applied." "I believe that this is the simplest mode of stating and explaining the law of variation: that some forms acquire something which their parents did not possess; and that those which acquire something additional have to pass through more numerous stages than their ancestors; and those which lose something pass through fewer stages than their ancestors; and these processes are expressed by the terms 'acceleration' and 'retardation.'"*

It is clear, however, that we have here something more than acceleration and retardation of development in the ordinary sense of these words. It would be, therefore, more convenient to use the term "acceleration" for the condensation of *the same series* of developmental changes into a shorter period of time; "retardation" for the lengthening of the period in which *the same series* of changes are effected; and "arrested development" for those cases in which the young are born in an immature or embryonic condition. Whether there is any distinct tendency, worthy of formulation as a law, for organisms to acquire, as a result of protracted embryonic development, definite characteristics which their ancestors did not possess, I think very questionable. If so, this will fall under the head of the origin of variations.

That acceleration, in the sense in which I have used the term, does occur as a variation is well known. "With our highly improved breeds of all kinds," says Darwin,† "the periods of maturity and reproduction have advanced with respect to the age of the animal; and in correspondence with this, the teeth are now developed earlier than formerly, so that, to the surprise of agriculturalists, the ancient rules for judging of the age of an animal by the state of its teeth are no longer trustworthy." "Disease is apt to come on earlier in the child than in the parent; the exceptions in the other direction being very much rarer." ‡ Professor Weismann contends that the time of reproduction has been accelerated through natural selection, since the shorter the time before reproduction, the less the number of possible accidents. We may, perhaps, see in the curious cases of

---

* Cope, " Origin of the Fittest," pp. 226, **125, and** 297.
† " Animals and Plants under Domestication," vol. ii. p. 313.
‡ Ibid. p. 56.

reproduction during an otherwise immature condition, extreme instances of acceleration. The axolotl habitually reproduces in the gilled, or immature condition. Some species of insects reproduce before they complete their metamorphoses. And the females of certain beetles (*Phengodini*) are described by Professor Riley as larviform.*

Precocity is variation in the direction of acceleration, and that condensed development which is familiar in the embryos of so many of the higher animals may be regarded as the result of variations constantly tending in the same direction. That there are fewer examples of retardation is probably due to the fact that nature has constantly favoured those that can do the same work equally well in a shorter time than their neighbours. But there can be no doubt that, accompanying that fosterage and protection which is of such marked import in the higher animals, there is also much retardation. And as bearing upon the supposed law of variation as formulated by Messrs. Hyatt and Cope, it should be noted that this retardation or *decreased* rate of growth leads to the production of the more advanced descendant.

The Inheritance of Variations.

Given the occurrence of variations in certain individuals of a species, we have the alternative logical possibilities of their being inherited or their not being inherited. The latter alternative seems at first sight to be in contradiction to the law of persistence. Sir Henry Holland, seeing this, remarked that the real subject of surprise is, not that a character should be inherited, but that any should ever fail to be inherited.† Intercrossing may diminish a character, and sooner or later practically obliterate it: annihilate it at once and in the first generation it cannot. This logical view, however, ceases to be binding if we admit,

---

* *Nature*, vol. xxxvi. p. 592.
† Quoted from " Medical Notes and Reflections," 1855, p. 267, by Darwin, " Animals and Plants under Domestication," vol. i. p. 446.

with Professor Weismann, that variations may be produced in the body without in any way affecting the germ. It is also vitally affected if we believe that the hen does not produce the egg, though she may, perhaps, modify the eggs inside her; for the modification of the hen (*i.e.* the variety in question) may not be of the right nature or of sufficient strength to impress itself upon the germinal matter of the egg. We may at once admit, then, that acquired variations need not be inherited.

Passing to innate variations—variations, that is to say, which are the outcome of normal development from the fertilized ovum—must they be inherited, at any rate, in some degree? It seems to me that they must, on the hypothesis that sexual generation involves simply the blending or commingling of the characters handed on in the ovum or the sperm. The only cases where this would *apparently* fail to hold good would be where the ovum and the sperm handed on exactly opposite tendencies— a variation in excess contributed by the male precisely counterbalancing a variation in the opposite direction contributed by the female parent. Even here the tendency is inherited, though it is counterbalanced. On the hypothesis of "organic combination" before alluded to (p. 150), variations might, however, in the union of ovum and sperm, be not only neutralized, but augmented. If the variation be, so to speak, a definite organic compound resulting from a fortunate combination of characters in ovum and sperm, it might either fail altogether, or be repeated in an enfeebled form, or augmented in the offspring, according as the new conditions of combination were unfavourable or favourable.

Whether innate variations ever actually fail to be inherited, even in an enfeebled form, it is very difficult to say; for if this, that, or the other variation fail to be thus inherited, it is difficult to exclude the possibility of its being an acquired variation not truly innate. Certainly variations seem sometimes to appear in one generation, and not to be inherited at all. And, as we have seen, Mr.

Romanes appeals to a gradual failure of heredity, apart from intercrossing, to explain the diminution of disused organs.

That a variation strongly developed in both parents is apt to be augmented in the offspring is commonly believed by breeders. Darwin was assured that to get a good jonquil-coloured canary it does not answer to pair two jonquils, as the colour then comes out too strong, or is even brown. Moreover,* " if two crested canaries are paired, the young birds rarely inherit this character; for in crested birds a narrow space of bare skin is left on the back of the head, where the feathers are upturned to form the crest, and, when both parents are thus characterized, the bareness becomes excessive, and the crest itself fails to be developed."

On the whole, it would seem that variations may either be neutralized or augmented in inheritance; but the determining causes are not well understood.

Another fact to be noticed with regard to the inheritance of variations is that some characters blend in the offspring, while others apparently fail to do so. Mr. Francis Galton,† speaking of human characters, gives the colour of the skin as an instance of the former, that of the eyes as an example of the latter. If a negro marries a white woman, the offspring are mulattoes. But the children of a light-eyed father and a dark-eyed mother are either light-eyed or dark-eyed. Their eyes do not present a blended tint. Among animals the colour of the hair or feathers is often a mean or blended tint; but not always. Darwin gives the case of the pairing of grey and white mice, the offspring of which are not whitish-grey, but piebald. If you cross a white and a black game bird, the offspring are either black or white, neither grey nor piebald. Sir R. Heron crossed white, black, brown, and fawn-coloured Angora rabbits, and never once got these colours mingled in the same animal, but often all four colours in the same litter. He also

---

* Darwin, "Animals and Plants under Domestication," vol. i. p. 465.
† "Natural Inheritance," p. 12.

crossed "solid-hoofed" and ordinary pigs. The offspring did not possess all four hoofs in an intermediate condition; but two feet were furnished with properly divided and two with united hoofs.* Professor Eimer† has noticed that, in the crossing of striped and unstriped varieties of the garden snail, *Helix hortensis*, the offspring are either striped or unstriped, not in an intermediate or faintly striped condition.

These facts are of no little importance. They tend to minimize, for some characters at least, the effects of intercrossing. The variations which present this trait may be likened to stable organic compounds, which may be inherited or not inherited, but which cannot be watered down by admixture and intercrossing. It is well known‡ that, in 1791, a ram-lamb was born in Massachusetts, with short, crooked legs and a long back, like a turn-spit dog. From this one lamb § the *otter*, or *ancon*, breed was raised. When sheep of this breed were crossed with other breeds, the lambs, with rare exceptions, perfectly resembled one parent or the other. Of twin lambs, even, one has been found to resemble one parent, and the second the other. All that the breeder has to do is to eliminate those which do not possess the required character. And very rarely do the lambs of ancon parents fail to be true-bred.

Now, it can scarcely fail that such sports occur in nature. And if they are stable compounds, they will not be readily swamped by intercrossing. It only requires some further isolation to convert the sporting individuals into a distinct and separate variety. Now, Darwin tells us that

* Darwin, "Animals and Plants under Domestication," vol. ii. p. 70.
† "Organic Evolution," Mr. Cunningham's translation, p. 76.
‡ Darwin, "Animals and Plants under Domestication," vol. i. p. 104.
§ Similarly, from a chance sport of a one-eared rabbit, Anderson formed a breed which steadily produced one-eared rabbits ("Animals and Plants under Domestication," vol. i. p. 456). This is an example of asymmetrical variation. Variations are generally, but not always, symmetrical. Superficial colour-variations are sometimes asymmetrical. Gasteropod molluscs are nearly always asymmetrically developed. Among insects, *Anisognathus* affords an example of the asymmetrical development of the mandible. Our right-handedness is a mark of asymmetry.

the ancons have been observed to keep together, separating themselves from the rest of the flock when put into enclosures with other sheep. Here, then, we have preferential mating as the further isolating factor. I feel disposed, therefore, to agree with Mr. Galton when he says,* "The theory of natural selection might dispense with a restriction for which it is difficult to see either the need or the justification, namely, that the course of evolution always proceeds by steps that are severally minute, and that become effective only through accumulation. That the steps *may* be small, and that they *must* be small, are very different views; it is only to the latter that I object, and only when the indefinite word 'small' is used in the sense of 'barely discernible,' or as small as compared with such large sports as are known to have been the origins of new races."

Connected, perhaps, with the phenomena we have just been considering is that of *prepotency*.† It is found that, when two individuals of the same race or of different races are crossed, one has a preponderant influence in determining the character of the offspring. Thus the famous bull Favourite is believed to have had a prepotent influence on the short-horn race; and the improved short-horns possess great power in impressing their likeness on other breeds. The phenomena are in some respects curiously variable. In fowls, silkiness of feathers seems to be at once bred out by intercrossing between silk-fowl and any other breed. But in the silky variety of the fan-tail pigeon this character seems prepotent; for, when the variety is crossed with any other small-sized race, the silkiness is invariably transmitted. One may fairly suppose that prepotent characters have unusual stability; but to what causes this stability is due we are at present ignorant.

Lastly, we have to consider the phenomenon of *latency*.

---

* "Natural Inheritance," p. 32.
† See "Animals and Plants under Domestication," vol. ii. p. 40, from which illustrations are taken.

This is the lying hid of characters and their subsequent emergence. We may distinguish three forms of latency.

1. Where characters lie hid till a certain period of life, and then normally emerge.

2. Where the characters normally lie hid throughout life, but are, under certain circumstances, abnormally developed.

3. Where the characters lie hid throughout life, but appear in the offspring or (sometimes distant) descendants.

Latency is often closely connected with correlated variations. Secondary sexual characters, for example, are correlated with the functional maturity or activity of the reproductive organs. They therefore lie hid until these organs are mature and ready for activity. When they are restricted to the male, they normally remain latent throughout the life of the female, but reappear in her male offspring. Under abnormal conditions, such as the removal of the essentially male organs, the secondary sexual characters correlated with them do not appear, or appear in a lessened and modified form. The males may even, under such circumstances, acquire female characters. Thus capons take to sitting, and will bring up young chickens. Conversely, females which have lost their ovaries through disease or from other causes sometimes acquire secondary sexual characters proper to the male. Characters thus normally latent abnormally emerge. Mr. Bland Sutton[*] gives a case of a hen golden pheasant which "presented the resplendent dress of the cock, but her plumage was not quite so brilliant; she had no spurs, and the iris was not encircled by the ring of white so conspicuous in the male." Her ovary was no larger than a split pea.

A curious instance of latent characters correlated with sex is seen in hive bees. The worker bee differs from the female in the rudimentary condition of the sexual organs, in size and form, and in the higher development of the sense-organs. But it is well known that, if a very young worker grub be fed on "royal jelly," she will develop into

---

[*] "Evolution and Disease," p. 169.

a perfect queen. Not only are the sexual organs stimulated to increased growth and functional activity, but the correlated size and condition of the sense-organs are likewise acquired. The characters of queen and worker are latent in the grub. According to the nature of the food it receives, the one set of characters or the other emerges. Professor Yung's tadpoles and Mrs. Treat's butterflies (*ante*, p. 59) afford similar instances.

We come now to those cases of latency in which this obvious correlation does not occur. They afford examples of reversion to more or less remote ancestral characters. In some cases the cause of such reversion—such unexpected emergence of characters, which have remained latent through several, perhaps many, generations—is quite unknown. In others, at any rate among domesticated animals, the determining condition of such reversion is the crossing of distinct breeds.

Darwin gives* an instance of reversion, on the authority of Mr. R. Walker. He bought a black bull, the son of a black cow with white legs, white belly, and part of the tail white; and in 1870 a calf, the gr-gr-gr-gr-grandchild of this cow, was born, coloured in the same very peculiar manner, all the intermediate offspring having been black. In man partial reversions are not infrequent. An additional pair of lumbar ribs is sometimes developed, and in such cases the fan-shaped tendons which are normally connected with the transverse processes of the vertebræ are replaced by functional levator muscles. Since it is probable that the ancestor of man had more than the twelve pairs of ribs that are normally present in the human species, we may, perhaps, fairly regard the supernumerary rib as a reversion. But it may be a new sport on old lines.

The occasional occurrence in Scotland of red grouse with a large amount of white in the winter plumage, especially on the under parts, is justly regarded by Mr. Wallace † as a good example of reversion or latency in

* "Animals and Plants under Domestication," vol. ii. p. 8.
† "Darwinism," p. 107.

wild birds. There can be little doubt that, as he suggests, the Scotch red grouse is derived from a form which, like the wide-ranging willow grouse, has white winter plumage. During the glacial epoch this would be an advantage. "But when the cold passed away, and our islands became permanently separated from the mainland, with a mild and equable climate, and very little snow in winter, the change to white at that season became hurtful, rendering the birds more conspicuous, instead of serving as a means of concealment." The red grouse has lost its white winter dress; but occasional reversions point to the ancestral habit.

That crossing tends to produce reversion is a fact familiar to breeders and fanciers, and one which is emphasized by Darwin. When pigeons are crossed, there is a strong tendency to revert to the slatey-blue tint and black bars of the ancestral rock-pigeon. There is always a tendency in sheep to revert to a black colour, and this tendency is emphasized when different breeds are crossed. The crossing of the several equine species (horse, ass, etc.) "tends in a marked manner to cause stripes to appear on various parts of the body, especially on the legs," and this *may be* a reversion to the condition of a striped and zebra-like ancestor. Professor Jaeger described a good case with pigs. "He crossed the Japanese, or masked breed, with the common German breed, and the offspring were intermediate in character. He then recrossed one of these mongrels with a pure Japanese, and in the litter thus produced one of the young resembled in all its characters a wild pig; it had a long snout and upright ears, and was striped on the back. It should be borne in mind that the young of the Japanese breed are not striped, and that they have a short muzzle and ears remarkably dependent."[*] Darwin crossed a black Spanish cock with a white silk hen. One of the offspring almost exactly resembled the *Gallus bankiva*, the remote ancestor of the parents.

Such cases would seem to show that in our domestic

[*] Darwin, "Animals and Plants under Domestication," vol. ii. pp. 17, 18.

breeds ancestral traits lie latent. The crossing of distinct varieties may either neutralize the variations artificially selected, and thus allow the ancestral characters which have been masked by them to reappear; or they may allow the elements of the ancestral traits, long held apart in separate breeds by domestication, to recombine with the consequent emergence of the normal characters of the wild species. But, in truth, any attempted explanations of the facts are little better than guess-work. There are the facts. And the importance of crossing as a determining condition in domesticated animals should make us cautious in applying reversion, as it occurs in such cases, to wild species which live under more stable conditions where crossing is of rare occurrence.

The Origin of Variations.

The subject of the origin of variations is a difficult one, one concerning which comparatively little is known, and one on which I am not able to throw much light.

Taking a simple animal cell as our starting-point, we have already seen that it performs, in primitive fashion, certain elementary and essential protoplasmic activities, and gives rise to certain products of cell-life. In the metazoa, which are co-ordinated aggregates of animal cells, together with some of their products, there is seen a division of labour and a differentiation of structure among the cells. We see, then, that variation among these related cells has led to differences in size, in form, in transparency, and in function; while the cell-products have been differentiated into those which are of lifelong value, such as bone, cartilage, connective tissue, horn, chitin, etc., together with a variety of colouring matters; those which are of temporary value, such as the digestive secretions, fat, etc.; and those which are valueless or noxious, such as carbonic acid gas and urea, which are excreted as soon as possible. Here are already a number of important and fundamental variations to be accounted for.

Let us notice that, wide as the variations are, they are to a large extent hedged in by physical, chemical, and organic limitations. We have already seen that the size of cells is to a large extent limited, because during growth mass tends to outrun surface; and because, while disruptive changes occur throughout the mass, nutriment and oxygen must be absorbed by the surface. This is a physical limitation. Since the products of cell-life and cell-activity are chemical products, it is clear that they can only be produced under the fixed limitations of chemical combination; and though in organic products these limitations are not so rigid as among inorganic substances, still that there are limitations no chemist is likely to question. The organic limitations are to the varied, but not very numerous, modes of protoplasmic activity.

Probably, even at the threshold of metazoan life, such variations did not affect only individual cells, but rather groups of cells. In other words, the differentiation was at once and primarily a tissue-differentiation. What do we know, however, about the primitive tissue-differentiation of the earliest metazoa? Hardly anything. We may fairly suppose that the first marked difference to appear was that between the outside and the inside. In the formation of an embryo this is the first differentiation we notice. From the beginning of segmentation or, in any case, very early, the outer-layer cells become marked off from the inner-layer cells. The next step was, perhaps, the formation of the mid-layer between the outer and inner. But how further differentiations were effected we really do not know, though we may guess a little. This, perhaps, we may fairly surmise—that fresh differentiations presupposed previous differentiations, and formed the basis of yet further differentiations. Thus calcified cartilage presupposes cartilage, and leads up to the formation of true bone. In all this, however, we are very much in the dark. We can watch, always with fresh wonder, the genesis of tissues in the development of the embryo; but we do not at present know much of the mode of their primitive genesis in the early days of organic evolution: how can we, then, pretend to understand their origins?

If we speculate at all on the matter, we are led to the view that the variations must be primarily due to the differential incidence of mechanical stresses and physical or chemical influences. It may be admitted that this is little more than saying that they are due to some physical cause. Still, this at least may be taken as certain for what it is worth—that the primitive tissue-differentiations are due to physical or chemical influences, direct or indirect, on the protoplasm of the cell. Here is one mode of the origin of variations.

I do not wish to reopen the question whether these variations originate in the germ or in the body. I content myself with indicating the difference, from this standpoint, between the two views. Take, for example, the end-organs of the special senses, which respond explosively to physical influences in ways we shall have to consider more fully in the next chapter. If we hold that variations originating in the body may be transmitted through the germ to the offspring, then we may say that these variations are the direct result of the incidence of the physical or molecular vibrations on the protoplasm. But if we believe, with Professor Weismann, that all variations originate in the germ, then the variations in the end-organs of the special senses, fitting them to be the recipients of special modes of influence, result from physical effects upon the germ of purely fortuitous origin, that is to say, wholly unrelated to the end in view. The rods and cones of the retina are due to purely chance variations, impressed by some chemical or physical causes completely unknown on the germinal protoplasmic substance. Those individuals which did not have these chance variations have been eliminated. It matters not that the rods and cones are believed to have reached their present excellence through many intermediate steps from much simpler beginnings. The fact remains that the origin of all these step-like variations was fortuitous, and not in any way the direct outcome of the physical influences which their products, the rods and cones, have become fitted to receive. I am not at present prepared to accept this theory of the germinal origin of all tissue-variations.

Whether use and disuse are to be regarded as sources of origin of variations is, again, a matter in which there is wide difference of opinion. But if we admit that any variations can take their origin in the body (as distinguished from the germ), then there is no *à priori* reason for rejecting use and disuse as factors. As such, we are, I think, justified, in the present state of our knowledge, in reckoning them, at all events, provisionally.

It is clear, however, that they are a proximate, not an ultimate, source of origin. I mean that the structures must be there before they can be either strengthened or weakened by use or disuse. They are at most a source of positive or negative variations of existing structures. They cannot be a direct source of origin of superficial variations. Gain or loss of colour; form-variations not correlated with organic variations;—these cannot be directly due to use or disuse. It is in the nervous and muscular systems and the glandular organs that use and disuse are mainly operative. When, however, organs are brought into relation, or fail to be brought into relation, to their appropriate stimuli, we speak of this, too, as use and disuse. We say, for example, that persistent disuse may impair the essential tissues of the recipient end-organs of the special senses, implying that these tissues require to be brought into continued relation to the appropriate stimuli in order that their efficiency be maintained. So, too, we say that the epidermis is thickened by use, meaning that it is brought into relation with certain mechanical stresses. Through correlation, too, the effects of use and disuse may be widespread. Thus increase in the size of a group of muscles may be correlated with increase in the size of the bones to which they are in relation. In fact, so knit together and co-ordinated is the organism into a unity, it is probable that hardly any variation could take place through use or disuse without modifying to some extent the whole organic being.

Once more, let it be clearly remembered that a large and important school of zoologists reject altogether use or disuse as a factor in variation. They believe that those germs are selected through natural selection in which there is an increased tendency to use or disuse of certain organs. In this, however, we are all agreed. The real question is what is the source of origin of this tendency. On the view of germinal origin, we are forced back on unknown physical or chemical influences in no wise related in origin (though, of course, related in result) with the use or disuse to which they give rise.

So far the main distinction between the two biological schools seems to be that the one, placing the origin of variation in the body-tissues, regards the variations as evoked in direct reaction to physical or chemical influences; while the other, placing the origin of variation in the germ, regards the variations as of fortuitous origin.

I do not use the phrase, "of fortuitous origin," as in any sense discrediting the theory. I am not attempting the cheap artifice of damning a view that does not happen to be my own with a phrase or a nickname. And I therefore hasten to point out what variations I do believe to have had a fortuitous origin. The phrase is often misunderstood, and they will serve to explain its meaning.

If the reader will kindly refer to the tables of variations in the bats' wings (Figs. 14-17), he will see that there are a great number of bones which vary in length and vary independently. And if he will also refer to Fig. 18, in which seven species of bats are compared, he will see that the differences arise from the increased length of one set of bones in one species and another set of bones in another species. Now, let us suppose that the long, swallow-like wing of the noctule, a high flyer with rapid wing-strokes, that catches insects in full flight, and the broad wings of the horse-shoe, a low flyer, flapping slowly, and, at any rate, sometimes catching insects on the ground, and covering them with its wings as with a net; let us suppose, I say, that to each species its special form of wing is an advantage. Among thousands of independent variations in the lengths of the bones there would be occasional combinations of variations, giving either increased length or increased breadth to the wing. In the noctule, the former would tend to be selected; in the horse-shoe, the latter. Thus the wing of the noctule would be lengthened, and that of the horse-shoe broadened, through the selection of fortuitous combinations of variations which chanced to be favourable. Now, each individual bone-variation is, we believe, due to some special cause; but the fortunate combination is fortuitous, due to what we term "mere chance."

Darwin believed that chance, in this sense, played a very important part in the origin of those favourable variations for which, as he said, natural selection is constantly and unceasingly on the watch. And there can be little question that Darwin was right.

We must now consider very briefly some of the proximate causes of variations. In most of these cases we cannot hope to unravel the nexus of causation. When a plexus of environing circumstances acts upon a highly organized living animal, the most we can do in the present state of knowledge is to note—we cannot hope to explain—the effects produced.

All readers of Darwin's works know well how insistent he was that the nature of the organism is more important than the nature of the environing conditions. "The organization or constitution of the being which is acted on," he says,* "is generally a much more important element than the nature of the changed conditions in determining the nature of the variation." And, again,† "We are thus driven to conclude that in most cases the conditions of life play a subordinate part in causing any particular modification; like that which a spark plays when a mass of combustible matter bursts into flame—the

---

* "Animals and Plants under Domestication," vol. ii. p. 201.

† Ibid. p. 282. The phenomena of the seasonal dimorphism of butterflies and moths show that changes of temperature (and perhaps moisture, etc.) determine very striking differences in these insects.

nature of the flame depending on the combustible matter, and not on the spark."

Recent investigations have certainly not lessened the force of Darwin's contention. From which there follows the corollary that the vital condition of the organism is a fact of importance. Darwin was led to believe that among domesticated animals and plants good nutritive conditions were favourable to variation. "Of all the causes which induce variability," he says,* "excess of food, whether or not changed in nature, is probably the most powerful." Darwin also held that the male is more variable than the female—a view that has been especially emphasized by Professor W. K. Brooks. Mr. Wallace, as we have already seen, regards the secondary sexual characters of male birds as the direct outcome of superabundant health and vigour. "There is," he says,† "in the adult male a surplus of strength, vitality, and growth-power which is able to expend itself in this way without injury." And Messrs. Geddes and Thomson contend‡ that "brilliancy of colour, exuberance of hair and feathers, activity of scent-glands, and even the development of weapons, are in origin and development outcrops of a male as opposed to a female constitution."

There is, I think, much truth in these several views thus brought into apposition. Vigour and vitality, predominant activity and the consequent disruptive changes, with their abundant by-products utilized in luxuriant outgrowths and brilliant colours, are probably important sources of variation. They afford the material for natural selection and sexual selection to deal with. These guide the variations in specific directions. For I am not prepared to press the theory of organic combination so far as to believe that this alone has served to give definiteness to the specific distinctions between secondary sexual characters, though it may have been to some extent a co-operating factor. This, however, is a question apart from that of

* "Animals and Plants under Domestication," vol. ii. p. 214.
† "Darwinism," p. 293.   ‡ "Evolution of Sex," p. 22.

origin. Superabundant vigour may well, I think, have been a source of *origin*, not only of secondary sexual characters, but of many other forms of variation.

And while these forms of variation may be the special prerogative of the male, we may perhaps see, in superabundant female vigour, a not less important source of developmental and embryonic variations in the offspring. The characteristic selfishness of the male applies his surplus vitality to the adornment of his own person; the characteristic self-sacrifice of the mother applies her surplus vitality to the good of her child. Here we may have the source and origin of those variations in the direction of fosterage and protection which we have seen to have such important and far-reaching consequences in the development of organic life. The storage of yolk in the ovum, the incubation of heavily yolked eggs, the self-sacrificing development in the womb, the elaboration of a supply of food-milk,—all these and other forms of fosterage may well have been the outcome of superabundant female vigour, the advantages of which are thus conferred upon the offspring.

We may now proceed to note, always remembering the paramount importance of the organism, some of the effects produced by changes in the environment.

The most striking and noteworthy feature about the effects of changes of climate and moisture, changes of salinity of the water in aquatic organisms, and changes of food-stuff, is that, when they produce any effect at all, they give rise to *definite* variations. Only one or two examples of each can here be cited. Mr. Merrifield,[*] experimenting with moths (*Selenia illunaria* and *S. illustraria*), finds that the variations of temperature to which the pupæ, and apparently also the larvæ, are subjected tend to produce "very striking differences in the moths." On the whole, cold "has a tendency, operating possibly by retardation, to produce or develop a darker hue in

[*] "Incidental Observations in Pedigree Moth-breeding," F. Merrifield. Transactions Entomological Society, 1889, pt. i. p. 79, *et seq*.

the perfect insect; if so, it may, perhaps, throw some light on the melanism so often remarked in north-country examples of widely distributed moths." Mr. Cockerell * regards moisture as the determining condition of a certain phase of melanism, especially among Lepidoptera. The same author states that the snail "*Helix nemoralis* was introduced from Europe into Lexington, Virginia, a few years ago. Under the new conditions it varied more than I have ever known it to do elsewhere, and up to the present date (1890) 125 varieties have been discovered there. Of these, no less than 67 are new, and unknown in Europe, the native country of the species." The effects of the salinity of the water on the brine-shrimp *Artemia* have already been mentioned. One species with certain characteristics was transformed into another species with other characteristics by gradually altering the saltness of the water. So, too, in the matter of food, the effects of feeding the caterpillars of a Texan species of *Saturnia* on a new food-plant were so marked that the moths which emerged were reckoned by entomologists as a new species.

The point, I repeat, to be especially noted about these cases and others which might be cited,† is that the variation produced is a definite variation. Very probably it is generally, or perhaps always, produced in the embryonic or larval period of life. In some cases the variation seems to be transmissible, though definite and satisfactory proofs of this are certainly wanting. Still, we may say that if the changed conditions be maintained, the resulting variation will also be maintained. Under these conditions, at least, the variation is a stable one. It is probable that, apart from preferential mating, the varieties thus produced will tend to breed together rather than to be crossed with the parent form or varieties living under different conditions. In this way varieties may sometimes arise by

---

* *Nature*, vol. xli. p. 393.

† See Professor Meldola's edition of Professor Weismann's "Studies in the Theory of Descent," and Mr. Cunningham's translation of Professor Eimer's "Organic Evolution."

definite and perhaps considerable leaps under the influence of changed conditions. We must not run the adage, *Natura nil facit per saltum*, too hard, nor interpret *saltum* in too narrow a sense.

It is true, and we may repeat the statement of the fact for the sake of emphasis, that we do not know how or why this or that particular variation should result from this or that change of climate, environment, or food-stuff; nor do we know why certain variations (such as that which produced the ancon breed of sheep) should be stable, while other variations are peculiarly unstable. But in this we are not worse off than we are in the study of inorganic nature. We do not know why calcite should crystallize in any particular one of its numerous varieties of crystalline form; we do not know why some of these are more stable than others. We may be able to point to some of the conditions, but we cannot be said to understand why arragonite should be produced under some circumstances, calcite under others; or why the same constituents should assume the form of augite in some rocks, and hornblende in other rocks. We are hedged in by ignorance; and perhaps one of our chief dangers, becoming with some people a besetting sin, is that of pretending to know more than we are at present in a position to know. Our very analogies by which we endeavour to make clear our meaning may often seem to imply an unwarrantable assumption of knowledge.

In the last chapter I used the term "organic combination," and drew a chemical analogy. I wished to indicate the particularity and the stability of certain variations, and the possibility of new departures through new combinations of variations, the new departure not being necessarily anything like a mean between the combining variations.* I trust that this will not be misunderstood as a new chemico-physical theory of organic forms. I have some fear lest I should be represented as maintaining that a giraffe or a peacock is a definite organic compound, with its proper

* See Darwin, "Animals and Plants under Domestication," vol. ii. p. 252.

organic form, in exactly the same way as a rhombohedron of calcite or a rhombic dodecahedron of garnet is a definite chemical compound, with its proper crystalline form. All that the analogy is intended to convey is that variations seem, under certain circumstances, to be definite and stable, and may possibly combine rather than commingle.

Summary and Conclusion.

It only remains to bring this chapter to a close with a few words of summary and conclusion.

The diversity of animal life must first be grasped. We believe that this diversity is the result of a process or processes of evolution. Evolution is the term applied to continuity of development. It involves adaptation; and adaptation to an unchanging environment may become more and more perfect. But the environment to which organisms are adapted also changes. Where the change is in the direction of complexity, we have elaboration; where it is in the direction of simplicity, we have degeneration. Of these elaboration is the more important. It involves both a tendency to differentiation giving rise to individuality, and a tendency to integration giving rise to association. Continued elaboration is progress; and this is opposed to degeneration.

The factors of evolution fall under two heads—origin and guidance. The origin of variations lies in mechanical stresses, and chemical or physical influences. Whether these act on the body (and are transmitted by inheritance) or only on the germ, is a question which divides biologists into two schools. In the latter case all variations are fortuitous; in the former the development of tissue-variations has been in direct response to the physical or chemical influences. There are, however, in any case fortuitous combinations of variations.

Whether use and disuse are factors of origin is also a debatable point. Those who believe that physical influences on the body are transmissible believe also that the effects of use and disuse are transmissible.

The vital vigour of the organism is a determining condition of importance. The vital vigour of males has favoured the origin of secondary sexual characters; that of females, the fostering and protection of young, and therefore the development in them of vital vigour.

The almost universally admitted factor in guidance is natural selection. But we must be careful not to use it as a mere formula.

Whether sexual selection is also a factor is still a matter of opinion. Without it the specific character and constancy of secondary sexual features are at present unexplained. If inherited use and disuse are admitted as factors in origin, they must also be admitted as important factors in guidance.

Questions of origin and guidance should, so far as is possible, be kept distinct. These terms, however, apply to the origin and guidance of variations. In the origin of species guidance is a factor, no doubt a most important factor. The title of Darwin's great work was, therefore, perfectly legitimate. And those who say that natural selection plays no part in the origin of species are, therefore, undoubtedly in error.

CHAPTER VII.

THE SENSES OF ANIMALS.

It is part of the essential nature of an animal to be receptive and responsive. The forces of nature rain their influence upon it; and it reacts to their influence in certain special ways. Other organisms surround it, compete with it, contend with it, strive to prey upon it, and occasionally lend it their aid. It has to adjust itself to this complex environment.

There are two kinds of organic response—one more or less permanent, the other temporary and transient. We have already seen something of the former, by which the tissues (the epidermis of the oarsman's hand, and the muscles of his arm) respond to the call made upon them. The response is here gradual, and the effects on the organism more or less enduring. This, however, is not the kind of response with which we have now to deal. What we have now to consider is that rapid response, transient, but of the utmost importance, by means of which the organism directly answers to certain changes in the environment by the performance of certain activities. The parts specially set aside and adapted to receive special modes of influence of the environment are the sense-organs. We human folk get so much pleasure from and through the employment of our sense-organs, that it is important to remember that the primary object of the process of reception of the influences from without was not the æsthetic one of ministering to the enjoyment of life by the recipient organism, but the essentially practical one of enabling that organism to respond to these influences. In other words, the *raison d'être* of the sense-organs is to set agoing suitable activities—activities in due response to the special stimuli.

In this chapter we shall consider the modes in which the special sense-organs are fitted to receive the influences of the environment, deferring to a future chapter the consideration of the resulting activities. For the present we take these activities for granted, observing them only in so far as they give us a clue to the sense-reaction by which they are originated. In this chapter, too, we shall deal, for the most part, with the physiological aspects of sensation. In all other organisms than ourselves, that is to say, than each one of us individually for himself, the psychological accompaniments of the physiological reactions of the sense-organs are matters of inference. Still, so closely and intimately associated are the physiological and the psychological aspects, that the exclusion of all reference to the latter would be impracticable, or, if practicable, unadvisable. What is practicable and advisable is to remember that, even if the two are mentioned in a breath, the physiological and the psychological belong to distinct orders of being.

In addition to the time-honoured "five senses," there are certain *organic sensations*, so called, which take their origin within the body. These are, for the most part, somewhat vague and indefinite. They do not arise immediately and in direct response to changes in the environment, but indicate conditions of the internal organs. Such are hunger, thirst, nausea, fatigue, and various forms of discomfort. Although they are of vital importance to the organism, prompting it to perform certain actions or to desist from others, they need not detain us here.

More definite than these, but still of internal origin, is the *muscular sense*. This, too, is of continual service to every active animal. By it information is given as to the energy of contraction of the muscles, and of the amount of movement effected—not to mention the rapidity and duration of the muscular effort. By it the position, or changes of position, of the motor-organs are indicated. It is obvious, therefore, that the sensations obtained in this way, some of which are exceedingly delicate, are an important guide to the organism in the putting forth of its activities. It is through the muscular sense that we maintain an upright position. It is through an educated and refined muscular sense that the juggler and the acrobat can perform their often surprising feats. Concerning the physiology of the muscular sense, we have at present no very definite knowledge. Some have held that we judge of muscular movements by the amount of effort required to initiate them; but it is much more probable that there are special sensory nerves, whose terminations are either in the muscles themselves or in the membranes which surround them.

We come now to the special senses. Of these we will take first the *sense of touch*. Through this sense we are made aware of bodies solid or liquid (or perhaps gaseous) which are actually in contact with the skin or its infoldings at the mouth, nostrils, etc. There are considerable differences in the sensitiveness of the skin in different parts of its surface; some parts, like the filmy membrane which covers the eye, being very sensitive, while others, like the horny skin that covers the heel of a man who is accustomed to much walking, are relatively callous. Different from this is the delicacy of the sense of touch. This delicacy is really the power of discrimination, and therefore involves some mental activity. But it is also dependent upon the distribution of the recipient end-organs of the nerve. The highest pitch of delicacy is reached in the tip of the tongue, which is about sixty times as delicate as the skin of the back. The power of discrimination is tested in the following way: The points of a pair of compasses are blunted, and with them the skin is lightly touched. When the points are close together, the sensation is of one object;

when they are more divergent, each point is felt as distinct from the other. On the thigh and in the middle of the back, two distinct points of contact are not felt unless the compass-tips are about $2\frac{1}{3}$ inches (67·7 millimetres) apart. When the divergence is 2 inches, they are felt as one. With the tip of the tongue, however, we can distinguish the two separate points when they are only $\frac{1}{25}$ of an inch (1·1 millimetre) apart. For the finger-tip the distance is about $\frac{1}{12}$ of an inch (2 millimetres); for the tip of the nose, about $\frac{1}{4}$ of an inch (6·8 millimetres); for the forehead, a little less than an inch (22·6 millimetres); and so on. Shut your eyes, and allow a friend to draw the compass with the points about $\frac{1}{2}$ an inch apart, from the forehead to the tip of your nose, or (setting the points about $\frac{1}{4}$ of an inch apart) from the ball of your thumb to the finger-tip. The increasing delicacy and power of discrimination is readily felt, and it is difficult to believe that the compasses are not being slowly opened.

It is beyond the purpose of this chapter to describe minutely the nature and structure of the nerve-ends in the sense-organs. This is a matter of minute anatomy, or histology. A full description of them as they occur in man will be found in any standard text-book of physiology; while Sir John Lubbock's "Senses of Animals" gives much information concerning, and many illustrations of, the minute structure of the sense-organs in the invertebrates. Here I can only touch very briefly on some of the more important points.

One of the larger nerves of the body (*e.g.* the sciatic nerve), consists of a bundle of nerve-threads collected from a considerable area; some of these (motor threads) end in muscles, others (sensory threads) in the skin or its neighbourhood. Each nerve-thread has a central axis-fibre, which is surrounded by a fatty, insulating medullary sheath, and this by a delicate primitive sheath. In some parts of the skin the sensory nerve-threads lose their medullary sheath, and end in very fine branches between the cells of the tissue. In other cases the cells near their termination are specially modified to form tactile cells, or tactile corpuscles, in contact with or surrounding the axis-fibre or its expansion (Fig. 23).

Hairs are delicate organs of touch, though, of course,

Fig. 23.—Tactile corpuscles.
1. In the beak of a goose. 2. In the finger of a man. 3. In the mesentery of a cat.

this is not their only function. They act as little levers embedded in the skin.

Turning now to the vertebrate animals other than man, we find in them a sense of touch closely analogous to our own. As in us, so in them, the specially mobile parts are eminently sensitive and delicate; for instance, the lips in many animals, such as the horse, and the finger-like organ at the end of the elephant's trunk. In some of them special hairs are largely developed as organs of touch, as in the whiskers of the cat and the long hairs on the rabbit's lip. With the aid of these the rabbit finds its way in the darkness of its burrow; and it is said that, deprived of these organs, the poor animal blunders about, and is unable to steer its course in the dark.

The wing of the bat is very sensitive to touch; and it is supposed that it is through this sense that the bat is able to direct its course in the darkness of caves. Miss Caroline Bolton thus describes an experimental trial of this power of the bat at which she was herself present. A room, about twenty feet by sixteen, was arranged with strings crossing each other in all directions so as to form a network with about sixteen inches space between the strands. To each string was attached a bell in such a way that the slightest touch would make it ring. One corner of the room was left free for those who were present at the experiment. A bat, measuring about one foot from the tip of one wing to that of the other, was let loose in the room when it was quite dark, "and it was distinctly heard flying about all over the room, but never once did it touch a string or stop flying. It several times came quite near to the spectators, so that they could feel the vibration of the air in their faces. The experiment was continued for half an hour. Then, when the door was opened and light let in, the bat stopped flying, and settled down in the darkest corner." Now, here it may be said that, although the room was dark to human spectators, there may have been light enough for a bat to see his way. The cruel experiments of Spalanzani, however, who put out the eyes of bats and obtained a similar result, seem to show that the animal is guided by some sense other than that of sight.

The crustaceans and many insects are covered with a dense armour, and it might be supposed that in them there could be no sense of touch. But this sense is by no means absent. Seated on the tough integument are delicate little hairs, to the base of which a nerve-fibril passes through a perforation in the integument. These are specially numerous in the antennæ of insects.

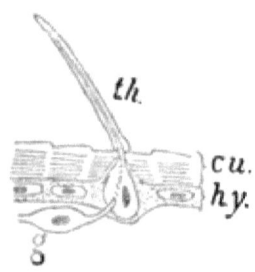

Fig. 24.—Touch-hair of insect.

t.h., touch-hair; *cu.*, cuticle; *h.y.*, hypodermis; *g.*, ganglion-cell connected with nerve passing into the cavity of the touch-hair (after Miall). The ganglion is often surrounded by several—eight or less—accessory cells, which are not figured here.

In yet lower organisms we know in some cases the manner in which they are sensitive to touch; but in a great number of cases, although observation shows that they are thus sensitive, we know nothing definite as to how the surface is specially fitted to receive the stimuli. Even the primitive amœba, however, is sensitive in the sense spoken of on p. 8; that is to say, it reacts under the influence of a stimulus.

Closely associated with the sense of touch is the

temperature-sense. Goldschneider and others have shown that on the skin of the human hand, for example, there are special points that are sensitive to heat and cold. Some of these little specialized areas are sensitive to cold; others are sensitive to heat; and neither of these seem to be sensitive to pressure. It therefore seems probable that special nerve-fibrils are set apart for the temperature-sense; but of the manner in which these fibrils terminate little or nothing is known.

Let us note that this temperature-sense, unlike the sense of touch, may make us aware of distant bodies. It is, then, what we may term a *telæsthetic* sense in contradistinction to a contact-sense. It is stimulated by a molecular throb; the throbbing body may be in contact, but it may be as distant as the sun, in which case the molecular pulsations are brought to us on waves of æther. Whether these waves act directly on the nerve end-organs, or indirectly on them through the warming of the skin-surface in which they terminate, we cannot say for certain. But if the hand be held before a heated stove and be sheltered from the heat by a screen, the removal of the screen, even for the fraction of a second, gives rise to a strong stimulation of the temperature-sense, though the skin-surface be not appreciably raised in temperature. Hence it is probable that the end-organs are stimulated directly, and not indirectly.

Concerning the temperature-sense in the lower animals, nothing definite is known. But it is impossible to see our familiar pets basking in the sunshine, or a butterfly sunning itself on a bright summer's day, without feeling confident that the temperature-sense is a channel of keen enjoyment. As before mentioned, however, this is not to be regarded as the primary end in sensation. The primary end is not life-enjoyment, but life-preservation. And we must regard the temperature-sense as developed in the first instance to enable the organism to escape from the ill effects of deleterious heat or cold, and to seek those temperature-conditions which are most helpful to the continued and healthful fulfilment of the process of life.

The *sense of taste* is called into play by certain soluble substances, or liquids, which must come in contact with the specialized nerve-endings. Under normal circumstances, the sense of taste is closely associated with that of smell, the result of the combination of the two special senses being a *flavour*. The *bouquet* of a choice wine, the flavour of a peach, involve both senses; quinine involves taste alone; and garlic and vanilla are nearly, if not quite, tasteless,—what we call their taste is in reality their action on the organ of smell.

It is difficult to classify tastes. Sweet, bitter, salt, alkaline, sour, acid, astringent, acrid,—these are the prominent and characteristic varieties.

This sense is generally localized in or near the mouth; in us mainly in the tongue. One manner, but not the only manner, in which the nerves in this region terminate is in the minute flask-shaped taste-buds, which have near one end, where they reach the surface, a funnel-shaped opening, the taste-pore. They are made up of elongated cells, some of which near the centre are spindle-shaped, and are called taste-cells. They are found chiefly round the large circumvallate papillæ; but in the rabbit and some other animals they are collected in the folds of a little ridged or pleated patch—the *papilla foliata*—on each side of the tongue near the cheek-teeth.

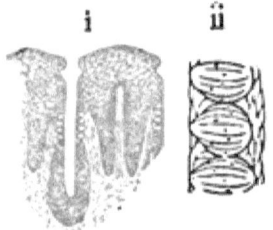

Fig. 25.—Taste-buds of rabbit.
i., section across part of the pleated patch (enlarged); ii., taste-buds further enlarged.

It is probable that the stimulation of the end-organs of taste is effected by the special mode of molecular vibration due to the chemical nature of the sapid substance. Mr. J. B. Haycroft, in a paper read before the Royal Society of Edinburgh,[*] suggests that "a group of salts of similar chemical properties have their

[*] See abstract in *Nature*, vol. xxxiv. p. 515.

molecules in a similar vibrating condition, giving rise to similar colours and similar tastes." "Thus the chlorides and sulphates of a series of similar elements—called a group of elements by Mendeljeff—have similar tastes."

The delicacy of the sense of taste in man has been the subject of investigation by Messrs. E. H. S. Bailey and E. L. Nichols.* They give the following table :—

I. Quinine—
  Male observers detected 1 part in 390,000 parts of water.
  Female   „   „   1   „   456,000   „   „
II. Cane-sugar—
  Male observers   „   1   „   199   „   „
  Female   „   „   1   „   204   „   „
III. Sulphuric acid—
  Male observers   „   1   „   2,080   „   „
  Female   „   „   1   „   3,280   „   „
IV. Bicarbonate of sodium—
  Male observers   „   1   „   98   „   „
  Female   „   „   1   „   126   „   „
V. Common salt—
  Male observers   „   1   „   2,240   „   „
  Female   „   „   1   „   1,980   „   „

The above figures represent means or averages of a great number of individuals. There was very considerable variation for some tastes. In the case of the bitter of quinine, the maximum delicacy was the detection of 1 part in 5,120,000 parts of water; the minimum 1 part in 456,000 parts of water. Except in the case of salt, the sense was more delicate in women than in men. It is not stated whether the men tested were smokers.

It does not seem necessary to say anything concerning the sense of taste in the lower mammalia.

In birds and reptiles the sense of taste does not appear to be highly developed. Parrots are, perhaps, better off in this respect than the majority of their class; and the ducks have special organs on the edges of the beak, which seem to minister to this sense. A python at the Zoological Gardens, partially blind owing to a change of skin, is said to have struck at an animal, but to have only succeeded

---

* See *Nature*, vol. xxxvii. p. 557.

in capturing its blanket. This, however, it constricted, and proceeded to swallow with abundant satisfaction.

It may here be mentioned that the scales and skin of many fishes are provided with sense-organs which very closely resemble the taste-buds of higher animals. They occur in the head and along the "lateral line" which runs down the side of the fish, and may be readily seen, for example, in the cod. Mr. Bateson's [*] careful observations at Plymouth gave, however, no indication of the possession of an olfactory or gustatory function, and their place in the sensory economy of the fish remains problematical. In or near the mouth similar end-organs are found to be somewhat variously developed in different fishes—on the palate and lips, on the gill-bars, more rarely on the tongue, and on the barbels of the rockling and the pout. How far any or all of these have a gustatory function remains to be proved.

Anglers and fishermen, however, from their everyday experience, and naturalists from special observations, do not doubt that fishes have a sense of taste. Professor Herdman's recent experiments on feeding fishes with nudibranchs [†] (naked molluscs) seem to show, for example, that the fishes concerned, including shannies, flat-fish, cod, rockling, and others, have a sense of taste leading them to reject these molluscs as nasty. They show, too, that some of the nudibranchs (*Doris*, *Ancula*, *Eolis*) are protected by warning coloration.

Our knowledge of the sense of taste among the lower (invertebrate) animals is imperfect, and is largely based rather on observation of their habits than on the evidence of anatomical structure. Here, again, comes in the difficulty of distinguishing between taste and smell. But even if the caterpillars which refuse to eat all but one or two special herbs, or the races of bloodsuckers which seem to have individual and special tastes, are guided in part by an olfactory sense, there is much evidence which seems

---

[*] "Sense-Organs and Perception of Fishes:" Journal of Marine Biological Association, New Series, vol. i. No. 3, p. 225.

[†] *Nature*, vol. xlii. p. 201.

to admit of no alternative explanation. Moisten, for example, the antennæ of a cockroach with a solution of Epsom salts or quinine, and watch him suck it off; or repeat F. Will's experiments on bees, tempting them with sugar, and then perfidiously substituting pounded alum. The way in which these little insects splutter and spit suggests that, whatever may be the psychological effect, the physiological effect is analogous to that produced in us by an exceedingly nasty taste. Here smell would seem to be excluded. Forel, moreover, mixed strychnine with honey, and offered it to his ants. The smell of the honey attracted them, but when they began to feed, the effect of the taste was at once evident.

The organs of taste in insects are probably certain minute pits, in each of which is a delicate taste-hair, which, in some cases, is perforated at the free end. They occur in the maxillæ and tongue in ants and bees, and on the proboscis of the fly.

In many of the invertebrates, the crayfish and the earthworm, for example—to take two instances from very different groups—observation seems to show that a sense of taste is developed, for they have marked and decided food-preferences. Nevertheless, the existence of special organs for this purpose has not been definitely proved.

The sense of taste no doubt ministers to the enjoyment of life. But, presumably, it has been developed in subservience to the process of nutrition. Primarily, taste was not an end in itself, but was to guide the organism in its selection of food that could be assimilated. Nice and nasty were at first, and still are to a large extent, synonymous with good-for-eating and not-good-for-eating. With unwonted substances, however, its testimony may be false. Sugar of lead is sweet, but fatal. Brought to a new country, cattle often eat, apparently with relish, poisonous plants. Still, under normal circumstances, the testimony of taste is reliable.

The *sense of smell* is, to a large extent, telæsthetic. It

is true that the stimulation of the end-organs is effected by actual contact with the odoriferous vapour. But since this vapour may be given off from an odoriferous body at some distance from the organism, such as a flower or a decomposing carcase, it is clear that the sense gives information of the existence of such bodies before they themselves come in contact with us. Primitively, we may suppose that it was developed in connection with that sense of taste with which, as we have seen, it is so closely associated. In this respect smell is a kind of anticipatory taste. But it has now other ends, apart from those which are purely æsthetic. In us it may serve as a warning of a pestilential atmosphere; in many organisms, such as the deer, it gives warning of the presence of enemies; in many again, and some insects among the number, it is the guiding sense in the search for mates.

The organ of smell in ourselves and in all the mammalia is the delicate membrane that covers the turbinal bones in the nose. It contains cells with a largish nucleus, around which the protoplasm is mainly collected. A filament passes from this to the surface, and ends in a fine hair or cilium (or a group of hairs or cilia in birds and amphibia); a second filament runs downwards into the deeper parts of the tissue, and may pass into a nerve-fibril.

In us and air-breathing creatures, the substance which excites the sensation of smell must be either gaseous or in a very fine state of division; but in water-breathers the substance exciting this sensation—or, in any case, one of anticipatory taste—may be in solution. The sensitiveness of the olfactory membrane is very remarkable. A grain of musk will scent a room for years, and yet have not sensibly lost in weight. Drs. Emil Fischer and Penzoldt found that our olfactory nerves are capable of detecting the $\frac{1}{460000}$ part of a milligramme of chlorophenol, and the $\frac{1}{460000000}$ part of a milligramme, or about one thirty-thousand-millionth of a grain, of mercaptan. It may be that to such substances our olfactory sensibility is especially delicate.

Not much is known concerning the manner in which the end-organs of smell are stimulated. As in the case of taste, it is probably a matter of molecular vibration; and Professor William Ramsay has suggested that the end-organs are stimulated by vibrations of a lower order than those which give rise to sensations of light and heat. He has also drawn attention to the fact that to produce a sensation of smell, the substance must have a molecular weight at least fifteen times that of hydrogen.

It is well known that the sense of smell is in some of the mammalia exceedingly acute. The dog can track his master through a crowded thoroughfare. The interesting experiments of Mr. Romanes [*] show that, under ordinary conditions of civilized life, the smell of boot-leather is a factor, and the dog tracks his master's boots. In one case, the boots were soaked in oil of aniseed, but this to us powerful scent did not overcome the normal odour of the master's boots. Mr. W. J. Russell, in a subsequent number of the same periodical, describes how his pug could find a small piece of biscuit by scent, and this odour of biscuit was not overmastered by a strong smell of eau-de-Cologne. Deer-stalkers know well how keen is the sense of smell in the antlered ruminants.

We must not, however, be too ready to conclude, from these observations, that the olfactory membrane is absolutely more sensitive in such animals than it is in man. It may well be that, though they are so keen to detect certain scents, they are dull to those which affect us powerfully. It is quite possible that the odour of aniseed or eau-de-Cologne is—possibly from the fact that their end-organs are not attuned to these special molecular vibrations—out of their range of smell. Their special interests in life have led to the cultivation of extreme sensibility to special tones of olfactory sensation. Under unusual circumstances, man may cultivate unwonted modes of utilizing the sense of smell. A boy, James Mitchell, who was born blind, deaf, and dumb, and who was mainly dependent on

[*] *Nature*, vol. xxxvi. p. 273.

the sense of smell for keeping up some connection with the external world, observed the presence of a stranger in the room, and formed his opinion of people from their characteristic smell. On the whole, therefore, we may, perhaps, conclude that the variations in sensitiveness are mainly relative to the needs of life.

In birds the sense of smell is but little developed, notwithstanding all that most interesting naturalist, Charles Waterton, wrote on the subject. Vultures seem unable to discover the presence of food which is hidden from their sight. Probably reptiles share with them this dulness of the sense of smell.

It has already been remarked that, in the case of aquatic animals, there is probably little distinction between taste and smell. It would be well, perhaps, to restrict the word "smell" to the stimuli produced by vapours or airborne particles, and to use the phrase "telæsthetic taste," or simply "taste," for those cases where the effects are produced through the medium of solution. In this case, however, the point to be specially noticed is that taste in aquatic animals becomes a telæsthetic sense, informing the organism of the presence of more or less distant food. Thus, if you stir with your finger the water in which leeches are living, they will soon flock to the spot, showing that the telæsthetic sense is associated with an appreciation of direction. If a stick be used to stir the water, they do not take any notice of it. Mr. W. Bateson* has shown that there are many fishes, among which are the dog-fish, skate, conger eel, rockling, loach, sole, and sterlet, which habitually seek their food by scent (telæsthetic taste), aided to some extent by touch, and but little, if at all, by sight. "None of these fishes ever starts in quest of food when it is first put into the tank, but waits for an interval, doubtless until the scent has been diffused through the water. Having perceived the scent of food, they swim vaguely about, and appear to seek it by examining the whole area pervaded by the scent, having seemingly no sense of the

---

* *Journal of Marine Biological Association*, New Series, vol. i. No. 3, p. 235.

direction whence it proceeds." I venture to think that further observation and experiment may show that such a sense of direction does in some cases exist. Some years ago I was fishing in Simon's Bay, at the Cape, with a long casting-line. The sea was unusually calm, and the water clear as crystal. Beneath me was a clear patch of granite, two or three yards across, surrounded by tangled seaweed. Evening was coming on, and I was just going to put up my tackle when I saw a long dark fish slowly sail into the open space and take up his position at one side. My line was out, baited, I think, with a piece of cuttle-fish, and I tried to draw it into the clear space, but only succeeded in bringing it to within a foot or so of the side furthest from the fish. There it got hitched in the weed; but the fish being still undisturbed, I awaited further developments. After two or three minutes the fish slowly turned, crossed the pool, and remained motionless for a few moments; then he proceeded straight to the bait; and in a few minutes I had landed a dog-fish between four and five feet long. I did not then know that the dog-fish sought its food mainly or solely by scent (taste); but in any case I do not think in this instance he could have seen the bait, hidden as it was amid the seaweed.

Although I am aware, and have already mentioned, that Mr. Bateson's observations do not support the view that the sense-organs of the lateral line minister to this telæsthetic sense, still I think that further observations and experiments may show that these sense-organs are "olfactory," and that the lateral development may be in relation to the appreciation of the direction in which the food lies. It is, however, a difficult matter to determine, and the few experiments I have made are so far inconclusive.

Much has been written concerning the sense of smell in insects. That they possess such a sense few will be disposed to doubt. The classical observations of Huber show that bees are affected by the smell of honey, and that the penetrating odour of fresh bee-poison will throw a whole hive into a state of commotion. He was of opinion that the impunity with which his assistant, Francis Burnens, performed his various operations on bees was due to the gentleness of his motions, and the habit of repressing his respiration, it being the odour transmitted by the breath to which the bees objected. Sir John Lubbock formed a little bridge of paper, and suspended over it a camel's-hair brush containing scent, and then put an ant at one end. She ran forward, but stopped dead short when she came to the scented brush. Dr. McCook introduced a pellet of blotting-paper saturated with eau-de-Cologne into the neighbourhood of some pavement-ants, who were engaged in a free fight. The effect was instantaneous; in a very few seconds the warriors had unclasped mandibles, relaxed their hold of their enemies' legs, antennæ, or bodies.

The correct localization of the sense of smell has been a matter of difficulty. Kirby and Spence localized it at the extremity of the "nose," between it and the upper lip. That the nose, they naïvely remark, corresponds with the so-named part in mammalia, both from its situation and often from its form, must be evident to every one who looks at an insect. Lehman, Cuvier, and others, misled by the fact that the organ of smell is in us localized at the entrance of the air-track, supposed that at or near the spiracles of insects were the organs of smell. Modern research tends more and more clearly to localize the sense of smell, as first suggested by Réaumur, in the feelers or antennæ, and in some cases also in the palps. If the antennæ of a cockroach be extirpated or coated with paraffin, he no longer rushes to food, and takes little notice of, and will sometimes even walk over, blotting-paper moistened with turpentine or benzoline, which a normal insect cannot approach without agitation. There can be little doubt that it is by means of its large branching antennæ that the male emperor moth (*Saturnia carpini*) is able to find its mate.* If a collector take a virgin female

---

* Mr. S. Klein mentions a similar fact in connection with *Bombyx quercus* (*Nature*, vol. xxxv. p. 282).

into a locality frequented by these moths, he will soon be surrounded by twenty or thirty males; but if the moth be not a virgin, he will at most see one or two males. The sense of smell is thus delicate enough to distinguish the fertilized from the unfertilized female, and has associated with it a sense of direction by which the insect is guided to the right spot. Carrion flies whose antennæ have been removed fail to discover putrid flesh; and E. Hasse has observed that male humble-bees whose antennæ have been removed cannot discover the females. The sensory elements are lodged in pits or cones, which may be filled with liquid, peculiar sensory rods or hairs being associated with the nerve-endings. Of these pits the queen-bee has, according to Mr. Cheshire, 1600, the worker 2400, and the drone nearly 19,000, on each antennæ. On the antennæ of the male cockchafer, Hauser estimates the number to be 39,000.

In the aquatic crayfish there are, besides the long antennæ, smaller antennules, each of which has two filaments, an inner and an outer. On the under surface of most of the joints of the outer filament there are two bunches of minute, curiously flattened organs, which were regarded by Leydig, their discoverer, as olfactory. Observation, too, seems to confirm the view that the sense of smell (or telæs-

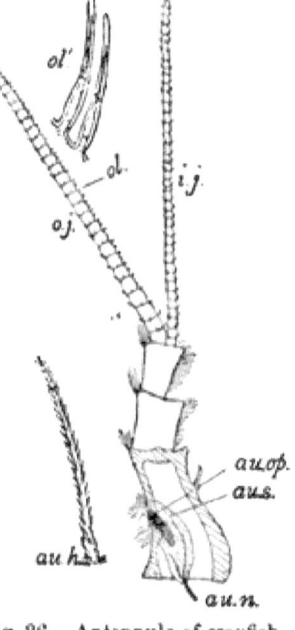

Fig. 26.—Antennule of crayfish.
i.j., inner joint; *o.j.*, outer joint; *ol.*, olfactory setæ; *ol'.*, the same, enlarged; *au.op.*, auditory opening in the basal division, which has been cut open to show *au.s.*, the auditory sac; *au.n.*, auditory nerve branching to the two ridges beset with auditory hairs; *au.h.*, auditory hair, enlarged. (After Howes.)

thetic taste) is located in the antennule. I tried on a crayfish the following experiment: When it was at rest at the bottom of its tank, I allowed a current of pure water (the water in which it lived) to flow from a pipette over its antennæ and antennules. The antennæ moved slowly, but the antennules remained motionless. I then took some water in which a cod's head had been boiled, and allowed some of this to stream over the antennæ and antennules. The former moved slightly as before, but the antennules were thrown into a rapid up-and-down jerky vibration, and shortly afterwards the crayfish began moving about the bottom of its tank. If only one antennule be thus stimulated, or stimulated to a higher degree than the other, the crayfish seems generally (but not always) to turn to that side in search of food. Mr. Bateson* has shown to how large an extent shrimps and prawns seek their food by smell, and states that a prawn, though blind, will often find his way back to his proper place, and stay in it.

In the snail the anterior pair of "horns," or tentacles, are said to be olfactory. Near the end of each is a large ganglion, or nerve-knot, from which fibres pass to the surface, in which there are said to be developed sensory knobs. Snails, however, from which these tentacles have been removed are apparently still possessed of a sense of smell. Certain lobed processes round the mouth have been regarded as the seat of olfactory sensation, but this is doubtful. In the foot of the snail, the part on which it glides, there is a hollow gland, and in this there are special cells, each of which gives off a delicate rod, enlarging at the free end into a ciliated knob. These are regarded as sensory and, it may be, olfactory. In shellfish like the mussel, in which the water is sucked in by an inhalent tube or siphon, and ejected through an exhalent siphon above it (see Fig. 2, p. 4), there is at the entrance of the incoming current a thin layer of elongated cells which are described as olfactory, and are in association with a special ganglion. Olfactory depressions have been described in some worms. But in a great number of the lower invertebrates very little or nothing is known concerning a sense of smell.

* Journal of Marine Biological Association, New Series, vol. i. No. 2, p. 211.

Hearing is a telæsthetic sense. Through it we become aware of certain vibratory states of more or less distant objects. The vibrations of these bodies are transferred to the air or other medium surrounding the body, and are transmitted through the air or other medium to the ear. The sound-waves traverse the air at a rate of 337 metres (1106 feet) in a second; but they travel about four times as fast in water. If the vibration is periodic or regular, the sound is called a tone; non-periodic or irregular sounds are noises. The pitch of a tone is determined by the number of vibrations in a second. The lowest or gravest tone most of us can hear is that where there are about 30 vibrations in a second; twice this number give us a tone of an octave higher; twice this again, another octave; and so on. In musical composition, tones from about 40 to about 4000 vibrations per second are employed. This is a range of somewhat over six octaves. But many of us are capable of hearing sounds over a range of about ten octaves, that is to say, from 30 to 30,000 vibrations per second. The upper limit of hearing is, however, very variable. Some people are deaf to tones of more than 15,000 or 20,000 vibrations per second.* Others may hear shrill tones of 40,000, or even in rare cases 50,000. I could as a boy hear the shrill squeak of a bat; now I am quite deaf to it. A friend of mine in South Africa was unable to hear the piping of the frogs in the pond, which was to me so loud as almost to drown the tones of his voice.

Apart from the pitch of a note is its quality. The same note struck on different instruments or sung by different persons has a different ring. This is determined by the number and intensity of overtones, or partials, which are associated with the fundamental tone. Suppose the deep fundamental tone of 33 vibrations be sounded; with it there may be associated overtones, eight or nine in number, all of which are simple multiples (twice, thrice, four times,

---

* A friend of mine informs me that his limit is about 17,500 per second, 20,000 being quite inaudible.

and so on) of the fundamental 33. The effects of these on the organ of hearing fuse or combine with the predominant effect of the fundamental tone. In harmonious chords, also, two or more fundamental tones, with their accompaniment of partials, blend in sensation so completely that it requires a keen musical ear and some training to analyze them into their component elements.

The delicacy of discrimination of tones is greatest in the mid-region of hearing; and there is much individual variation in accuracy of ear. I have made experiments on many individuals to test their powers in this respect. I found some who were unable, in the mid-region of hearing, to state which was the higher of two notes sounded on a violin, the tones of which were separated by a major third, and in one case by a fifth. With notes on the piano the discrimination was more delicate, and yet more delicate when the notes were sung. In such cases tone-discrimination is deficient; and between these and the musician, who is stated to be able to distinguish tones separated by only $\frac{1}{64}$ of a tone, there are many intermediate stages.

It is beyond my purpose to describe, in more than a very general way, the nature of the auditory apparatus of man. The vibrations of the air are received by the drum-membrane, which lies in the auditory passage. From this it is transmitted, by a chain of small bones, to the inner auditory apparatus. This consists of two small membranous sacs, with one of which three membranous looped tubes, the semicircular canals, are connected; with the other is connected a spiral tube, the cochlear canal. These membranous sacs and canals are filled with fluid, and are surrounded by the fluid which fills the bony cavity in which they lie. This bony cavity has two little windows, one oval and the other round, across each of which a membrane is stretched. The oval membrane is in connection with the chain of auditory bones; and when this is made to vibrate in and out, the membrane of the round window vibrates out and in. Thus the fluid around and within the membranous sacs and canals is set in vibration.

And the parts are so arranged that the vibrations, in passing from the oval to the round membrane, must run up one side and down the other side of the cochlear canal. As they run down they set in vibration a delicate membrane which is supported on beautiful arched rods (the

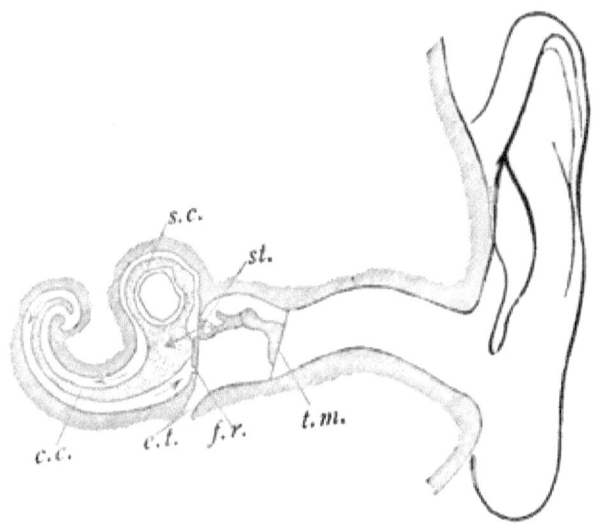

Fig. 27.—Diagram of ear.

t.m., tympanic membrane, to which is attached a chain of small bones stretching across the cavity of the drum, the innermost of which, *st.*, fits into the "oval window." The vibrations are transmitted up one side and down the other side of the cochlear canal, *c.c.*, and thus reach the "round window," *f.r.*; *s.c.* is one of the semicircular canals, the other two are omitted; *e.t.* is the Eustachian tube connecting the cavity of the drum with the mouth-cavity.

organs of Corti). And this membrane contains a number of special hair-cells, so called because they bear minute hair-like structures. These are the special end-organs of hearing. It has been suggested that the fibres of the membrane on the arched rods, which are of different lengths and may be stretched with differing degrees of tension, respond to vibrations of different pitch. Thus the hair-cells on that particular part of the membrane would be stimulated, and the note might be appreciated in its true position in the scale.

We must now pass on to consider the sense of hearing in animals. That the mammalia have this sense well developed is a matter of familiar observation, and in some of them, such as the horse and the deer, it is exceedingly acute. The form and movements of the external ear also enable many of the mammalia to collect and attend to sounds from special directions. The mammalia possess also the power of tone-discrimination, as is shown by the fact that our domesticated animals recognize different modulations of the human voice, and that wild creatures distinguish tones or noises of different quality. A Newfoundland dog, possessed by a friend of mine, always howled when the tenor D was struck on the piano, or sung. And Théophile Gautier reports that one of his cats could not endure the note G, and always put a reproving and silencing paw on the mouth of any one who sang it.

In birds the sense of hearing is not only very sensitive, but the power of discrimination is exceedingly delicate. No one who has watched a thrush listening for worms can doubt that her ear is highly sensitive. The astonishing accuracy with which many birds imitate, not only the song of other birds, but such unwonted sounds as the clink of glasses or the ring of quoits, shows that the delicacy in discrimination has reached a high level of development. In birds, however, the cochlear canal has not the same development that it has in mammals, and there are no arched rods—no organs of Corti.

Nothing special is to be noted concerning the sense of hearing in the reptiles, amphibia, and fishes. In all (with the exception of the lowly lancelet) the auditory organ is developed. We shall, however, presently see reason to question whether the possession of an "auditory organ," with well-developed semicircular canals, necessarily indicates the power of hearing. And Mr. Bateson's recent experiments at Plymouth* seem to indicate that fishes are not so sensitive in this respect as anglers† are wont to believe. "The sound made by pebbles rattling inside an opaque glass tube does not attract or alarm pollack; neither are they affected by the sharp sound made by letting a hanging

---

* Journal of Marine Biological Association, New Series, vol. i. No. 3, p. 251.

† Of course, anglers will say that what may be true for pollack and other coarse and vulgar sea-fish does not apply to King Salmon or Prince Trout.

stone tap against an opaque glass plate standing vertically in the water." Carp at Potsdam are, indeed, said to come to be fed at the sound of a bell. But Mr. Bateson well remarks that this " can scarcely be taken to prove that the sound of the bell was heard by them, unless it be clearly proven that the person about to feed them was hidden from their sight." There is clearly room for further observation and experiment in this matter.

Turning to the invertebrata, we find, even in creatures as low down in the scale of life as jelly-fish, around the margin of the umbrella in certain medusæ, simple auditory organs. In some cases they are pits containing otoliths (minute calcareous or other bodies, which are supposed to be set a-dance by the sound-vibrations); in others there is a closed sac with one or more otoliths; in others, again, they are modified tentacles, partially or completely enclosed in a hood. All these are generally regarded as auditory, there being specially modified cells of the nature of hair-cells. We shall see, however, that another interpretation of organs containing otoliths is at any rate possible. For the present, we will follow the usual interpretation, and regard them as auditory.

Vesicular organs containing otoliths are found near the cerebral ganglia in some of the worms and their relations. But the common earthworm, though it appears to be sensitive to sound, does not appear to have any such organs.

Molluscan shell-fish are generally provided with auditory organs. In the fresh-water mussel it is found in the muscular foot. It can be more readily seen in the *Cyclas*, if the transparent foot of this small mollusc be examined under the microscope. It is a small sac containing an otolith. Mr. Bateson found that the mollusc *Anomia* "can be made to shut its shell by smearing the finger on the glass of the tank so as to make a creaking sound. The animals shut themelves thus when the object on which they were fixed was hung in the water by a thread." In the snail and its allies the auditory sac is found in close connection with the nerve-collar that surrounds the gullet.

In the cuttle-fishes it is found embedded in the cartilage of the head.

In the lobster or crayfish the auditory organs are found at the base of the smaller feelers or antennules. They are little sacs formed by an infolding of the external integument (see Fig. 26, p. 259). Beautifully feathered auditory hairs project into the sac along specialized ridges, and the sac in many cases contains grains of sand which play the part of otoliths. Hensen seems to have proved that shrimps collect the grains of sand and place them in the auditory sac for this purpose. The curious shrimp-like *Mysis* has two beautiful auditory sacs in its tail. These are provided with auditory hairs. Hensen watched these under the microscope while a musical scale was sounded, and found that the special hairs responded each to a certain note. When this particular note was sounded the hair was thrown into such violent vibration as to become invisible, but by other notes it was unaffected.

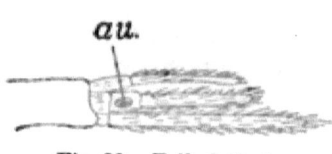

Fig. 28.—Tail of *Mysis*. au., auditory organ.

Passing now to insects, we may first note that grasshoppers and crickets have an auditory organ on the front leg. These are provided with tympanic membranes, and the breathing-tubes, or tracheæ, are so arranged that the pressure of the air is equalized on the two sides of the membrane—just as in us and other vertebrates the same end is effected by a tube which runs from the interior of the drum of the ear to the mouth-cavity (see Fig. 27). In the organ within the leg there is a group of cells, followed by a row of similar cells which diminish regularly in size from above downwards. Each is in connection with a nerve-fibril, and contains a delicate auditory rod. It has been suggested that the diminution in size of the cells may have reference to the appreciation of different notes, but nothing definite is known on the matter. Ants, too, have

Fig. 29.—Leg of grasshopper. ty., tympanic membrane.

an auditory organ, as shown by Sir John Lubbock, in the tibia of the front leg. But in locusts it is situated on the first segment of the abdomen. In flies there are a number of vesicles, generally regarded as auditory (but by some as olfactory), at the base of the rudimentary hind wings—the so-called halteres, or balancers.

Observation seems to point to the fact that in most insects the sense of hearing is lodged in the feelers, or antennæ. Kirby made the following observation on a little moth: "I made," he says, "a quiet, not loud, but distinct noise; the antenna nearest to me immediately moved towards me. I repeated the noise at least a dozen times, and it was followed every time by the same motion of that organ, till at length the insect, being alarmed, became more agitated and violent in its motions." Hicks wrote, in 1859, "Whoever has observed a tranquilly proceeding capricorn beetle which is suddenly surprised by a loud sound, will have seen how immovably outward it spreads its antennæ, and holds them porrect, as it were, with great attention, as long as it listens." The same observer described certain highly specialized organs in the antennæ of the hymenoptera (ants, bees, and wasps), which he thus describes: "They consist," he says, "of a small pit leading into a delicate tube, which, bending towards the base, dilates into an elongated sac having its end inverted." Of these remarkable organs, Sir John Lubbock says there are about twelve in the terminal segment, and he has suggested that they may serve as microscopic stethoscopes.

Mayer, experimenting with the feathered antenna of the male mosquito, found that some of the hairs were thrown into vigorous vibration when a note with 512 vibrations per second was sounded. And Sir John Lubbock, who quotes this observation, adds,* "It is interesting that the hum of the female gnat corresponds nearly to this note, and would consequently set the hairs in vibration." The same writer continues, "Moreover, those auditory hairs are most affected which are at right angles to the

* "Senses of Animals," p. 117.

direction from which the sound comes. Hence, from the position of the antennæ and the hairs, a sound would act most intensely if it is directly in front of the head. Suppose, then, a male gnat hears the hum of a female at some distance. Perhaps the sound affects one antenna more than the other. He turns his head until the two antennæ are equally affected, and is thus able to direct his flight straight towards the female."

It is difficult to determine the range of hearing in the lower organisms. But it is quite possible, nay, very probable, that the superior limit of auditory sensation is much more extended in insects than it is in man. We know that many insects, such as the cicadas, the crickets and grasshoppers, many beetles, the death's-head moth, the death-watch, and others, make, in one way or another, sounds audible to us. But there may be many insect-sounds—we may not call them voices—which, though beyond our limits of hearing, are nevertheless audible to insects. At the other end of the scale, on the other hand, slow pulsations may be appreciated—for example, by aquatic creatures—by means of what we term the auditory organs, in a way that is not analogous to the sensation of sound in us. It may be noted that auditory organs are dotted about the body somewhat promiscuously in the various invertebrates. We have seen that auditory organs, or what are generally believed to be such, are found in the foot of bivalves, in the antennules of lobsters, in the fore legs of crickets and ants, in the abdomen of locusts, in the balancers of flies, and in the tail of *Mysis*. But when we come to consider the matter, there is no reason why the organ of hearing should be in any special part of the body. The waves of sound rain in upon the organism from all sides. There is no great advantage in having the organs of hearing in the line of progression, as with sight, where the rays come in right lines; nor in having them in close association with the mouth, as in the case of the organ of smell.

Closely connected with the organ of hearing in vertebrates is the organ of another and but recently recognized sense.

In briefly describing the auditory apparatus in man, mention was made of three curved membranous loops, the so-called semicircular canals. A few more words must now be said about them and the membranous sac with which they are connected.

The sac lies in a somewhat irregular cavity in a bone at the side of the head, in the walls of which are five openings leading into curved tunnels in the bone in which lie the membranous loops. The planes in which the three semicircular canals lie are nearly at right angles to each other, and they are called respectively the horizontal, the superior, and the posterior. The two latter unite at one end before they reach the sac; hence there are five, and not six, openings into the cavity. At one end of each semicircular canal is a swelling, or ampulla, in each of which is a ridge, or crest, abundantly supplied with hair-cells. And in a little recess in the sac there is, occupying its floor, its front wall, and part of its outer wall, a patch of hair-cells covered by a gelatinous material with numerous small crystalline otoliths. The only other point that calls for notice is that the membranous sac does not fit closely in the bony cavity in which it lies, while the diameter of the membranous semicircular canals is considerably less than that of their bony tunnels, except at the ampullæ, or swellings, where they fit pretty closely. Both the bony cavity and the membranous labyrinth (as it is called) are filled with fluid.

From its close connection with the organ of hearing, this apparatus was for long regarded as in some way auditory in its function, and it was surmised that it enabled us to perceive the direction from which the sound came. But how it could do so was not clear. In 1820 M. Flourens made the observation that the injury or division of a membranous canal gave rise in the patient to rotatory movements of the animal round an axis at right angles to the plane of the divided canal; and he, therefore, suggested that the canals might be concerned in the co-ordination of movement. They are now regarded as the organs of a sense of rotation or acceleration.

That we have such a sense of rotation has been proved experimentally.* Let a man, blindfolded, sit on a smooth-running turn-table. When it begins to rotate he feels that he is being moved round, but if the rotation be continued at the same rate, this feeling quickly dies away. If the rotation be increased, he again feels as if he were being moved round, but this again soon dies away. Further increase gives a fresh sensation, which in turn subsides, and the man may then be spinning round rapidly, and be perfectly unconscious of the fact. He is only aware that he has been gently turned round a little two or three times. Now let the speed of rotation be slackened. He has a sensation of being gently turned round a little in the opposite direction. Each time the speed is lessened he has this sense of being turned the reverse way. From these experiments we see that what we are conscious of is change of rate of rotation, or, in technical language, acceleration, positive or negative.

From Professor Crum Brown's paper in *Nature* I transcribe, with some verbal modifications, his account of how the semicircular canals enable us to feel these changes of

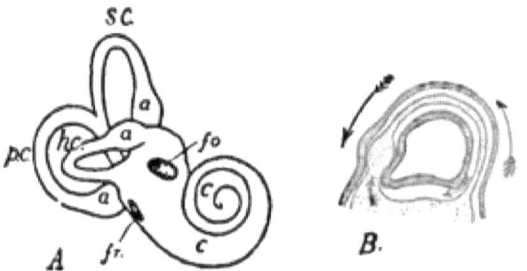

Fig. 30.—Diagram of semicircular canals.

A. bony labyrinth of human ear (after Sömmering). *c, c.,* the cochlea; *s.c.*, superior semicircular canal; *p.c.*, posterior semicircular canal; *h.c.*, horizontal semicircular canal; *a, a, a,* their swellings, or ampullæ; *f.o., f.r.*, fenestra ovalis and rotunda (oval and round windows) in the vestibule.

B. Diagram of semicircular canal to illustrate effect of rotation. The large arrows indicate the direction of the rotation. The small arrow to the left indicates the resulting flow of the inner fluid into the ampulla; that to the right, the flow of the outer fluid into the vestibule.

motion. Let us consider the action of one canal. If the head be rotated about a line at right angles to the plane of the canal, with the ampulla leading, there will be a tendency

* See a very interesting and lucid paper by Professor Crum Brown, whose name is intimately connected with this subject, in *Nature,* vol. xl. p. 449.

for the fluid within the sac to flow into the ampulla, and for the fluid around the semicircular canal to flow into the cavity in which the sac lies. These movements will conspire to stretch the membranous ampulla, and thus to stimulate the hair-cells. This stretching will not take place in that canal if the rotation be in the reverse direction. But on the opposite side of the head is another canal in the same plane, but turned the other way. In the reversed rotation the ampulla in this canal will lead, and its hair-cells will be stimulated. Thus by means of the two canals on either side of the head in the same plane, rotation in either direction can be appreciated. And since there are two other pairs of semicircular canals in two other planes, rotation in any direction will be recognized by means of one or more of the six canals.

It is thus by means of the semicircular canals that we can appreciate acceleration of rotatory motion.[*] But we can also appreciate acceleration of movements of translation —forwards or backwards, up or down. And Professor Mach has suggested that it is through the stimulation of the hair-cells in the patch in the sac itself (the so-called *macula acustica*) that we are able to appreciate these changes. The otoliths, held loosely and lightly in position by the gelatinous substance in which they are embedded, may, through their inertia, aid in the stimulation of the sense-hairs.

And this naturally suggests the question whether those sense-organs in the invertebrates which contain otoliths may not be regarded with more probability as organs for the appreciation of changes of motion than as auditory organs. This for some years has been my own belief. I have always felt a difficulty in understanding how the otoliths are set a-dance by auditory vibrations. But their inertia would materially aid in the appreciation of changes of motion. In some forms the otoliths are held in suspension in a gelatinous material. In others—the molluscs,

---

[*] It is interesting to note that in the blind-fish (*Amblyopsis spelæus*) the semicircular canals are, according to Wyman, unusually large.

for example—the otolith (which is generally single) is retained in a free position by ciliary action. In aquatic creatures an organ for the appreciation of changes of motion might be of more service than an auditory organ. And if one be permitted to speculate, one may surmise that the sense of hearing may be a refinement of the sense through which changes of motion are appreciated. First would come a sense of movements of the organism in the medium through the stimulation of the sense-hairs by the relative motion of the otolith; then these sense-hairs, with increased delicacy, might appreciate shocks in the medium; and, eventually, those more delicate shocks which we know as auditory waves. In this way we might account for the fact that in the vertebrates the same organ, through different parts of its structure, appreciates both change of motion and auditory vibrations. And thus the organs in the invertebrata which are generally regarded as auditory, and for which has been suggested the function of reacting to changes of motion, may, in truth, subserve both purposes—may be organs in which the differentiation I have hinted at is taking place.

Sight, like hearing, is a telæsthetic sense. Through it we become aware of certain vibratory states of more or less distant objects. The medium by means of which these vibrations are transmitted is not, as in the case of hearing, the air, but the æther which pervades all space. The rate of transmission is about 186,000 miles in a second. That which answers in vision to pitch in hearing is colour. The lowest, or gravest, light-tone to which we are sensitive is deep red, where the number of vibrations per second is about 370 billions (370,000,000,000,000). The highest, or most acute, light-tone is violet, with about 833 billion vibrations in a second. If white light be passed through a prism, the rays are classified according to their vibration-periods, and are spread out in a spectrum, or band of rainbow colours. But different individuals vary, as we shall presently see, in their sensibility to the lowest and the highest vibrations. Some people are, moreover, relatively or absolutely insensible to certain colours, generally either red or green. Such persons are said to be colour-blind. When the rainbow colours are combined in due proportion, or when pairs or sets of them are combined in certain ways, white light is produced.

We saw that in the case of sound-waves, when the number of vibrations in a second is doubled, the sound is raised in pitch by an octave. Using this term in an analogous way for colour-tones, we may say the range in average vision is about one octave—that is, from about 400 billion to about 800 billion vibrations in a second. But, though these are the limits in human vision, we know of the existence of many octaves of radiant energy physically in continuity with the light-vibrations. Photography has made us acquainted with ultra-violet vibrations up to about 1600 billions per second—an octave above the violet. And Professor Langley's observations with the bolometer indicate the existence of waves with as low a vibration-period as one billion per second, and even here, in all probability, the limit has not been reached. To the vibrations more rapid than those that are concerned in the sensation of violet, the human organism is apparently in no manner sensitive. But to infra-red vibrations down to about thirty billions per second the nerves of the skin respond through the temperature-sense. We shall have to return to these limits of sensation at the close of this chapter.

The human eye is a nearly spherical organ, capable of tolerably free movements of rotation in its socket. What we may call the outer case, which is white and opaque elsewhere, is quite transparent in front. Through this transparent window may be seen the coloured iris, in the centre of which is a circular aperture, the pupil. The size of the pupil changes with the amount of light—it dilates or contracts, according as the light is less or more intense. Just behind it, and still in the front part of the eye, is the transparent lens, the convexity of the anterior surface of

T

which can be altered in the accommodation of the organ for near or far vision. The space between the lens and

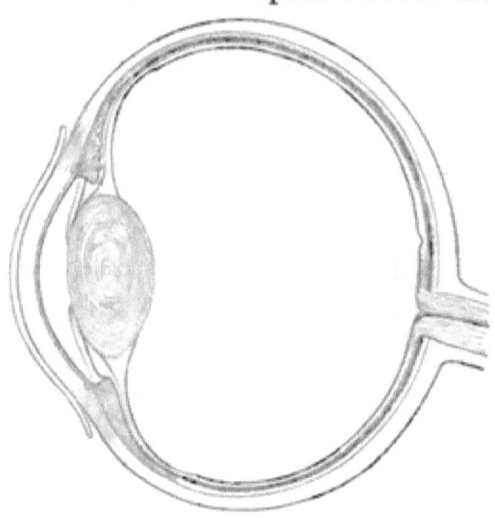

Fig. 31.—The human eye.
Horizontal section, to show general structure.

iris and the corneal window of the eye is filled with a watery fluid. Behind the lens there is a transparent, semi-fluid, jelly-like material, filling the rest of the chamber of the eye. At the back of the eye is spread out the sensitive membrane —the retina. The structure of this membrane is very complicated, and cannot be described here. It is, however, indicated in Fig. 32. For our present purpose it is sufficient to note that here are the end-organs of the optic nerve; that these consist of a number of delicate rods and cones; and that these rods and cones do not face in the direction from which the light comes, but face towards the back of the eyeball, where a pigmented substance is developed. The rays of light are thus focussed through the retina on to this pigmented substance; the ends of the rods and cones are stimulated; and the stimulation is handed on, augmented in

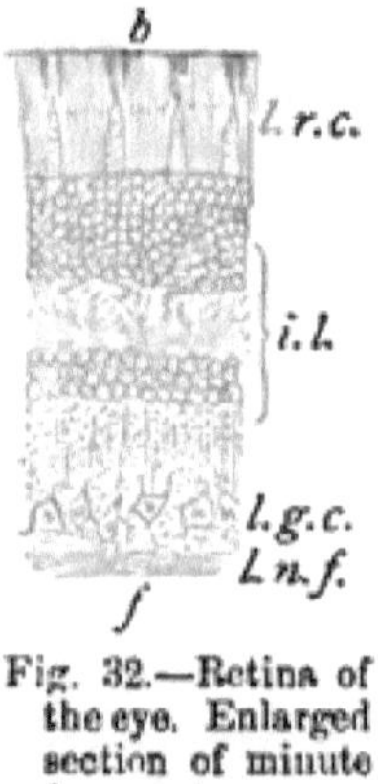

Fig. 32.—Retina of the eye. Enlarged section of minute fragment.
b., back of retina next the outer coat; *l.r.c.*, layer of rods and cones; *i.l.*, intermediate layers; *l.g.c.*, layer of ganglion-cells; *l.n.f.*, layer of nerve-fibres; *f.*, front of retina, the surface turned towards the pupil.

certain intermediate ganglia, to the delicate transparent nerve-fibres in the front of the retina. These collect to a certain spot, where they pass through the retina to form the optic nerve. Where they pass through the retina there can, of course, be no rods and cones. And in this spot there is no power of vision. It is the blind spot. The reality of its existence can easily be proved. Make a dot on a piece of writing-paper, and about three inches to the left of it place a threepenny or sixpenny bit. Close the right eye, and look with the left eye at the dot. The sixpenny bit will also be seen, but not distinctly. Keep the eye fixed on the dot, and move the head slowly away from the paper. At a distance of about ten inches the coin will completely disappear from view. Its image then falls on the blind spot.

The organ of vision, then, in us consists of an essential sensory membrane, the retina, with its delicate rods and cones; and an accessory apparatus for focussing an inverted image on to the sensitive surface of the retina. The surface is not, however, equally sensitive, or, in any case, does not give an equal power of discrimination, throughout its whole extent. This is seen in the experiment above described. When we look at the dot we see the coin, but not distinctly. The area of clear and distinct vision is, in fact, very small, constituting the yellow spot about $\frac{1}{12}$ of an inch (2 millimetres) long, and $\frac{1}{30}$ of an inch ($\cdot$8 millimetre) broad. And even within this small area there is a still more restricted area of most acute sensibility only $\frac{1}{120}$ of an inch ($\cdot$2 millimetre) in diameter. Nevertheless, within this minute area there are some two thousand cones, the rods being here absent. In carefully examining an object we allow this area of acute vision to range over it. Hence the extreme value of that delicate mobility which the eye possesses—a mobility that is accompanied by muscular sensations of great nicety.

We saw that the sense of touch in the tongue is sufficiently delicate to enable us to recognize, as two, points of contact separated by $\frac{1}{25}$ of an inch (1$\cdot$1 millimetre). What, in similar terms, is the delicacy of sight?

At what distance apart, on the most delicate part of the retina, can two points of stimulation be recognized as distinct from each other? If the points of stimulation be not less than $\frac{1}{6000}$ of an inch (·004 millimetre) apart, they can be distinguished as two. Below this they fuse into one. The diameter of the end of a single cone in the yellow spot is also about $\frac{1}{5000}$ of an inch (·0045 millimetre).

With regard to the mode in which the stimulation of the retinal elements is effected, we have no complete knowledge. Certain observations of Boll and Kühne, however, show that when an animal is killed in the dark the retina has a peculiar purple colour which is at once destroyed if the retina be exposed to light. If a rabbit be killed at the moment when the image, say, of a window, is formed on the retina, and the membrane at once plunged in a solution of alum, the image may be fixed, and an "optogram" of the window may be seen on the retina. The discharge of the colour of the retinal purple may be regarded as the sign of a chemical change effected by the impact of the light-vibrations. But in the yellow spot there seems to be no visual purple. It is, indeed, developed only in the rods, not in the cones. Here, probably, chemical or metabolic changes occur without the obvious sign of the bleaching of retinal purple. In the dusk-loving owl the retinal purple is well developed, but in the bat it is said to be absent.

We saw that in the case of hearing the auditory organ is fitted to respond to air-borne vibrations varying from about thirty to thirty thousand per second. And though the details of the process are at present not well understood, it is believed that certain parts of the recipient surface are fitted to respond to low tones, other parts to intermediate tones, and yet others to high tones. Thus the reception is serial. If there be two pianos near each other, accurately in tune, any note struck on one will set the corresponding note vibrating in the other.* The auditory organ may be likened to this second piano. Special parts respond to special tones.

* The dampers must, of course, be lifted by depressing the loud pedal.

Now, in the case of vision, the conditions are different. The reception cannot be serial. As I range my eye over a flower-bed, I bring the area of distinct vision on to a number of different colours, and these are seen to be distinct, though they are received on the same part of the retinal surface. It might, perhaps, be suggested that special cones were set apart for each shade of colour. But there are only some two thousand cones in the central area of most acute vision, and Lyons silk-manufacturers prepare pattern cards containing as many shades of coloured silks. So that there would be only one cone to each colour. And Herschel thought that the workers on the mosaics of the Vatican could distinguish at least thirty thousand different shades of colour! There are also many phenomena of colour-blending which show that colour-reception cannot in any sense be serial.

How, then, are we to account for our wide range of colour-sensation? Just as the blending by the artist on his palette of a limited number of pigments gives him the wide range of colour seen on his canvas, so the blending of a few colour-tones may give us the many shades we are able to distinguish. The smallest number of fundamental colour-tones which will fairly well account for the phenomena of colour-vision, is three. And these three are red, green, and blue or violet. These are the three so-called primary colours. All others are produced from these elements by blending.

To explain our ability to appreciate differences of colour, then, it is supposed, on the hypothesis of Young and Von Helmholtz, that three kinds of nerve-fibres exist in the retina, the stimulation of which gives respectively, red, green, and violet in consciousness. Professor McKendrick, interpreting Von Helmholtz, gives * the following scheme:—

"1. Red excites strongly the fibres sensitive to red, and feebly the other two.

"2. Yellow excites moderately the fibres sensitive to red and green, feebly the violet.

* "Special Physiology," p. 636.

"3. Green excites strongly the fibres sensitive to green, feebly the other two.

"4. Blue excites moderately the fibres sensitive to green and violet, feebly the red.

"5. Violet excites strongly the fibres sensitive to violet, feebly the other two.

"6. When the excitation is nearly equal for the three kinds of fibres, the sensation is white."

This theory cannot be regarded as more than a provisional hypothesis. Still, by its means we can explain many colour-phenomena. It is well known, for example, that if we gaze steadily at a red object, and then look aside at a grey surface, an after-image of the object will be seen of a blue colour. According to the theory, the red fibres have been tired and cannot so readily answer to stimulation. Over this part of the retina, therefore, the effect of grey light is to stimulate normally the fibres sensitive to green and violet, but only slightly those sensitive to red, owing to their tired condition. The result will be, as we see from the above scheme (4), the sensation of blue. Colour-blind people, on this view, are those in whom one set of the fibres, generally the red or the green, are lacking or ill developed.

We may, perhaps, with advantage restate this theory in terms of chemical change, or metabolism. On this view three kinds of "explosives" are developed in the retinal cones; for it is seemingly the cones, rather than the rods, which are concerned in colour-vision. All three explosive substances are unstable; but one, which we may call R., is especially unstable for the longer waves of the spectrum; another, G., for the waves of mid-period; a third, V., for those of smallest wave-length.

Suppose that R. only were developed. If, then, we were to look at a band of light spread out in spectrum wavelengths, we should see a band* of monochromatic r. light. Its centre would be bright, and here would be the maximum instability of R. On either side it would fade away. The

---

* A band and not a line, because R. is unstable to the impact of a considerable range of light-vibrations.

lateral edges of the spectrum would be the limits of the instability of R. If G. only were developed, we should see only a band of monochromatic *g*. light. Its centre would not coincide with that for R., but would lie in a region of smaller wave-length. Here would be the maximum instability for G. On either side the green would fade away. Its lateral edges would mark the limits of the instability of G. But though their centres would not coincide, the R. band and the G. band would to a large extent overlap. Similarly with the band for V. It, too, would have its centre of maximum instability and its lateral edges of lessening instability. Its centre would lie in a region of yet smaller wave-length than that for G. And the *v*. band would overlap the green and the red.

Normally, all three bands are developed, and their blended overlapping gives the colours of the rainbow. For this reason the monochromatic bands *r*., *g*., and *v*. are unknown to us in experience. All the colour-tints we know are blended tints. What we call full-red light causes strong disruptive change in R., but decomposes slightly G., and probably also, but in much less degree, V.

Whether R., G., and V. are all three present in each cone, or whether they are each developed in separate cones, we do not know for certain. Nor are we certain that there are separate nerve-fibres for the transmission of stimuli due to R., G., and V.

When we look steadily at a red object we cause the disruption of R.; and since it takes some time for the reformation and reconstitution of this explosive substance, on turning the eye to a grey surface, G. and V. are alone, or in preponderating proportions, caused to undergo disruption. Hence the phenomena of complementary after-images. It is not merely a matter of the tiring of certain nerve-fibres, but a using-up of the explosive material in certain of the cones.

What is called *colour-blindness* is probably due to one of several abnormal conditions. It is *possible* that in some cases R., G., or V. may be entirely absent. More frequently they are in abnormal proportions. They probably vary in their sensitiveness, and not improbably in the wave-period to which they show the maximum response.

To test the variation, if any, in the limits of instability for R. and V., or in any case in the limits of colour-vision at the red end and at the violet end of the spectrum, in apparently normal individuals, my friend and colleague, Mr. A. P. Chattock, made, at my suggestion, a number of observations on some of the students of the University College, Bristol, to whom my best thanks are due for their kind willingness to be submitted to experiment. The instrument used * was a single-prism spectro-goniometer.

In the accompanying diagram (Fig. 33) the results of some of these observations are graphically shown. The middle part of the spectrum, between the wave-lengths 420 and 740 millionths of a millimetre, is omitted, only the red end and the violet end being shown. The observations on thirty-four individuals, seventeen men and seventeen women, all under thirty years of age, are given for both eyes. The left-hand vertical line of each pair stands for the right eye in each case. To the left of the table are placed the wave-lengths in millionths of a millimetre.

Take, for example, the first pair of vertical lines. The

---

* Mr. Chattock has kindly supplied me with the following note:—

"Readings at the violet end were taken at the extremity of the lavender rays, at the point where the faint band of lavender light seemed to end off about half-way across the field of view (the cross-wires being invisible).

"At the red end the cross-wires were always visible, and were in each case set to the point where the top horizontal edge of the spectrum lost its definition.

"Other things equal, the 'red' readings should be more reliable than the violet, therefore, from the greater definiteness of the point observed, and the means of observing it. But against this has to be set off the fact that the extreme violet rays were spread out by the prism used more than eight times as much as the red rays.

"In any case, the wide differences observed in the 'red' readings are much greater than could have been due to misunderstanding or careless observation —as shown by setting the instrument to maximum and minimum readings, and noting the very obvious difference between them apparent to a normal eye. The same conclusion is rather borne out by the closer (average) agreement between the two eyes of the same individual than between those of different persons.

"The source of light was the central portion of an ordinary Argand burner."

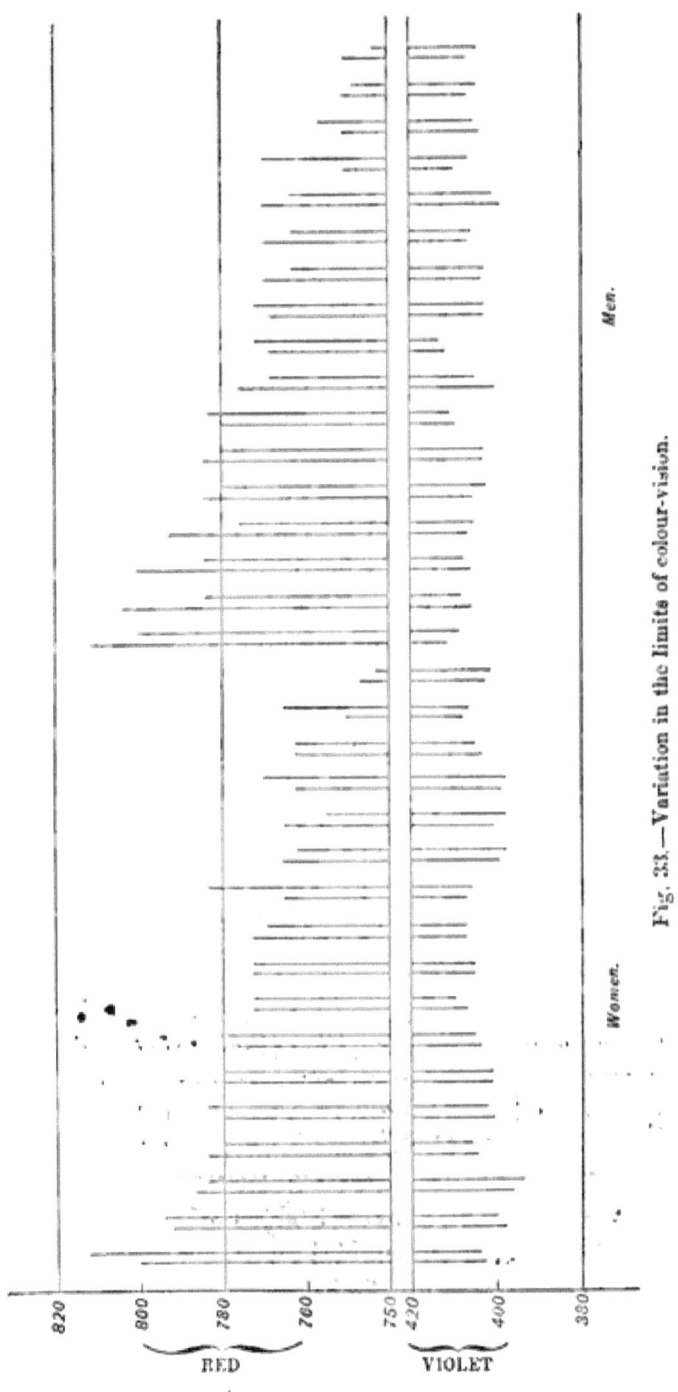

Fig. 33.—Variation in the limits of colour-vision.

individual whose colour-range they represent could detect red light in the spectrum up to 800 millionths of a millimetre wave-length for the right eye, and up to 811 for the left; and could detect violet light down to 403 and 404. Beyond these limits all was dark. But the last individual in the series, while his range in the violet was about the same, could only detect red light up to 743 and 750 millionths of a millimetre. His spectrum was so much shorter.

It is seen that there is more variation at the red end than at the violet end of the spectrum, and this notwithstanding that the violet rays are more spread out by the prism than the red rays. It is seen that the two eyes are often markedly different. This is not due to inaccuracy of observation, for certain individuals in which this occurred were tested several times with similar results. It is seen that the variations at the red end and the violet end are often independent, and that the absolute length of the visible spectrum differs in different individuals.

The following table presents these observations and a few others in another light:—

TABLE OF MAXIMA AND MINIMA IN WAVE-LENGTHS, EXPRESSED IN MILLIONTHS OF A MILLIMETRE.

	Violet.			Red.			No. of Individuals.
	Highest.	Mean.	Lowest.	Highest.	Mean.	Lowest.	
Women under 30 ..	410·0	402·75	394·0	811	772·85	743	17
Men ,, ,, ..	413·0	405·0	399·0	811	772·8	743	17
Women over 30 ..	410·5	406·65	401·5	792	767·8	743	7
Men ,, ,, ..	407·0	404·5	402·5	787	773·7	758	3
N { right eye		406			687		
{ left eye		407			717		

The individual N showed signs of colour-blindness, and is therefore not included in the table, but entered separately. He was unable to recognize the C line of the hydrogen spectrum (wave-length 656), which was brilliantly obvious to the normal eye.

These observations [*] need further confirmation and extension. We intend to continue the investigation each session. They are, however, sufficient to show that in some individuals R. undergoes disruptive change on the impact of light-waves which have no noticeable effect on the retina of other individuals.

It is impossible here to do more than just touch the fringe of the difficult subject of colour-vision. And the only further fact that can here be noticed is that trichromatic colour-vision is apparently in us limited to the yellow-spot and its immediate neighbourhood. Around this is an area which is said to be bichromatic—all of us being, for this area, more or less green-blind. In the peripheral area around this, colour is indistinguishable, and we are only sensitive to light and shade. So far as the structure of the retina is concerned, we may notice in this connection that in the central region of most complete trichromatic vision there are cones only; around the yellow spot each cone is surrounded by a circle of rods; and further out into the peripheral region by two, three, or more circles of rods.

Concerning the sense of sight in the lower mammals little need be said. In many cases the acuteness of vision is remarkable. Mr. Romanes's experiments on Sally, the bald-headed chimpanzee at Regent's Park, led him to conclude that she was colour-blind, but I question whether the experiments described quite justify this conclusion. Sir John Lubbock was unable to teach his intelligent dog Van to distinguish between coloured cards; but the failure was as complete when the cards were marked respectively with one, two, or three dark bands. We are not justified, therefore, in ascribing the failure to colour-blindness. The real failure, probably, was in each case to make the animal understand what was wanted. Bulls are, at any rate,

---

[*] The variations above indicated throw light on a fact to which Lord Rayleigh has directed attention. The yellow of the spectrum may be matched by a blending of spectral red and spectral green; but the proportions in which these spectral colours must be mixed differ for different individuals. The complementary colours for different individuals are also not precisely the same.

credited with strong colour-antipathies, and insect-eating mammals are probably not defective in the colour-sense.

It is said that nocturnal animals, such as mice, bats, and hedgehogs, have no retinal cones; and if the cones are associated with colour-vision, they may not improbably be unable to distinguish colours. Some moles are blind (*e.g.* the Cape golden mole). But the common European mole, though the eyes are exceedingly minute ($\frac{1}{25}$ of an inch in diameter), has the organ fairly developed, and is even said not to be very short-sighted. It is protected by long hairs when the animal is burrowing, and is only used when it comes to the surface of the ground.

It is probably in birds that vision reaches its maximum of acuteness. A tame jackdaw will show signs of uneasiness when seemingly nothing is visible in the sky. Presently, far up, a mere speck in the blue, a hawk will come within the range of far-sighted human vision. Steadily watch the speck as the hawk soars past, until it ceases to be visible; the jackdaw will still keep casting his eye anxiously upward for some little time. He may be only watching for the possible reappearance of the hawk. But just as he saw it before man could see it, so, probably he still watches it after, to man's sight, it has become invisible. So, too, for nearer minute objects, the swift, as it wheels through the summer air, presumably sees the minute insects which constitute its food. And every one must have noticed how domestic fowls will pick out from among the sand-grains almost infinitesimal crumbs.

It is probable that the area of acute vision is much more widely diffused over the retina of birds than it is with us. In any case, the cones are more uniformly and more abundantly distributed over the general retinal surface.

An exceedingly interesting and important peculiarity in the retina of birds, which they share with some reptiles and fishes, is the development, in the cones, of coloured globules. "The retinæ of many birds, especially of the finch, the pigeon, and the domestic fowl, have been carefully examined by Dr. Waelchli, who finds that near the centre green is the predominant colour of the cones, while among the green cones red and orange ones are somewhat sparingly interspersed, and are nearly always arranged alternately—a red cone between two orange ones, and *vice versâ*. In a surrounding portion, called by Dr. Waelchli the red zone, the red and orange cones are arranged in chains, and are larger and more numerous than near the yellow spot; the green ones are of smaller size, and fill up the interspaces. Near the periphery the cones are scattered, the three colours about equally numerous and of equal size, while a few colourless cones are also seen. Dr. Waelchli examined the optical properties of the coloured cones by means of the micro-spectroscope, and found, as the colours would lead us to suppose, that they transmitted only the corresponding portions of the spectrum; and it would almost seem, excepting for the few colourless cones at the peripheral part of the retina, that the birds examined must have been unable to see blue, the whole of which would be absorbed by their colour-globules." *

These facts are of exceeding interest. They seem to show that for these birds the retinal explosives are not the same as for us. They are R., O., and G. Moreover, the colour-globules will have the effect of excluding the phenomena of overlapping. For each kind of cone the spectrum must be limited to the narrow spectral band transmissible through the associated colour-globule. If these facts be so, it is not too much to say that the colour-vision of birds must be so utterly different from that of human beings, that, being human beings, we are and must remain unable to conceive its nature. The factors being different, and the blending of the factors by overlap being, by specially developed structures, lessened or excluded, the whole set of resulting phenomena must be different from ours. And this is a fact of the utmost importance when we consider the phenomena of sexual selection among birds, and those theories of coloration in insects which involve a colour-sense in birds.

* "Colour-Vision and Colour-Blindness," R. Brudenell Carter (*Nature*, vol. xlii. p. 56).

Concerning the sense of sight in reptiles and in amphibians, little need here be said. At near distances some of them undoubtedly have great accuracy of vision. This is, perhaps, best seen in the chamæleon. In this curious animal the eyes are conical, and each moves freely, independently of the other. The eyelids encase the organ, except for a minute opening, looking like a small ink-spot at the blunted apex of the cone. The animal catches the insects on which it feeds by darting on to them its long elastic tongue and slinging them back into the mouth, glued to its sticky tip. Its aim is unerring, but it never strikes until both eyes come to rest on the prey, and great accuracy of vision must accompany the great accuracy of aim. Frogs and toads capture their prey in a somewhat similar way; and a great number of reptiles and amphibians are absolutely dependent for their subsistence on the acuteness and accuracy of their vision, which is, however, on the whole, markedly inferior to that of birds.

In fishes, from their aquatic habit, the lens and dioptric apparatus are specially modified, in accordance with the denser medium in which they live; and one curious fish, the Surinam sprat, is stated to have the upper part of the lens suited for aerial, and the lower part for aquatic vision.

Mr. Bateson[*] has made some interesting observations on the sense of sight in fishes. He finds that in the great majority of fishes the shape and size of the pupil do not alter materially in accordance with the intensity of the light. The chief exceptions are among the Elasmobranchs (dog-fishes and skates). In the torpedo the lower limb of the iris rises so as almost to close the pupil, leaving a horizontal slit at the upper part of the eye. In the rough dog-fish, the angel-fish, and the nurse-hound, the pupil closes by day, forming merely an oblique slit. In the skate a fern-like process descends from the upper limb of

---

[*] Journal of Marine Biological Association, New Series, vol. i. Nos. 2 and 3. His experiments with regard to the colour-sense in fishes gave, for the most part, negative results.

the iris. The contraction in these cases does not seem to take place rapidly as in land vertebrates, but slowly and gradually.

Among diurnal fishes belonging to the group of the bony fishes (Teleosteans), the turbot, the brill, and the weever have a semicircular flap from the upper edge of the iris, which partially covers the pupil by day, but is almost wholly retracted at night.

None of the fishes observed by Mr. Bateson appears to distinguish food (worms) at a greater horizontal distance than about four feet, and for most of them the vertical limit seemed to be about three feet; but the plaice at the bottom of the tank perceived worms when at the surface of the water, being about five feet above them. Most of them exhibited little power of seeing an object below them. But though the distance of clear vision seems to be so short for small objects in the water, many of these fish (plaice, mullet, bream) notice a man on the other side of the room, distant about fifteen feet from the window of the tank. The sight of some fishes, such as the wrasses (*Labridæ*), is admirably adapted for vision at very close quarters. "I have often seen," says Mr. Bateson, "a large wrasse search the sand for shrimps, turning sideways, and looking with either eye independently, like a chamæleon. Its vision is so good that it can see a shrimp with certainty when the whole body is buried in grey sand excepting the antennæ and antenna-plates. It should be borne in mind that, if the sand be fine, a shrimp will bury itself absolutely, digging with its swimmerets, kicking the sand forwards with its chelæ, finally raking the sand over its back, and gently levelling it with its antennæ; but if the least bit be exposed, the wrasses will find it in spite of its protective coloration."

Although it is probably not functional in any existing form, mention must here be made of the median or pineal eye. On the head of the common slow-worm, or blind-worm, there is a dark patch surrounding a brighter spot.

This is the remnant of a median eye. It has been found in varying states of degeneration in many reptiles (Fig. 34), and in a yet more vestigial form in some fishes and amphibia. It is connected with a curious structure, associated with the brain of all vertebrates, and called the pineal gland. Descartes thought that this was the seat of the soul; but modern investigation shows it to be a structure which has resulted from the degeneration of that part of the brain which was connected with the median eye. There is some reason to suppose that, in ancient life-forms, like the Ichthyosaurus, and Plesiosaurus, and the Labyrinthodont amphibians, it was large and functional. In any case, there is a large hole in the skull (Fig. 35) through which the nervous connection with the brain may have been established. The structure of the eye is not similar to that of the lateral eye, but more like that of some of the invertebrates.

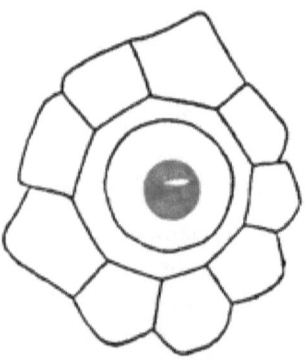

Fig. 34.—Pineal eye.

Modified eye-scale of a small lizard, *Varanus benekalensis*. (After Baldwin Spencer.)

To these invertebrates we must now turn.

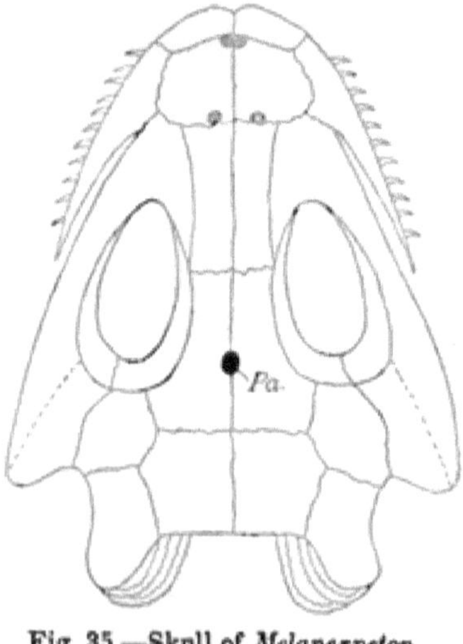

Insects have eyes of two kinds. If we examine with a lens the head of a bee, we shall see, on either side, the large compound or facetted eye; but in addition to these there is on the forehead or vertex a triangle of three small, bright, simple eyes, or ocelli. These ocelli, or eyelets, differ, in different insects, as to the details of their structure; but in general

Fig. 35.—Skull of *Melanerpeton*.

A Labyrinthodont amphibian from the Permian of Bohemia (after Fritsch). × 4. *Pa.*, the parietal foramen.

they consist of a lens produced by the thickening of the integumentary layer which is at the same time rendered transparent. Behind this lies the so-called vitreous body, composed of transparent cells, and then follows the retina, in which there are a number of rods, the recipient ends of which are turned towards the rays of light, and not away from them as in the vertebrate. Spiders have from six to eight ocelli, arranged in a pattern on the top of the head. Facetted eyes are not found in them.

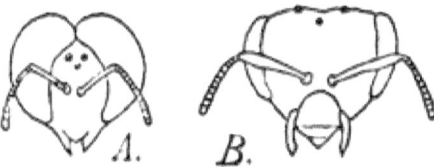

Fig. 36.—Eyes and eyelets of bee.
A. Drone. *B.* Worker.

These facetted eyes, which are found in both insects and crustacea, have apparently a more complex structure than the ocelli. Externally—in the bee, for example—the surface is seen to be divided up into a great number of hexagonal areas, each of which is called a facet, and forms (in some insects, but not in all) a little lens. Of these the queen bee has on each side nearly five thousand; the worker some six thousand; and the drone upwards of twelve thousand; while a dragon-fly (*Æschna*) is stated to have twenty thousand. Beneath each facet (in transverse section, Fig. 37) is a crystalline cone, its base applied to the lens, its apex embraced by a group of elongated cells, in the midst of which is a nerve-rod which is stated to be in direct connection with the fibres of the optic nerve. Dark pigment is developed around the crystalline cones. And retinal purple is said to be present in the cells which underlie it.

With regard to these facetted eyes there has been much discussion. The question is—Is each facetted organ an eye, or is it an aggregate of eyes? To this question the older naturalists answered confidently—An aggregate. A simple experiment seems to warrant this conclusion. If the facetted surface be cleared of its internal structures (the crystalline cones, etc.) and placed under the microscope, each lens may, at a suitable distance of the object-glass, be made to give a separate image of such an object as a candle reflected in the mirror of the microscope. If each lens thus gives an image, is not each the focussing apparatus of a single eye? But a somewhat more difficult experiment points in another direction. If the facetted cornea be removed *with the crystalline cones still attached* (Grenacher was able to do it with a moth's eye), and placed under the microscope, when the instrument is focussed at the point of the cone (where the nerve-rod comes), a spot of light, and not an image, is seen. No image can be seen unless the microscope be focussed for the centre of the cone; and here there are no structures capable of receiving it and transmitting corresponding waves of change to the "brain."

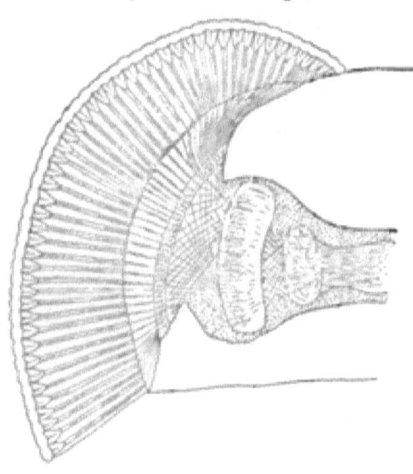

Fig. 37.—Eye of fly. Transverse section through head. (After Hickson.)

But what, it may be asked, can be the purpose of an eye-structure which gives, not an image, but merely a spot of light? The answer to this question can only be found when it is remembered that there are thousands of these facets and cones giving thousands of spots of light. The somewhat divergent cones and facets of the insect's eye (Fig. 37) embrace, as a whole, an extended field of vision; each has its special point in that field; and each conveys to the nerve-rod which lies beneath it a stimulation in accordance with the brightness, or intensity, or quality of that special point of the field to which it is directed. The external field of vision is thus reproduced in miniature mosaic at the points of the crystalline cones—thus there is produced by the juxtaposition of contiguous points a stippled image. And it must be remembered that, even in human vision, the stimulation is not that of a continuum, but is stippled with the fine stippling of the ends of the rods and cones. In insect-vision the stippling is far coarser, and the image is produced on different principles.

In the vertebrate the image is produced by a lens; in the insect's eye, by the elongated cones. How this is effected will be readily seen with the aid of the diagram. At $a\,b$ are a number of transparent rods, separated by pigmented material absorbent of light. They represent the crystalline cones. At $c\,d$ is an arrow placed in front of them; at $e\,f$ is a screen placed behind them. Rays of light start in all directions from any point, c, of the arrow; but of these only that which passes straight down one of the transparent rods reaches the screen. Those which pass obliquely into other rods are absorbed by the pigmented material. Similarly with rays starting from any other point of the arrow. Only those which, in each case, pass straight down one of the rods reach the screen. Thus there is produced a reduced stippled image, $c'\,d''$, of the arrow.

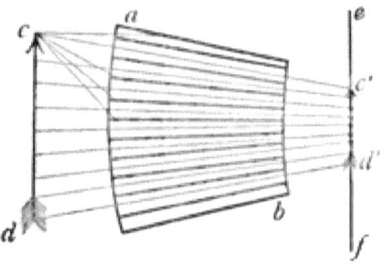

Fig. 38.—Diagram of mosaic vision.

There has been a good deal of discussion as to the relative functions of the ocelli and the facetted eyes of insects. The view generally held is that the ocelli are specially useful in dark places and for near vision; while the facetted eyes are for more distant sight and for the ascertainment of space-relations. How the two sets of impressions are correlated and co-ordinated in insect-consciousness, who can say?*

The interesting observations of Sir John Lubbock seem to show that insects can distinguish between different colours. "Amongst other experiments," he says,† "I

---

* We must remember how largely the antennæ are used when an insect is finding its way about. Watch, for example, a wasp as it climbs over your plate. If the antennæ be removed, it seems to stumble about blindly. The antennæ seem almost to take the place of eyes at close quarters.

† "Senses of Animals," p. 194.

brought a bee to some honey which I placed on a slip of glass laid on blue paper, and about three feet off I placed a similar drop of honey on orange paper. With a drop of honey before her a bee takes two or three minutes to fill herself, then flies away, stores up the honey, and returns for more. My hives were about two hundred yards from the window, and the bees were absent about three minutes or even less. After the bee had returned twice, I transposed the papers; but she returned to the honey on the blue paper. I allowed her to continue this for some time, and then again transposed the papers. She returned to the old spot, and was just going to alight, when she observed the change of colour, pulled herself up, and without a moment's hesitation darted off to the blue. No one who saw her at that moment could have the slightest doubt about her perceiving the difference between the two colours."

Passing now to the crustacea, we find in them eyes of the same type as in insects; but in the higher crustacea ocelli are absent. In the crabs and lobsters the eyes are seated on little movable pedestals; in the former the crystalline cones are very long, in the latter they are short. There can be little doubt that vision is by no means wanting in acuteness in an animal which, like the lobster, can dart into a small hole in the rocks with unerring aim from a considerable distance. The experiments of Sir John Lubbock have shown that the little water-flea (*Daphnia*) can distinguish differences of colour, yellows and greens being preferred to blues or reds.

Among the molluscs there are great differences in the power of sight. Most bivalves, like the mussel, are blind. Interesting stages in the development of the eye may be seen in such forms as the limpet, *Trochus* and *Murex*. The limpet has simply an optic pit, the *Trochus* a pit nearly closed at the orifice and filled with a vitreous mass, and the *Murex* a spherical organ completely closed in with a definite lens. The snail has a well-developed eye on the hinder and longer horn or tentacle. But it does not seem to be aware of the presence of an object until it is brought within a quarter of an inch or less of the tentacle. In all probability the eye does little more than enable the snail to distinguish between light and dark. And the same may be said of the eye of many of the molluscs. In some, however, the cuttle-fishes and their allies, the eye is so highly developed that it has been compared with that of the vertebrate. There is an iris with a contractile pupil. And the ganglion with which it is connected forms a large part of the so-called brain. The powers of accurate vision in these higher forms are probably considerable.

It is interesting to note that whereas in the cuttle-fishes and most molluscs, the rods of the retina are turned towards the light, in *Pecten*, *Onchidium* (a kind of slug), and some others, they are, as in vertebrates, turned from the light. In *Pecten* the nerve to supply the retina bends round its edge at one side. But in *Onchidium* it pierces the retina as in vertebrates.

In worms, eyes are sometimes present, sometimes absent. In star-fishes and their allies they often occur. In medusæ (jelly-fish) they are sometimes found on the margin of the umbrella. Even in lowly organisms, like the infusoria, eye-spots not unfrequently occur. We must remember, however, that, in these lower forms of life, the organs spoken of as eyes or eye-spots merely enable the possessor to distinguish light from darkness.

Even when eyes or eye-spots are not developed, the organism seems to be in some cases sensitive to light— using the word "sensitive," once more, in its merely physical acceptation. The earthworm, for example, though it has no eyes, is distinctly sensitive to light; and the same has been shown to be the case with other eyeless organisms. Graber holds that his experiments demonstrate that the eyeless earthworm can distingush between different colours —in other words, is differentially sensitive to light-waves of different vibration-period—preferring red to blue or green, and green to blue. And the same observer has shown that animals provided with eyes—the newt, for example—can distinguish between light and darkness by the general surface of the skin. M. Dubois, by a number of experiments on the blind *Proteus* of the grottoes of Carniola, has shown that the sensitiveness of its skin to light is about half that of its rudimentary eyes; and, further, that this sensibility varies with the colour of the light employed, being greatest for yellow light.[*]

We have not been able to do more than make a rapid survey of the sense of sight as it seems to be developed in the invertebrates and lower animals. The visual organs differ, not only in structure, but in principle. We may, I think, distinguish four types.

1. Organs for the mere appreciation of light or darkness (shadow), exemplified by pigment-spots, with or without concentrating apparatus.

2. Organs for the appreciation of the direction of light or shadow, with or without a lens. The simple retinal eyes of gasteropods, and perhaps in some cases the ocelli of insects, probably belong to this class.

3. True eyes, or organs in which a retinal image is formed, through the instrumentality of a lens, as in vertebrates and cephalopods.

4. The facetted eyes of insects, in which a stippled image is formed, on the principle of mosaic vision.

Unfortunately, all these are called indiscriminately eyes, or organs of vision. An infusorian or a snail is said to see. But the terms "eye," "vision," "sight," imply that final excellence to which only the higher animals, each on its own line, have attained.

This final excellence probably has its basis and earliest inception in the fact that the functional activity of protoplasm is heightened in the presence of ætherial vibrations. If, then, we imagine, as a starting-point, a primitive transparent organism with a general susceptibility to the influence of light-vibrations, the formation within its tissues of pigment-granules absorbent of light will render the spots where they occur specially sensitive to the

[*] See *Nature*, vol. xli. p. 407.

ætherial vibrations. Special refraction-globules would also act as minute lenses, focussing the light, and thus concentrating it upon certain spots.

In many of the lower animals we find such organs, belonging to our first category, and constituting either eye-spots of pigmented material or simple lenses covering a pigmented area. If we call these eyes, we must remember that in all probability they have no power of what we call vision—only a power of distinguishing light from dark. Where, however, there exists beneath the lens a so-called retina, that is, a layer of rod-like endings of a nerve, it might, at first sight, be thought that there, at any rate, we have true vision. But in all probability, in a great number of cases the retinal rods are simply for the purpose of rendering the organism sensitive, not only to the presence of light, but to its direction. Light straight ahead (a) stimulates the middle rods; from one side (b, c) it is focussed on the rods of the opposite side of the retina; and similarly for intermediate positions. The presence of a retinal layer is thus no infallible sign of a power of vision as apart from mere sensibility to light. Indeed, in a great number of cases, from the convexity and position of the lens, the formation of an image is impossible. Only when it can be shown that a more or less definite image can be focussed on the retina, or can be formed on the principle of mosaic vision, can we justly surmise that a power of true vision is present. I doubt whether this can be shown to be unquestionably the case in any forms but the higher arthropods, the cuttle-fishes and their allies, and the vertebrates.

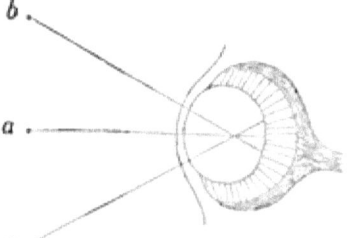

Fig. 39.—Direction-retina.
Simple retina for distinguishing the direction of the source of light or of shadow.

There is one more point for consideration before we leave the sense of sight—Are the limits of vision the same in the lower forms of life as they are in man? or, to put the question in a more satisfactory form—Are the limits of sensibility to light-vibrations the same in them as in us? M. Paul Bert concluded that they are. But Sir John Lubbock has, I think, conclusively shown that they are not. For the full evidence the reader is referred to his "Senses of Animals."[*] His experiments on ants, with which those of M. Forel are in complete accordance, satisfied him that these little animals are sensitive to the ultra-violet rays which lie beyond the range of our vision. Other experiments with fresh-water fleas (*Daphnia*) showed that they have colour-preferences, green and yellow being the favourite colours.

The daphnias were placed in a shallow wooden trough, divided by movable partitions of glass into divisions. Over this was thrown a spectrum of rainbow colours. The partitions were removed, and the daphnias allowed to collect in the differently illuminated parts of the trough. The partitions were then inserted, and the number of crustaceans in each division counted. The following numbers resulted from five such experiments:—

Dark.	Violet.	Blue.	Green.	Yellow.	Red.
0	3	18	170	36	23

Special experiments seem to show that their limits of vision at the red end of the spectrum coincide approximately with ours; but at the violet end their spectrum is longer than ours. Sir John covered up the visible spectrum, so as to render it dark, and gave the daphnias the option of collecting in this dark space or in the ultra-violet. To human eyes both were alike dark. But not so to the daphnian eye; for while only 14 collected in the covered part, 286 were found in the ultra-violet. The width of the violet visible to man was two inches. Sir John divided the ultra-violet into three spaces of two inches each. Of the 286 daphnias, 261 were in the space nearest the violet, 25 in the next space, and none in the furthest of the three spaces. From which it would seem that, though these little creatures are sensitive to light of higher vibration-

[*] Chap. x. p. 202.

period than that which affects the human eye, their limits do not very far exceed ours. We have seen that human beings differ not a little in their limits of violet-susceptibility. We may presume that Sir John Lubbock and those who assisted him in these experiments were normal in this respect. But it is possible that some individuals could have perceived a faint purple where there was darkness to them, and that the majority of the 261 daphnias were collected in the region just beyond the partition between ultra-violet and darkened violet. Still, there is no cause for doubting the general conclusion that daphnias are sensible to ultra-violet rays beyond the limits of human vision.

Sir John Lubbock has an interesting chapter on problematical organs of sense. In the antennæ of ants and

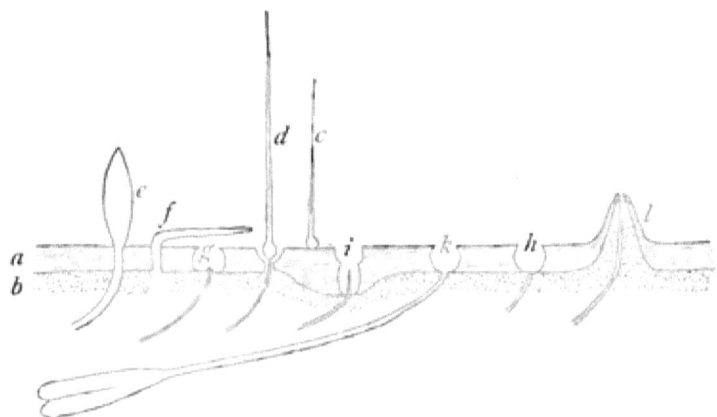

Fig. 40.—Antennary structures of hymenoptera. (After Lubbock.)

a., cuticle; *b.*, hypodermis; *c.*, ordinary hair; *d.*, tactile hair; *e.*, cone; *f.*, depressed hair lying over *g.* cup with rudimentary hair at the base; *h.*, simple cup; *i.*, champagne-cork-like organ of Forel; *k.*, flask-like organ; *l.*, papilla, with a rudimentary hair at the apex.

bees there are modified hairs and pits in the integument (at least eight different types, according to Sir John Lubbock), the sensory nature of which is undoubted. But what the sensory nature in each case may be is more or less problematical. Many worms have sense-hairs or bristles of the use of which we are ignorant. Some organs described as tactile or olfactory in the lower invertebrates are so described on a somewhat slender basis of evidence. The sense-value of the bright marginal beads of sea-anemones is unknown. Even in animals as high in the scale of life as fishes, there is a complete set of sense-organs—the muciparous canals, in the head and along the lateral line down the side, the function of which we can only guess. By some they are regarded as olfactory; by others, as fitted to respond to vibrations or shocks of greater wave-length than the auditory organ can appreciate; by others, as of importance for the equilibration or balancing of the fish.

It will thus be seen that, apart from the possibility of unknown receptive organs as completely hidden from anatomical and microscopic scrutiny as the end-organs of our temperature-sense, there are in the lower animals organs which may be fitted to receive modes of influence to which we human folk are not attuned.

And what are the physical possibilities? We have seen that, through the telæsthetic senses—hearing, vision, and the temperature-sense—we are made aware of the vibrations of distant bodies, the effects of which are borne to us on waves of air or of æther. The limits of hearing with us are between thirty and about forty thousand (or perhaps, in very rare cases, fifty thousand) vibrations per second. But these are by no means the limits of vibrations of the same class. By experiments with sensitive flames,[*] Lord Rayleigh has detected vibrations of fifty-six thousand per second; and Mr. W. F. Barrett has shown that a sensitive flame two feet long is sensitive to vibrations beyond the limit of his own hearing and that of several of his friends who were present at the experiment. We have some reason to suppose that vibrations too rapid to be audible by man are audible by insects, but not much is known with regard to the exact limits.

The following table shows what is known concerning

[*] The observations are not yet published, and I have to thank Lord Rayleigh for his courtesy in allowing me to make use of this fact.

the æther-vibrations. The figures are those given by Professor Langley:—

Quality of radiations.	Wave-lengths in thousandths of a millimetre.	Number of vibrations per second in billions.	Effects on man.
Limit of photography, artificial source	0·185	160	none known
Limit of photography, solar source	0·295		" "
Limit of violet to normal eyes	0·36	833 } vision.
Limit of red to normal eyes	0·81	370 }
Probable inferior limit of temperature-sensations	9·25*	30	temperature-sense
Longest waves hitherto recognized with bolometer	30·0	1	none known

From this table it will be seen that, apart from the possible extension of sight beyond human limits, there are possibilities of another sense for the ultra-violet actinic vibrations as different from sight as is the infra-red temperature-sense. Moreover, the temperature-sense for us has no scale; there is nothing corresponding to pitch in sound or colour in sight. It may not be so with lower organisms. Insects, for example, may be sensitive to tones of heat. The bee may enjoy a symphony of solar radiance. I am not saying that it is so; I am merely suggesting possibilities which we have not sufficient knowledge to authoritatively deny. We have no right to impose the limits of human sensation on the entire organic world. Insects may have "permanent possibilities of sensation" denied to us.

Even within our limits there may be, as we have already seen, great and inconceivable differences. We saw

---

* Professor Langley finds that the maximum effect with a radiating source at 170° C. is at about 5·0 thousandths of a millimetre wave-length.
 „ 100° C. „ „ 7·5 „ „ „
 „ 0° C. „ „ 11·0 „ „ „

We are sensitive to radiations from a body at 100° C. But when the temperature falls below the normal temperature of the body we are not sensitive to heat-vibrations, but to loss of heat from the surface exposed. The limit of sensibility to heat-vibrations, therefore, probably lies between 7·5 and 11 thousandths of a millimetre. I have taken about 9·25 as the limit.

that our own colour-sensations are probably due to the blending and overlapping in different proportions of three primitive monochromatic bands, but that in all probability in birds the bands are different, and overlapping is largely prevented. Their colour-phenomena must be inconceivably different from ours. And what shall we say of the colour-vision of invertebrates? Are we justified in supposing that for them, as for us, R., G., and V. are the unstable explosives, and that they are present in the same proportions as with us? If not, their colour-world cannot be the same as ours. Of the same order it probably is. And all that we can hope to do is to show, as has been shown, that colours which differently affect us affect them also differently.

In conclusion, we may return to the point from which we set out. The organism is fitted to respond to certain influences of the external world. The organs for the reception of these influences are the sense-organs. When they are stimulated waves of change are transmitted inwards to the great nerve-centres; they are there co-ordinated, and issue thence to muscles or glands. Thus the organism is fitted to respond to the influences from without. The activities of organisms are in response to stimulation.

We have seen that the cells of the organic tissues are like little packets of explosives, and that the changes which occur in the organism may be likened to their explosion and the setting free of the energy stored up in them. The end-organs of the special senses may be regarded as charged with explosives of extreme sensitiveness. Some are fired by a touch; the molecular vibrations of sapid or odorous particles explode others; yet others are fired by the coarser vibrations of sound; others, once more, by the energy of the ætherial waves. The visual purple is a highly unstable chemical compound of this kind; expose it for a moment to light, and it topples over to a new molecular arrangement, the colour being at the same time discharged. If the retina has been removed from the body, this is all that happens. But if (in the frog) it be replaced on the choroid layer from which it has been stripped, the visual purple is reformed. The explosive is thus reconstructed and the sensibility is restored. Thus, as fast as the explosives are fired off by sense-stimuli, so fast in normal life are they reconstituted and the sensibility restored. Meanwhile the explosion at the end-organs has fired the train of explosives in the nerve, and created molecular explosive disturbances in the brain. Thence the explosive waves pass down other nerves to muscles or glands, and, giving rise therein to further explosions, take effect in the activities of the organism.

We shall have to consider these activities hereafter. We must now turn to the psychical or mental accompaniments of the explosive disturbances in the brain or other aggregated mass of nerve-cells.

CHAPTER VIII.

MENTAL PROCESSES IN MAN.

I HAVE already drawn attention to the fact that the primary end and object of the reception of the influences (*stimuli*) of the external world, or environment, is to enable the organism to answer or respond to these special modes of influence, or stimuli. In other words, their purpose is to set agoing certain activities. Now, in the unicellular organism, where both the reception and the response are effected by one and the same cell, the activities are for the most part simple, though even among these protozoa there are some which show no little complexity of response. Where, however, the organism is composed of a number of cells, in which a differentiation of structure and a specialization of function have been effected, certain cells are set apart as *recipients*, while other cells are set apart to respond (*respondents*). There is thus the necessity of a channel of communication between the two. Hence yet other cells (*transmitters*), arranged end to end, form a line of connection and communication between the group of receiving cells and the group of responding cells, and constitute what we term a *nerve*. That which is transmitted may still be called a stimulus, each cell being stimulated in turn by its neighbour. Thus a stimulus must be first received and then transmitted.

But little observation is required to convince us of the fact that, in the higher creatures, a very simple stimulus may give rise to a very complex response. A light pinprick will cause a vigorous leap in a healthy frog—a leap that involves a most intricate, accurate, and complex co-ordination of muscular activities. And anatomical investigation shows us that in such creatures there is always, in the course of the channel of communication or transmission, a group of closely connected cells, which play the part of co-ordinants. In the vertebrate animals these co-ordinants are collected in the brain and spinal cord. In the insects, crustaceans, and worms they are arranged in a knotted chain running close to the under surface of the body. To this central nervous system, as it is called, nerves (afferent nerves) run inwards from the recipient organs. From it nerves (efferent nerves) run outwards to the organs of response. And in it the transmitted stimuli, brought in by the afferent nerves, are modified, through intervention of the co-ordinants, into stimuli carried out by the efferent nerves. A simple stimulus may create a great commotion among the co-ordinants of the central nervous system, and give rise to many and complex stimuli going out to the muscles and other organs of response. How this is effected is one of the many wonders of the animal mechanism. We believe that the connection and co-ordinations have gradually been established during a long process of development and evolution, reaching back far into the past. How, we can at present scarcely guess.

We must picture to ourselves, then, in the animal organism, a multitude of nerve-fibres running inwards from all the end-organs of the special senses, from the muscles, and from the internal organs, and all converging on the central nervous system. And we must picture to ourselves a multitude of nerve-fibres passing outwards from the central system, and diverging to supply the muscles, glands, and other organs which are to respond to the stimulation from without. We must picture the fibres coming from or going to related parts or organs collecting together to form nerves and nerve-trunks, which are, however, only bundles of isolated nerve-fibres. And, lastly, we must picture the central nerve-system itself co-ordinating and organizing the stimuli brought into it by afferent nerves, from the organs of special sense, and handing over the resultants by efferent nerves to the organs of special activities. So far we have purely physiological effects, many of which occur with surprising accuracy and precision when an organism is in a state of unconsciousness. Place your finger in the palm of a sleeping child, and the fingers will close over it without the child awaking to consciousness. If, in a frog, the brain of which has been extirpated, the side be touched with a drop of acid, the leg of that side will be drawn up, and the foot will be used to wipe away the acid. And if that leg be held and prevented from reaching the side, the other leg will be brought round so as to try and bring the foot within reach of the irritated spot. The actions are, however, in all probability, purely physiological, and are performed in complete absence of consciousness.

When we turn from the physiological to the psychological aspect of the question, we enter a new world, the world of consciousness, wherein the impressions received by the recipient organs (no longer regarded as mere stimuli, but as the elements of consciousness) are co-ordinated and organized, and are built up into those sensations and perceptions through which the objects of the external world take origin and shape. It is with this process that we have now to deal; and we will deal with it first in man.

The first fact to notice is that, apart from sense-stimuli received and exciting consciousness, we have also the revival of past impressions. This revival is the germ of memory. What exactly is the physical basis of memory, how the effects of stimuli in consciousness come to be registered, we do not know. It is clearly a matter that falls under the general law of persistence; but in what organic manner we are largely ignorant. Still, there can be no question of the fact that, quite apart from impressions due to immediate influences of the environment *now* acting on our recipient organs, we have also revivals of bygone influences of the environment—shadows or afterimages of previous modes of influence. Without this process of registration and revival, stimuli could never give rise to sensations and perceptions such as we know them. Without it experience would be impossible.

We may say, then, that impressions (resulting from stimuli) and their revival in memory are the bricks of the house of knowledge; and these are built up through experience into what we call the world of things around us. There may be and is a certain amount of mortar, supplied by the builder, in addition to the elementary bricks. But without the bricks no house of knowledge could be built. Let us now examine the bricks and the building.

From what we have already learnt in the chapter on "The Senses of Animals," it is clear that the impressions and their revivals in memory have differences in quality. Here, on the very threshold of the subject, we must pause. They have differences of quality. But in consciousness these differences must be distinguished. And this involves their recognition and discrimination, presupposing, therefore, a corresponding faculty, however simple, on the part of the recipient. Without cognition and recognition (twin sisters, born in the same hour) we can never get beyond mere impressions; which may, indeed, be differentiated physically, as different stimuli due to diverse action of the environment, but are psychically undifferentiated. This recognition and discrimination is thus the primary activity of the recipient mind. Here is already some of the mortar supplied by the builder. Memory is absolutely essential to the process. The sense-impression of external origin gives rise to an impression of similarity or dissimilarity, which is part of the internal reaction to the external stimulus. Thus impressions are raised to the level of *sensations*. A sensation is an impression that has been discriminated from others, and recognized as being of such and such a nature. The impressions of the sense-organs as we know them are thus not mere impressions, but impressions

x

raised to the level of sensations, in so far as they are recognized and discriminated.

Let us now glance at some of the differences in quality recognized in sensation. First, we have the broadly distinguished groups of touches and pressures, temperature-sensations, tastes, smells, sounds, sights, muscular sensations, and organic sensations from internal parts of the body. And then, within each of these groups, there are the more or less delicate and distinct shades of quality, well exemplified in vision by the different colour-sensations, in hearing by notes of different pitch, and in smell by the varieties of scents and odours. Many of those sensations, moreover, which are apparently simple, are in reality compound. There are differences of quality in the note A as sounded on a violin, a piano, and a flute; and these differences are due to different admixtures of overtones, which fuse with the fundamental tone and alter its timbre. So, too, with vision. The sensation given by a white disc is a compound sensation, due to waves of different period, which separately would give sensations of colour. Sensations, then, differ in quality.

They also differ in quantity or intensity. This needs little illustration. As evening falls, the sight-sensations derived from the surrounding objects grow more and more feeble. They may remain the same in quality, but the quantity or intensity gradually diminishes. So, too, in music, the pianos and fortes give us differences in intensity of sound-sensations.

Sensations also differ in duration. The stimulation may be either prolonged or instantaneous. Two or more sensations may, moreover, be simultaneous or successive. Just as they may be either similar or different in quality and in intensity, so they may be either simultaneous or successive in time. Simultaneous sensations are best exemplified in vision and through touch; successive sensations are given most clearly by the sense of hearing, through which we recognize a sequence of sounds.

And then, again, sensations not only differ in time, but

they seem also to differ in place. A sensation of touch may be referred to different parts of the body—the hand, the foot, or the forehead. But here we open up an important question—Where do we feel a sensation, such as, for example, that of pressure on the skin? Common sense answers, without hesitation, that we feel it at the particular part of the body which is affected by the external stimulus. I feel the pen with which I write with my finger-tips. And common sense is perfectly right from its own point of view. But it is a well-known fact that a person whose leg has been amputated experiences at times tickling and uneasiness in the absent member. This is due to irritation of the nerve-ends in the stump of the limb. But the sensations are referred outwards to the normal source of origin of impressions, the effects of which were carried inwards by the nerve affected. We shall have to consider hereafter the nature of the relation between physiological and psychological processes—the connection of mind and body. Assuming for the present that psychical processes have a physical basis in physiological processes, the fact given above and others of like implication seem to show that the sensation has for its physiological basis some nerve-change in the central nervous system—in us, no doubt, in the brain. Of course, it must be remembered that the sensation, as felt, is a mental fact (using the word "mental" in its broadest sense, as belonging to the psychical as opposed to the physiological series). But it would seem that the physiological accompaniment of this mental fact is some nerve-change in the brain. This nerve-change is caused by a stimulus having its origin in the end-organ of the afferent nerve, and we naturally refer the impression outwards to the place of its source of origin under ordinary and normal conditions. In other words, we *localize* it. That is what common sense means when it says that we feel pressure at the finger-tips.

To account for this process of localization, it is supposed that every sensation, apart from its special quality as a touch, a taste, or a smell, has a more or less defined spatial quality, or local sign, dependent upon the part of the body to which the stimulus is applied. These local signs have, doubtless, in the long run, been established by experience—if under this term we may include a more or less unconscious process, the outcome of evolution. But they are so rapidly established in the individual, that we are forced to conclude that we inherit very highly developed aptitudes for localization.

The refinement of localization is very different in the different senses. In smell and taste there seems no more than a general localization in the organ affected—the nose or the mouth. In hearing there is not much more, unless we regard the discrimination of pitch as a mode of localization. In touch (and temperature) the refinement is much higher, but it varies with the part of the body affected.

If the back be touched by two points less than two inches and a third apart, the sensation will be that of a single point; the finger-tips, however, can distinguish two points separated by less than one-tenth of an inch; and the tip of the tongue is still more refined in its power of discrimination, distinguishing as two, points separated by less than the twenty-fifth part of an inch. So that the tongue is about sixty times as refined in its discrimination as the skin of the back. Moreover, the delicacy of localization may be cultivated, so that in some cases the refinement may, by practice, be doubled.

When we come to sight, the refinement of localization reaches its maximum, the local signs in the retina showing the highest stage of differentiation, the distance on the retina between two points distinguishable by local signs being, according to Helmholtz, not much more than $\frac{1}{6000}$ of an inch (·0044 millimetre), which nearly corresponds with the space between two cones in the yellow spot.

We must remember that the presentations of sense are in all cases given in a stippled form, that is, by the stimulation of a number of separate and distinct points. In vision the stippling is very fine, owing to the minute size and close setting of the retinal cones. In the case of hearing, the stippling, if we may so extend the use of this term, is also very fine, as is shown by the fact that musicians can, according to Weber, distinguish notes separated in the scale of sounds by only one-sixtieth part of a musical tone. In touch the stippling is comparatively coarse. But in all cases there is a stippling; and yet from these stippled sensations the mind in all cases elaborates a continuum. The visual image is continuous, notwithstanding the retinal stippling and the existence of the blind spot. When we lay our hands on a smooth table we fill in the interstices between the sensational points, and feel the surface as continuous. In all cases out of the stippled sense-stimuli we form a continuum.

The next thing that we have to note is that it is not so much the sensation itself, as that which gives origin to it, that we habitually refer outwards to the recipient end of the afferent fibre. In referring a sensation of touch to a certain part of the skin, it is of something touching us that we seem to be immediately conscious. We refer the stimulus to an object in the external world, which we localize, and which we believe to have given rise to the sensation.

This, however, is more clearly seen in the case of vision. When we look through the window and see an object such as a house before us, we do not habitually localize the sensation in a certain part of the retina, but we refer the object to a particular position more or less distant in the world around us. This projection of the object outwards in a right line from the eyes is really a marvellous process, though the wonder of it is lost in its familiarity. It is the outcome of the experience of hundreds of generations. And the experience is not gained through vision alone, but through this in combination with other senses and activities. We see an object, but we have to go to it before we can touch it. It is not in contact with us, but distant from us. Its outness and distance is a matter of what is termed the geometry of the senses; and this geometry has been elaborated through many generations of organized beings, from data given by sight, touch, and the muscular sense. It is true that I can now estimate the distance of the house without going to it; but my eyes go to it, and I can feel them go. The panes of my window are separated by iron bars. As I look from them to the distant house and back to them again, I can feel my eyes going from one to the other. The lens of the eye is adjusted for near or far distance by the action of a ciliary muscle, through which its anterior surface can be flattened, returning again by its own elasticity to the more convex form when the muscle ceases to act. Each eye, moreover, is moved in its orbit by six eye-muscles, and in normal vision the two eyes act as one organ. For near distances they converge; for far distances there is less convergence. Through the muscular sense, which is here extraordinarily delicate, we can feel the amount of accommodation and convergence; and thus we can feel the eyes going to or coming from a near and a distant object. Of course, we are aided in judging or estimating distances by the apparent size of the object when the real size is known, by the clearness of its outlines in a slightly hazy atmosphere, and so forth. But apart from such judgments, it would probably be impossible to perceive that an object is near or distant in the absence of muscles of accommodation and convergence affording the data of the muscular sense. Not only the distance of two objects from the eye, but their distance apart, can be measured by the aid of the muscular sense as we move the eyes from one to the other. And in us this is so delicate that, according to Weber, a distinct muscular sensation is attached to a displacement of a sensitive point of the yellow spot through less than $\frac{1}{6000}$ of an inch.

Now, if it be true that the consciousness aroused by objects around us, through sensation, is an accompaniment of certain physiological changes in the brain, it is clear that the localization of their points of origin in special parts of the skin, and the outward projection of the objects exciting vision, is an act of the mind quite distinct from the mere passive response in consciousness which we call an impression, and more complex than that mental activity which, through discrimination and recognition, converts the bare impression into a sensation. It is, in fact, part of that mental process which is called *perception*.* Sensation has nothing to do with the objects around us as such; it is by perception that we are aware of their existence. Let us now follow the process of perception a little further, always remembering that it involves certain activities of the mind.

These activities are too often ignored. We often speak of the senses as the avenues of knowledge, and John Bunyan, likening the soul to a citadel, spoke of the five gateways of knowledge, Eye-gate, Ear-gate, Mouth-gate, Smell-gate, and Feel-gate. Hence arises a vague notion that through the eye-gate, for example, a sort of picture of the external object somehow enters the mind. And this idea is no doubt fostered by the fact that an inverted image of the object is formed on the retina, though how the inverted image is turned right way up again in passing into the mind bothers some people not a little.†

---

* I use this term in a broad sense, as the process involved in the formation of what I shall term *constructs*.

† And I may add it is not an easy matter to explain to those who have not considered such questions. It is a matter of the correlation of the testimony of the sense-organs. A boy stands before me. I go to him and touch him, and pass my hands downwards from head to foot. Then I stand a little way off and look at him. His image on my retina is inverted. But as I run my eye over him I direct my eye downwards to his feet and upwards to his head. I am not conscious that the stimuli are running *upwards* along the retinal image. Thus my eye-muscles and my other muscular and tactile sensations seem to tell me that he is one way upwards. The image on my retina tells me, though I am not conscious of the fact, that he is the other way upwards. But he cannot be both! The testimony of one sense has to give way. One standard or the other has to be adopted. Practically that of touch and the muscular sensations is unconsciously selected, and sight-sensations are habitually interpreted in terms of this standard. So long as the two are sufficiently accurately correlated, the practical requirements of the case are met. And it is well known that it is not difficult, with a little practice, to establish a new correlation. This is indeed done every day by the microscopist, for whom the images are all reversed by his instrument. He very soon learns, however, that to move the object, as seen, to the left, he must push it to the right. A new correlation is rapidly and correctly established.

A much closer analogy is this: Something stands without and knocks at the doorway of sense, and from the nature of the knocks we learn somewhat concerning that which knocks. In other words, at the bidding of certain stimuli from without we construct that mental product which we call the object of sense. It is of these mental constructions—"*constructs*"* I will call them for convenience—that I have now to speak.

In a fruiterer's shop on the opposite side of a street I see an orange. That is to say, certain cones of the retina of my eye are stimulated by light-waves of a yellow quality, and at the bidding of these stimuli I construct the object which I call an orange. That object is distant, roundish, yellow, resisting and yet somewhat soft, with a peculiar smell, and possessed of a taste of its own. Now, it is obvious that I cannot see all these qualities of the orange, as we call them. I construct the object on reception of certain light-waves which are focussed on the retina of my eye. If I go to the orange, however, I can test the correctness of my construct by the senses of touch, smell, and taste. But what led me to construct an object with these qualities? Experience has taught me that these qualities are grouped together in special ways in an orange. I constructed that particular object through what is termed the principle of association. I have learnt that these qualities are grouped together in certain relations to each other, and when I actually receive sight-stimuli of a certain quality, grouped in certain ways, they immediately call up the memories of the associated qualities. That which is actually received is a mere suggestion, the rest is suggested in memory through association. The object might be suggested through other senses. I come into a dining-room after dessert, and the object is suggested through smell. Or my little son says, "Open your mouth and shut your eyes, and see what the fairies will send you;" and an orange is suggested by taste. In all these cases

---

* I use this term because the word "percept" is used in different senses by different writers, *e.g.* by Mr. Mivart and Mr. Romanes.

the object is constructed at the bidding of certain sensations, which suggest to my mind the associated qualities. The object is a *construct*.

And here let us notice that we ascribe the form, the resistance, the taste, the smell, to the object. We do not say or think, " Sight-sensations inform me that there is something which I call an orange, and which is capable of exciting in me sensations of touch, taste, and smell ; " but we say, " There is an orange, which *has* such and such a taste, smell, and feel." In other words, we refer these sensations, related in certain ways, outwards to the object, and name them qualities of the object that we see. But remember, that we do not necessarily or normally say or think anything about it. We just inevitably construct the object, what we build in to the construct depending upon association through experience.

At this stage, perhaps, Common Sense steps in, and, shaking his head, says, with characteristic bluntness, " Nonsense ; you'll never persuade me that the things I see and feel around me are nothing but fictions of my own mind. I don't construct them, as you call it ; there they are for me to see and feel and taste if I will." Now, Common Sense is a sturdy, hard-headed individual, with whom I desire to keep on friendly terms. And I therefore hasten to explain that I most fully agree with every word that he says. The orange that I see before me is not a mere fiction of my mind. I can, if I will, take it up, feel it, smell it, and taste it. If it will satisfy Common Sense, I will say that it is the idea of the orange that I construct. Only I think that Common Sense, who has a horror of roundabout and indirect statements, will not like my saying, " I am receiving certain visual sensations related in certain ways, which lead me to construct an idea of an orange." He will prefer my saying simply, " I see an orange." Since what he wants me to call our ideas of things answer point for point to the things as they actually exist for us human-folk, it is not only more satisfactory but more correct to merge the two in one, and speak directly and simply of the object. The object is a thing I construct. That it is real may be proved by submitting it to the test of all the senses that I have.

And what do I mean by "real"? I mean that what it is for me it is also for you and any other normally constituted human being. This is, in truth, the only common-sense criterion of objective reality. Some people are colour-blind, and tell us that a rose is not red, but green. We reply that it is really red, but that, through a defect of sight, they cannot distinguish its redness. Here we take the normal human being as a standard for objective reality. For him the rose is red. And this is the only practical criterion that we have. This, however, does not satisfy some people, who think that the objects around them have the same reality, independent of man, that they have for us human-folk. Annihilate, they say, every human being —nay, all life—and the objects will remain as they are, and retain the same reality. Yes, the same reality; which means that if just one fortunate fellow escaped annihilation, he would find them all just as they were. And this nobody doubts. Nevertheless, it is (to me, at least) inconceivable that things independently of us are what they appear to us. Think of what we learnt about the sensations. They all arose in stimulations of the end-organs of special sense. Thence the explosive waves of change passed inwards to the brain, and somewhere therein gave rise to mental products. These mental products, the accompaniments of nerve-changes, can in no sense be like the outside something which gave rise to them. They are symbols of that outside something. And it is these symbols that we build up into objects. Hence I said that it is not only more satisfactory and convenient, but more correct, to speak directly of the object as constructed, and not our idea of the object. The mental product *is* the object for us, not only for me, but for you and all normal human beings, since the object is the same for all of us. And hence, also, I said that the analogy of gateways, through which pictures of objects gain access to the mind, was false and misleading, and that a truer analogy is that something stands without and knocks at the doorway of sense, and that from the nature of the knocks we learn somewhat concerning that which knocks. The person inside can never open the door to see what manner of thing it is which knocks. But he can build up a most cunning symbolism of knocks which shall suffice for all practical purposes. In other words, the object-world, symbolic though it is, which you and I and the rest of us construct at the bidding of something without us (the existence of which I assume), is amply sufficient for all our practical needs, and constitutes the only practical reality for human-folk.

I am well aware that there are many people who cannot bring themselves to believe in, or even to listen without impatience to, the view that the world we see around us is a world of phenomena. It is absurd, they say, to tell us that yonder tulip, as an object, is in any sense dependent on our perception of it. There it is, and there it would have been had man never been created. Can one conceive that the new species of fossil, which was only yesterday disentombed from the strata in which it has lain buried for long ages, is dependent on man's observation for its qualities as an object? To say that it was "constructed" by the lucky geologist who was fortunate enough first to set eyes on it is sheer nonsense. Its shelly substance protected a bivalve mollusc millions of years before man appeared upon the earth. When we see the orange in the fruiterer's shop, the sight of it merely reminds us of its other qualities—its taste, its smell, its weight, and the rest, which are essentially its own, and no endowments of ours—nowise bestowed upon it by us.

I have no hope of convincing, and not much desire to convince, one who thus objects. I would merely ask him how and when he stepped outside his own consciousness to ascertain that these things are so. Does he believe that consciousness is an accompaniment of certain nervous processes in the grey cortex of the brain? If so, let him tell us how these conscious accompaniments resemble (not merely symbolize, but *resemble*) tulips and oranges and fossil molluscs. If not, let him propound his new theory of consciousness.

Let it not be supposed that I am denying the existence, and the richly diversified existence, of the external world. We are fully justified, I think, in believing that, corresponding to the diversity of mental symbolism, there is a rich diversity of external existence. But its nature I hold that we can never know. The objects that we see are the joint products of two factors—the external existence and the percipient mind. We cannot eliminate the latter factor so as to see what the external factor is like without it. Those who, like Professor Mivart,* say that we can eliminate the percipient factor, and that the external world without it is just the same as it is with it, are content to reduce the human mind, in the matter of perception, to the level of a piece of looking-glass.

There are some people who seek to get behind phenomena by an appeal to evolution. It will not do nowadays, they say, to make the human mind a starting-point in these considerations; for the human mind is the product of evolution, and throughout that evolution has been step by step moulded to the external world. The external world has, therefore, the prior existence, and to it our perceptions have to conform. All this is quite true; but it is beside the point. Mind has, throughout the process of evolution, been moulded to the external world; our perceptions do conform to outside existences. But they conform, not in exact resemblance, but in mental symbolism. They do not copy, but they correspond to, external existences. It is just because, throughout the long ages of evolution, mind has lived and worked in this

* "Let the perception be considered to be made up of $x + y$; x being the ego, or self, and y the object. The mind has the power of supplying its own $-x$, and so we get (through the imagination of the mind and the object) $x + y - x$, or y pure and simple" (Mivart, "On Truth," p. 135). Mr. Mivart devotes a whole section of this work to the defence of ordinary common-sense realism. The above assertion seems to contain the essence of his teaching in the matter.

symbolic world that common sense is unable to shake off the conviction that this is the only possible world, and exists as such independently of mental processes. The world of phenomena *is* the world in which we, as conscious beings, live and move. No one denies it. But it is none the less a symbolic world; none the less a world which mind has constructed in the sense that it is an inalienable factor in its being.

Each of us, when we perceive an object, repeats and summarizes the constructive process which it has been the end of mental evolution to compass. Hence it is that, at the bidding of a simple impression, percepts or constructs take origin and shape in the mind. In taking possession of this faculty in the early years of life, we are entering upon a rich ancestral heritage. But if what I have been urging has truth, what we call objects are human constructs, and cannot by any manipulation be converted into anything else.

I will now take another and more complex case of construction, which will bring out some other facts about what I have termed "constructs." I hear in the street a piercing howl, which suggests a dog in pain. Rising from my seat and going to the window, I see a white terrier with a black patch over the left eye limping down the road on three legs. Now, what was the nature of the construct framed at the bidding of the piercing howl? A dog in pain. But what dog? The nature of the howl suggested a small dog; but there was nothing further to particularize him. The construct was, therefore, exceedingly vague and ill defined, and was not rendered definite and particular till I went to the window, and saw that it was a white terrier with a black patch over the eye. The howl, moreover, suggested certain activities of the dog. The construct was not merely a passive, inanimate object, like the orange, but an object capable of performing, and actually performing, certain actions. Here, again, we can only say that it is through experience that special activities are associated with certain objects. Just as the construct orange is capable of exciting sensations of taste, so the construct dog is capable of doing certain things and performing certain actions, that is, of affecting us in certain further ways.

But, further, the howl suggested a dog in pain. No amount of sensations entering into any manner of relations could give me that element of the construct. I can neither see, touch, taste, smell, nor hear pain in another being. Pain is entirely subjective and known only to the sufferer. But I have been a sufferer. I have experienced pain and pleasure. And just as my experiences, individual and ancestral, lead me to project into inanimate objects certain qualities, the products of my sensations, so do my experiences, individual and ancestral, lead me to project into certain animals feelings analogous to those I have myself experienced. This is sometimes described as an inference. But if we call this an inference, then we must, I think, call the taste, smell, and feel of the orange I see before me inferences. In both cases the inference, if we so call it, enters at once into the immediate construct.

And when I went to the window and saw the dog limping down the street, I saw also a small boy, with arm drawn back, in the act of throwing a stone. In other words, I saw the objects in the scene before me standing in certain relations to each other. I concluded that the boy had thrown a stone at the dog and was about to throw another. In other words, I saw the scene before me as part of a sequence of events.

One more example I will give to bring out another and important feature in the mental process. Strolling before breakfast in early spring in my friend's garden, there is borne to me on the morning air a whiff of violet fragrance. Not only does this lead me to construct violets, but it reminds me of a scene in my childhood with which the scent of these flowers was closely associated. Not only is the object constructed, but a scene with which their fragrant odour has been associated is *reconstructed* in memory. The violets are immediate constructs or presentations of sense; the remembered scene is a *reconstruct* or representation in memory. So, too, when I heard a piercing howl in the street, the dog I constructed was a vague presentation of sense; but the street in which I instinctively placed him was a reconstruct or representation in memory. The difference between a construct or presentation of sense, and a reconstruct or representation in memory, is that the former is directly suggested through the immediate action of some quality or activity of the object, while the latter is indirectly suggested through some intermediate agency.

Before proceeding further, let us review the conclusions we have thus far reached. Through the action of certain surroundings on our sensitive organization, we receive certain impressions, and among these impressions and others revived in memory we recognize certain similarities or differences in quality, in intensity, in order of sequence, and in source of origin. The sensations which thus originate are mental facts in no sense resembling their causes, but representing them in mental symbolism. The consciousness of similarity or difference is no part of the impression, but a further mental fact arising out of the impression, and with it giving origin to sensation. It deals with the relation of impressions among each other and to the recipient. It involves recognition and discrimination. Its basis is laid in memory. The sensations are instantly localized, referred to objects, and projected outwards, mainly through the instrumentality of the muscular sense. The mental symbolism is thus built into the objects around us, and constructs are formed. But into the tissue of these constructs are woven, not only the sensations immediately received, but much that is only suggested through association as the outcome of past experience, individual and ancestral. The constructs and their associated reconstructs are thus endowed with qualities which have practical reality, since they are not for me only, but for you and for mankind. They are, therefore, in a sense independent of *me*, but nowise independent of *man*.*

* If it be said that the object does exist independently of man, though not

Some of the constructs are endowed with activities, and some with feelings akin to our own. Finally, in the field of vision which we construct or reconstruct, the objects are seen to stand in relationship to each other, and the scene as a whole is perceived to be part of an orderly sequence of events.

We have already got a long way beyond the impressions with which we started; and yet, if I may trust my own experience, such construction as I have described is direct and immediate. A child of four or five would not only construct as much, but might not improbably go a long way further, and say, "Naughty boy to throw a stone at poor doggie!" It is, I say, direct and immediate, and it implies a wonderful amount of mental activity. Some people seem to imagine that in the simpler forms of perception, as when I see an orange on the table, the mind is as passive as the sensitive plate in a photographer's camera. This surely is not so. It is a false and shallow psychology which teaches it. Just as a light pin-prick may set agoing complex physical activities in the frog, so may comparatively simple visual sensations give rise to complex mental activities in construction and reconstruction. It is to emphasize this mental activity that I have persistently used the terms "construct" and "construction." And I wish to emphasize it still further by saying that without the active and constructive mind no such process of construction or reconstruction is possible or (I speak for myself) conceivable. We might just as well suppose that the frog could leap away on stimulation of a pin-prick in the absence of its complex bodily organization, as that sensation could give rise to construction and reconstruction in the absence of a highly organized mind.

We have seen that when a howl suggested the construct dog, that construct was vague and undefined; but when I

in the phenomenal guise under which we know it, I would reply—Not so; for it is to the existence *under this phenomenal guise* that we apply the word "object." In philosophical language, the existence, stripped of its phenomenal aspect, is called the *Ding an sich*. Its essential character is its independence of man; and hence its unknowability.

went to the window and saw the terrier, the construct became particularized and defined. This seems to me the normal order of development: first the vague, general, and indefinite; then the particular, special, and defined. That which is immediately suggested at the bidding of sensations received is always more or less general; it only becomes specialized on further examination physical or mental—first a dog or an orange; then this dog or this orange. The more unfamiliar the object, the more vague and indefinite the construct. The more familiar the object, and the further our examination of it is carried, the more particular and defined the construct. I would, therefore, mark two stages in the process of construction: first, the formation of constructs by immediate association, more or less vague, indefinite, and ill defined; and, secondly, the definition of constructs by examination, by which they are rendered more definite, particular, and special, and supplemented by intelligent inferences.

I need not stay here to point out the immense importance of this process of defining and particularizing constructs, or the length to which it may be carried; nor need I pause to indicate how, through memory and association, representative or reconstructive elements crowd in to link or weave the constructs into more or less vivid and brilliant scenes. But I have next to notice that out of this intelligent examination arises a new, distinct mental process, *the analysis of constructs.*

This process involves the paying of special attention to certain qualities of objects, to the intentional exclusion of other qualities. When I cease to examine an orange as a construct, and pay attention to its colour or its taste to the exclusion of other properties, with the purpose of comparing this colour or taste with other colours and tastes, I am making a step in analysis. So, too, when I consider the form of an orange for the purpose of comparing it with the form of the earth, I am making a step in analysis. And, again, when I consider the howl of the dog with the object of comparing it with other sounds, I am making a step in

Y

analysis. We may call the process by which we select a certain quality, and consider it by itself to the neglect of other qualities, *isolation*, and the products of the process we may term *isolates*.*

This process could not be initiated till a large body of constructive and reconstructive experience had been gained. But once initiated, there is no end to the process. We pick to pieces all the phenomena of nature, all the qualities and relationships of objects, the activities and functions of animals, the mental phenomena of which we are conscious in ourselves. We isolate the qualities, relationships, feelings; and we name the isolates we obtain. Hence arises all our science, all our higher thought. In the terms which we apply to our isolates consists the richness of our language.

We *name* the isolates; that is, we apply to each an arbitrary symbol to stand for the isolated quality or relation. All words (except the obviously onomatopoetic, such as "bow-wow," "cuckoo," etc.) are arbitrary symbols associated with objects, or qualities, or relations, or other phenomena. And abstract names of isolates are, so to speak, the pegs on which we hang the qualities we have separated by analysis and isolation, while class-names are pegs upon which we can hang a group of similars reached by the process of isolation; for all classing and grouping of objects, or qualities, or relations involves, so far as the process is a conscious one, the principle of analysis. In classing objects, we group them in reference to certain characters which they have in common, disregarding certain other characters in which they differ. We group together, for example, sights, or sounds, or smells, and distinguish them from each other and from tastes and touches. And then we go further, and class all these together as sensations having certain characteristics in common whereby they are distinguished from perceptions of relation and so forth.

* I avoid, for the present, the use of the terms "abstraction" and "abstract idea" because they are employed in different senses by different authors.

Perhaps it may be objected that classification comes much earlier in the mental process than I am now putting it. It may be said that the recognition of a sensation as a touch, or a smell, or a sound involves a classification of sensations in these categories, and that the simple perception of an orange involves the placing of the object in this class of bodies. And, undoubtedly, we have here the germs of the process. Sensation and perception give us the materials for classification; the perception of similarity and difference gives us the *sine quâ non* of the process. Nevertheless, although there may be an earlier unconscious grouping of phenomena, it is only when the mind is specially directed to these materials, with the object of grouping them according to their similarities, that we can speak of classification proper—conscious and intentional classification, as opposed to unconscious grouping. And this involves the intentional selection of the points of similarity, and discarding or neglecting the points of difference. It involves the process of analysis or isolation. There is a vast difference between the perceptual recognition of objects as similar, and conceptual classification on grounds of similarity. Just as the recognition of a sensation as now and not then, or here and not there, or as due to something outside us, gives us the germs from which, on ultimate analysis, our ideas of time, space, and causation are reached; so does the recognition of these sensations as of this kind and not that give us the germ from which, on analysis, the process of classification may arise. True, conscious, scientific classification is late in development.

And here let us notice that the conclusions we have reached in this chapter are the outcome of analysis and classification. The sensations with which we started are isolates. In considering their quality, intensity, sequence, we were isolating and classifying these special modes of their existence. Localization and outward projection involved isolation. We simply see the orange before us. To understand and explain how we come to see it as we do see it involves a somewhat subtle analysis. We perceive it to be yellow, round, resistant; and then, isolating these qualities, we reach conceptions of yellowness, roundness, and resistance, quite apart from oranges. Throughout our description the terms we used were very largely terms denoting classified isolates.

Lastly, having enormously increased our knowledge by this process of isolation, we proceed to build in the knowledge thus gained to the structure of our constructs. This is the third and last stage in construction. The first stage is the formation of indefinite constructs by immediate association; the second is the definition of constructs by examination; and the third is the completion of constructs by synthesis.

And the further this process of analysis and isolation is carried, the more we are, so to speak, floated off from the immediate objects of sense into the higher regions of abstract thought. Furthermore, by recombining our isolates in new modes and under new relations, we reach the splendid results of constructive imagination.

In the brief description which I have now given of our mental processes, I have for the most part avoided certain terms which are current in the science of psychology. It will be well here to say a few words concerning these words and their use. The process of *sensation* is sometimes defined as the mere reception of a sense-stimulus. But it is more convenient, and more in accordance with common usage, to call the simple result of a stimulus an impression, and to apply the term "sensation" to the discrimination and recognition of the impressions as of such and such a quality. Sensation, then, is the reception and discrimination of impressions which result from certain modes of influence (stimuli) brought to bear on our organization. Viewed in this way, therefore, even sensation involves a distinct reaction of the mind; it implies the first stage of mental activity. But when the sensations are given objective significance, when they suggest the existence of an object-world without us, they enter the field of *percep-*

tion. Here the discriminated sense-impression is, to use the words of Mr. Sully, "supplemented by an accompaniment or escort of revived sensations, the whole aggregate of actual and revived sensations being solidified or integrated into the form of a *percept;* that is, an apparently immediate apprehension or cognition of an object now present in a particular locality or region of space."* Throughout the whole process of the formation of constructs by immediate association, and their definition by examination, we were dealing with perception and percepts. But when we reach the stage when particular qualities were isolated, then we enter the field of *conception.* The isolates are *concepts.* Class-names, reached through processes involving isolation, stand for concepts. And completed constructions, involving synthesis of the results of analysis, contain conceptual elements. The word "concept," however, is used in different senses by different authors. Mr. Sully says,† for example, "A concept, otherwise called a general notion, or a general idea, is the representation in our minds answering to a general name, such as 'soldier,' 'man,' 'animal.' . . . Thus the concept 'soldier' is connected in my mind with the representations of various individual soldiers known to me. When I use the word 'soldier,' . . . what is in my mind is a kind of composite image formed by the fusion or coalescence of many images of single objects, in which individual differences are blurred, and only the common features stand out distinctly. . . . This may be called a typical or generic image." But Noiré, quoted by Professor Max Müller,‡ taking another illustration, says, "All trees hitherto seen by me leave in my imagination a mixed image, a kind of ideal presentation of a tree. Quite different from this is my concept, which is never an image." I follow Noiré; and I hold that the image, in so far as it is an image, whether simple or composite,§ is a percept;

---

* "Outlines of Psychology," p. 153.   † Ibid. p. 339.
‡ "Science of Thought," p. 453.
§ For compound or generic ideas "not consciously fixed and signed by

but that, in so far as there enter into the idea of the soldier or the tree elements which have been isolated by analysis, just in so far does the word "soldier" or "tree" stand for a concept. How far a word stands for a percept, and how far there enter conceptual elements, depends to a large extent on the level of intelligence of the hearer. The moment educated and intellectual folk begin to think

means of an abstract name," Mr. Romanes ("Mental Evolution in Man," p. 36) has suggested the term "recept." In the photographic psychology which he adopts, the percept is an individual and particular photograph, the recept a generalized or composite photograph. "The word 'recept,'" he says, "is seen to be appropriate to the class of ideas in question, because, in receiving such ideas, the mind is passive." This, it will be observed, is in opposition to the teaching of this chapter, in which the activity of the mind in perception has been insisted on. Mr. Romanes's recepts answer in part to what I have termed *constructs*, which, as we have seen, are, as a rule, from the first general rather than particular, and in part to concepts reached through analysis. Mr. Romanes, for example, speaks of ideas of principles (*e.g.* the principle of the screw) and ideas of qualities (*e.g.* good-for-eating and not-good-for-eating) as recepts (p. 60). On the other hand, Mr. Mivart ("The Origin of Human Reason," p. 59; see also his work "On Truth") terms such generic affections "sensuous universals." It may be well to append Mr. Romanes's and Mr. Mivart's tabular statements.

Mr. Romanes.

IDEAS	General, abstract, or notional	=	Concepts.
	Complex, compound, or mixed	=	Recepts, or generic ideas.
	Simple, particular, or concrete	=	Memories of percepts.

Mr. Mivart.

IDEAS	General or true universals	=	Concepts.
	Particular or individual	=	Percepts.
SENSITIVE COGNITIVE AFFECTIONS	Groups of actual experiences combined with sensuous reminiscences	=	Sensuous universals, or recepts.
	Groups of simply juxtaposed actual experiences	=	Sense-perceptions, or sencepts.

In Mr. Mivart's terminology, the representations of the lower group are "mental images" or "phantasmata." The term "consciousness" is by him restricted to the higher region of ideas, the term "consentience" being applied to the faculty by which cognitive affections are felt, unified, and grouped without consciousness. *There is a difference in kind*, according to Mr. Mivart, between "consentience" and "consciousness;" and the former could therefore never develop into the latter, nor the latter be evolved from the former. For this reason (because of the philosophy it is intended to carry with it) I shall not employ the word "consentience," which would otherwise be a useful term.

about their words, or the objects for which they stand, conceptual elements are sure to crowd in.

There is one more feature of these mental processes in man, and that by no means the least important, that remains for brief consideration. I began by saying that the primary end and object of the reception of the influences of the external world, or environment, is to enable the organism to answer to them in activity. We saw that the sight of an orange suggests, through association, its taste; and that the validity of the association could be verified by going to the orange and tasting it. We saw, too, that when I heard a dog howl in the street, and, going to the window, saw a small boy with a stone in his hand, I concluded that he was going to throw it at the dog. What I wish now to elicit is that out of perceptions through association there arise certain expectations, and that the activities of organisms are moulded in accordance with these expectations.

It is clear that these expectations or anticipations belong partly to the presentative or constructive order, and partly to the reconstructive or representative order. They are in some cases directly suggested by the presentations of sense; they are also built up out of representations which have become associated with the constructs in memory and through experience. But what we have here especially to notice about them is that, in the latter case, they involve more or less distinctly the element which we, in the language of our developed thought, call causation. There is a sequence of events, and the perception of certain of these gives rise, through association and experience, to an expectation of certain succeeding phenomena. Expectations are, therefore, the outcome of the linked nature of phenomena. And when we come eventually to think about the phenomena, and how they are linked together into a chain (successional) or web (coexistent), we reach the conception of causation as the connecting thread. In early stages of the mental process, such a conception does not emerge. Nevertheless, the phenomena are perceived as linked or woven. And the mental process by which we pass from any perceived event or existence to other preceding, concomitant, or subsequent events or existences linked or woven with it in the chain or web of phenomena, we call *inference*.* When, for example, I find a footprint in the sand, I infer that a man has passed that way; and when the clouds are heaped up heavy and black, I infer that a storm is about to burst upon us.

Concerning inference, of which I shall have more to say in the next chapter, I have now to note that it is of two kinds: first, perceptual inference, or inference from direct experience; secondly, conceptual inference, or inference based on experience, but reached through the exercise of the reasoning faculties. The latter involves the process of analysis or isolation; the former does not. There is a marked difference between the two. Perceptual inferences are the outcome of practical experience, but do not go beyond such practical experience. Conceptual inferences are also based on experience, but they predict occurrences never before experienced. Perceptual inferences, again, deal with matters practically; but conceptual thought explains them.

The expectation of a storm when the thunder-clouds are heavy is a case of perceptual inference. It is the outcome of a long-established association, and is not reached by a process of reasoning involving an analysis of the phenomena. But if, though the sky is clear, a west wind and a rapidly falling barometer lead me to predict rain, the inference is conceptual, and gained by me or for me by a process of reasoning; for the barometer was the outcome of the analysis of phenomena. In the mind of the rough sailor-lad, however, the fall of the mercury and the succeeding storm may be connected by mere perceptual

* We do not speak of the filling in the complement of a percept (the construction of the object at the bidding of a simple impression) as a matter of conscious inference. I do not consciously *infer* that yonder moss-rose is scented. Scent is an integral part of the construct. From the appearance of the rose, I may, however, infer that a rose-chafer has disturbed its petals. The complement of the percept, if inferred at all, is unconsciously inferred.

inference, the phenomena being simply associated together. If, however, there is any attempt at explanation, correct or incorrect, there is so far a conceptual element. In a little fishing-village on our south coast, a benevolent lady presented the fishermen with a Fitzroy barometer. I happened shortly after to remark to one of the men that the summer had been unusually stormy. "Yes, sir," he said, "it has. But then, you see, the weather hasn't no chance against that new glass." Here there was an attempted explanation of the phenomena. The falling glass was conceived as somehow causing bad weather.

It is hard to draw the line between perceptual and conceptual inferences, or rather to say, in this or that case, to which class the inference belongs, because man, through language, lives in a conceptual atmosphere. Moreover, the same result may, in different cases, be reached by perceptual or by conceptual inference. A child who had seen a great number of ascending balloons might, on seeing a balloon, expect it to ascend by a perceptual inference; but a man, knowing that the balloon was full of a gas lighter than air, might expect it to ascend through the exercise of conceptual inference. And just as in adult civilized life our constructs have more and more conceptual elements built into them, so do our inferences become more and more reasoned. It is probable that in an adult Englishman every inference has a larger or smaller dose of the conceptual element.

With the development of language we state our inferences in the form of propositions, and call them judgments. "Every proposition," says Mr. Sully,[*] "is made up of two principal parts: (1) the subject, or the name of that about which something is asserted; (2) the predicate, or the name of that which is asserted. Thus, when we affirm, 'This knife is blunt,' we affirm or predicate the fact of being blunt of a certain subject, namely, 'this knife.' Similarly, when we say, 'Air corrodes,' we assert or predicate the power of corroding of

[*] "Outlines of Psychology," p. 392.

the subject 'air.'" The proposition always involves conceptual elements; for the predicate of a proposition is always an abstract idea or general notion.

Propositions so formed may then become links in a chain of reasoning. "To reason is," says Mr. Sully,[*] "to pass from a certain judgment or certain judgments to a new one." And so passing on from judgment to judgment, we may ascend to the higher levels of abstract thought. According to Mr. Sully's definition, therefore, we start from a judgment or judgments in the process of reasoning. The formation of a judgment (conceptual inference) is, however, the first step in a continuous process; and I propose, under this term, "reason,"[†] to include this first step also. The formation of a conceptual inference I regard as the first stage of reason. Any mental process involving conceptual inference I shall call *rational*.

In contradistinction to this, I shall use the term "intelligence" for the processes by which perceptual inferences are reached. An intelligent act is an act performed as the outcome of merely perceptual inference. A rational act is the outcome of an inference which contains a conceptual element.

[*] "Outlines of Psychology." p. 414.
[†] Mr. Romanes adopts a different use of the terms "reason" and "rational," to which allusion will be made in the next chapter.

CHAPTER IX.

MENTAL PROCESSES IN ANIMALS: THEIR POWERS OF PERCEPTION AND INTELLIGENCE.

Two things I have been especially anxious to bring out prominently in the foregoing chapter: first, that the world we see around us is a joint product of two factors—the outward existence, on the one hand, and our active mind on the other; and secondly, that our mental processes and products fall under two categories—on the one hand, perception, giving rise to percepts, perceptual inferences, and intelligence, and on the other, conception (involving the analysis of phenomena), giving rise to concepts, conceptual inferences, and reason.

Now, I am anxious that the former—to take that first—should be laid hold of and really grasped as an indubitable fact. It is implied in the word "phenomena," that is to say, appearances. We can only know the world as it appears to us; and the world is for us what it appears. There is nothing here in conflict with common sense; the practical reality of phenomena is altered no whit. Suppose philosophy tries to get behind phenomena, so as to get a peep at the world beyond. Suppose Carlyle tells us that "All visible things are emblems; what thou seest is not there on its own account; strictly taken, is not there [as such] at all; matter exists only spiritually, and to represent some idea and *body* it forth." Has he altered the reality of the phenomena themselves? Not in the smallest degree. Suppose the materialist gives us his analysis of phenomena. Are not the phenomena he analyzes still the same, still equally real? No matter how far he analyzes phenomena, behind phenomena he cannot get. The materialist resolves all phenomena into matter in motion or into energy, and says that these are the only real existences. But they are no more real (they are a good deal less real to most of us) than the phenomena with which he started. How can the results of analysis be more real than that which is analyzed? Moreover, the matter and energy are still phenomena, and involve, as such, the percipient mind. Do what you will, you cannot get rid of the mental factor in phenomena.

It is possible that my use of the word "construct," my saying that the object is a thing which each of us constructs at the suggestion of certain sense-stimuli, may lead some to suppose that the process is in some sense an arbitrary one. This, however, would be a misconception. The process under normal conditions is just as inevitable as is, under normal conditions, the fall of a stone to the ground. The law of construction for human-folk is as much a law of nature as the law of gravitation. Both laws are condensed statements of the facts of the case. There is nothing arbitrary, lawless, or unnatural in the one or the other; the phrase merely emphasizes the essential presence of the mental factor.

If this principle be once thoroughly grasped, it will be seen how shallow and misleading is the view that the world is just reflected in consciousness unchanged as in a mirror, or faithfully photographed as on a sensitive plate. This is to reduce the human mind, which is surely no whit *less* complex than the human body, to the condition of a mere passive recipient instead of a vital and active agent in the construction of man's world.

The next point we have to consider is why we believe, as you and I practically do believe, that the world of phenomena exists as such, not merely for you and for me, but for man. Is it not because we believe in the practical unity of mankind? Is it not because we believe that, greatly as the conceptual and intellectual superstructure may differ in different individuals, the perceptual basis and foundation are practically identical? The senses and sense-organs give, in all normal individuals, sense-data, which differ only within comparatively narrow limits; and though the intellectual and moral world of the Bushman and the North Australian may differ profoundly from those of Shakespeare and Pascal, the perceptual world is, we have every reason to suppose, within these narrow limits, the same. This we may fairly believe; but even so there must be, nay, we know that there are, very great differences in the interpretation of the perceptual world. The individual cannot divest himself of the intellectual and conceptual part of his nature. We, for whom phenomena are more or less conditioned by science, find it difficult to think ourselves into the position of the savage, whose perceptual world is conditioned by crude superstition. The elements of his perceptual world are the same as ours, but the light of knowledge in which we view them is, for him, very dim. When we try to realize his world we find it exceedingly difficult.

And when we come to the lower animals—even those nearest us in the scale of life—the difficulties are enormously increased. The sense-data are probably much the same, but they are combined in different proportions. Olfactory sensation must, one would suppose, be built into the constructs of the dog and the deer to an extent which we cannot at all realize. And then, as Mr. P. G. Hamerton has well said, we have to take into account the immensity of the ignorance of animals. That ignorance, in combination with perfect perceptual clearness (ignorance and mental clearness are quite compatible) and with inconceivably strong instincts, produces a creature whose mental states we can never accurately understand.

I am tempted here to give the instance Mr. Hamerton quotes * in illustration of the ignorance of animals.

"The following account of the behaviour of a cow," he says, "gives a glimpse of the real nature of the animal. These long-tailed cows, say Messrs. Huc and Gabet, are so

---

* "Chapters on Animals," p. 9.

restive and difficult to milk, that to keep them at all quiet the herdsman has to give them a calf to lick meanwhile. But for this device, not a single drop of milk can be obtained from them. One day a Llama herdsman, who lived in the same house as ourselves, came with a long dismal face to announce that his cow had calved during the night, and that, unfortunately, the calf was dying. It died in the course of the day. The Llama forthwith skinned the poor beast and stuffed it with hay. This proceeding surprised us at first, for the Llama had by no means the air of a man likely to give himself the luxury of a cabinet of natural history. When the operation was completed, we found that the hay-calf had neither feet nor head; whereupon it occurred to us that, after all, it was perhaps a pillow that the Llama contemplated. We were in error, but the error was not dissipated till the next morning, when our herdsman went to milk his cow. Seeing him issue forth, the pail in one hand and the hay-calf under the other arm, the fancy occurred to us to follow him. His first proceeding was to put the hay-calf down before the cow. He then turned to milk the cow herself. The mamma at first opened enormous eyes at her beloved infant; by degrees she stooped her head towards it, then smelt at it, sneezed three or four times, and at last proceeded to lick it with the most delightful tenderness. This spectacle grated against our sensibilities; it seemed to us that he who first invented this parody upon one of the most touching incidents in nature must have been a man without a heart. A somewhat burlesque circumstance occurred one day to modify the indignation with which this treachery inspired us. By dint of caressing and licking her little calf, the tender parent one fine morning unripped it. The hay issued from within, and the cow, manifesting not the slightest surprise nor agitation, proceeded tranquilly to devour the unexpected provender."

Are we surprised at the want of surprise on the part of the cow? Why should we be? What knows she of anatomy or of physiology? If she could think at all about the matter, she would, no doubt, have expected her calf to be composed of condensed milk. But failing that, why not hay? She had presumably some little experience of *putting* hay inside. Why not *find* hay inside; and, finding hay, why not enjoy the good provender thus provided? But clearly we must not expect the brutes to possess knowledge to which they cannot attain about matters which in no wise concern their daily life.

"In our estimates of the characters of animals," continues Mr. Hamerton, in his comments on this anecdote, "we always commit one of two mistakes—either we conclude that the beasts have great knowledge because they are so clever, or else we fancy that they must be stupid because they are so ignorant." "The main difficulty in conceiving the mental states of animals," says the same observer, "is that the moment we think of them as *human*, we are lost." Yes, but the pity of it is that we cannot think of them in any other terms than those of human consciousness. The only world of constructs that we know is the world constructed by man.

"To Newton and to Newton's dog, Diamond," said Carlyle, "what a different pair of universes! while the painting in the optical retina of both was most likely the same." Different, indeed; if we can be permitted, without extravagance, to speak of the universe as existing at all for Diamond, or allowed, except in hyperbole, to set side by side a conception of ultimate generality, like the universe, the summation of all conceptions, and "the painting in the optical retina." Carlyle's meaning is, however, clear enough. Given two different minds and the same facts, how different are the products! In the construct formed on sight of the simplest object, we give far more than we receive; and what we give is a special resultant of inheritance and individual acquisition. No two of us give quite the same in amount or in quality. It is not too much to say that for no two human beings is the world we live in quite the same. And if this be so of human-folk, how different must be the world of man from the world of the dog—the world of Newton from the world of Diamond!

And we must remember that it is not merely that the same world is differently mirrored in different minds, but that they are two different worlds. If there is any truth in what I have urged in the last chapter, we *construct* the world that we see. The sensations are, as we have seen, mental facts, in no sense resembling their causes, but representing them in mental symbolism. Percepts are the elaborated products of this mental symbolism. The question, then, is not—How does the world mirror itself in the mind of the dog? but rather—How far does the symbolic world of the dog resemble the symbolic world of man? How far is his symbolism the same as ours? Only by fully grasping the fact that the external world of objects does not exist independently of us (though something exists which we thus symbolize), shall we realize the greatness of the difficulty which stands in the path of the student of animal psychology. So long as we are content to accept John Bunyan's crude analogy of the gateways of sense, the difficulty is comparatively small. There is the outside world self-existent and independent; a knowledge of it comes into the mind through the five gateways of sense—a picture of it through the eye-gate, and so on. The dog has also five similar gateways. The world for him is, therefore, much the same as for us. But this is not a true analogy. The world we see around us is a joint product of an external existence, the independent nature of which we can never know, and the human mind. It is something we construct in mental symbolism. How far does the dog construct a similar world? The answer to this question must, as it seems to me, be largely speculative.

And what help have we towards answering it? That afforded by the theory of organic evolution. If we accept that theory, and accept also the view that mental or psychical products are the inseparable concomitants of certain organic or physiological processes, then we have a basis from which to start. That basis I adopt.

Unfortunately, we have at present but little particular knowledge of the correlation of psychical and physiological processes. We cannot, by the dissection of a brain, draw much in the way of valid and detailed inference as to the nature of the psychical processes which accompany its physiological action. Fortunately, however, on the other hand, there are certain physical manifestations which do aid us, and that not a little, in drawing inferences from the physical to the mental. For organisms exhibit certain activities, and from these activities we can infer to some extent the character of the mental processes by which they are prompted. We are wont, in observing the actions of our fellow-men, to draw conclusions (often, alas! erroneous) as to the mental processes which accompany them. We are ourselves active, and we are immediately conscious of the modes of consciousness which accompany our actions. Thus the activities of organisms give us some clue to their mental processes, and it is through observation of their physical activities that we gain nearly all that is of particular value concerning the mental activities of animals. These activities we shall have to consider more fully in a future chapter. In the present chapter we shall consider them only so far as they give us information concerning the perceptual world (or worlds) of animals, and the nature of the inferences which we may suppose animals to draw from the phenomena which fall within their observation.

I think that, from the fundamental identity of life-stuff, or protoplasm, in all forms of animal life, and from the observed similarity of nerves and nerve-cells when nervous tissue has been developed, and again from the essential resemblance of life-processes in all animal organisms, we are justified in believing that mental or conscious processes, when they emerge, are essentially similar in kind. Exactly when they do emerge in the ascending branches of the great tree of animal life it is exceedingly difficult, if not quite impossible, to determine. And it is, I fancy, quite impossible for us so to divest ourselves of the complexity of human consciousness as to imagine what the simplicity of the emergent consciousness in very lowly organisms is like. But I think that we may fairly believe that some dim form of discrimination is the germ from which the spreading tree of mind shall develop.*

I assume, then, that, granting the theory of evolution, the early stages of the process of construction—discrimination, localization, and outward projection—are the same in kind throughout the whole range of animal life, wherever we are justified in surmising that psychical processes occur, and the power of registration and revival in memory has been established. As will be gathered, however, from what I have already said, I hold that the nature of the constructs produced is and must be for us human-folk, since we are human-folk, to a large extent a matter of speculation. Remembering this, then, endeavouring never to lose sight of it for a moment, let us consider what we may fairly surmise concerning the constructs and the process of construction in animals.

There can be no question that the animals nearest us in the scale of life—the higher mammalia—form constructs analogous to, if not closely resembling, ours. I do not think the resemblance can be in any sense close, seeing to how large an extent our constructs are literally our *handiwork*. For though in many animals the tongue and lips are delicate organs of touch—not to mention the trunk of the elephant—and though in the monkeys and many rodents the hands are used for grasping, still we have no reason to suppose that in any other mammal the geometrical sense of touch plays so determining a part in the formation of constructs as in man. On the other hand, in the dog and the deer, for example, not only must the marvellously acute sense of smell have a far higher sug-

---

* Or perhaps we may say, in the language of analogy, that when the germinal psychoplasm of some dim form of organic memory is fertilized by the union therewith of the more active male element of discrimination, a process of segmentation of the psychoplasm sets in by which, in process of differentiation, the tissues and organs of the mind are eventually developed.

gestive value, but smells and odours must, one would suppose, be built into the constructs in a far larger proportion. But although their constructs may not closely resemble ours, the constructs of animals may, I believe, be fairly regarded as closely analogous to our own. And as with us, so with them, a comparatively simple and meagre suggestion may give rise, through association in experience, to the construction of a complex object. And again, as with us, so with them, the suggested construct may be very vague and indefinite.

A dog, for example, is lying asleep upon the mat, and hears an unfamiliar step in the porch without. There can be no question that this suggests the construct man. But from the very nature of the case, this must be vague and indefinite. So, too, when a chamois, bounding across the snow-fields, stops suddenly when he scents the distant footprints of the mountaineer, the construct that he forms cannot be in any way particularized—no more particularized than is to me the sheep that I hear bleating in the meadow behind yonder wall.

And no one is likely to question the fact that animals habitually proceed from this first stage—the formation of constructs by immediate association—to the second stage of construction—the defining of constructs by examination. In many of the deer tribe, notably the prong-horn of America, this tendency is so strongly developed that they may be lured to their destruction by setting up a strange and unfamiliar object which, as we put it, may excite their curiosity. A strange noise or appearance will make a dog uneasy until he has by examination satisfied himself of the nature of that which produces it. Of this an instance fell under my observation a few days ago. My cat was asleep on a chair, and my little son was blowing a toy horn. The cat, without moving, mewed uneasily. I told my boy to continue blowing. The cat grew more uneasy, and at last got up, stretched herself, and turned towards the source of discomfort. She stood looking at my boy for a minute as he blew. Then curling herself up, she went to sleep again, and no amount of blowing disturbed her further. Similarly, Mr. Romanes's dog was cowed at the sound of apples being shot on to the floor of a loft above the stable; but when he was taken to the place, and saw what gave rise to the sound, he ceased to be disquieted by it. Every one must have seen animals defining their constructs by examination. A monkey will spend hours in the examination of an old bottle or a bit of looking-glass. At the Zoological Gardens connected with the National Museum at Washington, a monkey was observed with a female opossum on his knee. He had discovered the slit-like opening of the marsupial pouch, and took out first one and then another of the young, looked them over carefully, and replaced them without injury.*

There may possibly be some difference of opinion as to whether animals are able to infuse into their constructs of other animals the element of feeling. One would, perhaps, fain believe that the beasts of prey were wholly unaware of the pain they inflict on other organisms. But I question whether any close observer of animals could hold this view. Even if it were supposed that when two dogs fight they are blind to the pain they are inflicting on each other, their mock-fighting seems to imply a consciousness of the pain they might inflict, but avoid inflicting. And many of us have presumably had experiences analogous to the following: A favourite terrier of mine was once brought home to me so severely gashed in the abdominal region that I felt it necessary to sew up the wound. In his pain the poor dog turned round and seized my hand, but he checked himself before the teeth had closed upon me tightly, and piteously licked my hand. For myself, I cannot doubt that animals project into each other the shadows of the feelings of which they are themselves conscious.

The fact that dogs may be deceived by pictures † shows that they may be led through the sense of sight to form

* *Nature*, vol. xxxviii. p. 257.
† For examples, see Romanes's "Animal Intelligence," p. 455.

false constructs, that is to say, constructs which examination shows to be false. Through my friend and colleague, Mr. A. P. Chattock, I am able to give a case in point. I quote from a letter received by Mr. Chattock: "Your father asks me to tell you about our old spaniel Dash and the picture. I remember it well, though it must be somewhere about half a century ago. We had just unpacked and placed on the old square pianoforte, which then stood at the end of the dining-room, the well-known print of Landseer's 'A Distinguished Member of the Humane Society.' When Dash came into the room and caught sight of it, he rushed forward, and jumped on the chair which stood near, and then on the pianoforte in a moment, and then turned away with an expression, as it seemed to us, of supreme disgust."

I think we may say, then, that the higher animals are able to proceed a long way in the formation and definition of highly complex constructs analogous to, but probably differing somewhat from, those which we form ourselves. These constructs, moreover, through association with re-constructs or representations, link themselves in trains, so that a sensation or group of sensations may suggest a series of reconstructs or a series of remembered phenomena. We here approach the question of inferences, of which more anon. But in this connection passing reference may be made to the phenomena of dreaming. Dogs and some other animals undoubtedly seem to dream.

The nature of dreaming may, perhaps, be best illustrated by a rough analogy. Professor Clifford likened the human consciousness to a rope made up of a great number of occasionally interlacing strands. Let us picture such a rope floating in water. Much of it is submerged; only the upper part is visible at the surface. This upper part is like the series of mental phenomena of which we are distinctly conscious. Below this lie other series in the half-submerged state of subconsciousness. Deeper still lie unconscious physiological processes capable of emerging into the shadow of subconsciousness or the light of distinct consciousness. Now picture this rope gradually slipping round as it floats, so that now one part, now another, sees the light. This is analogous to the musing state, when we allow our thoughts to wander unchecked by any effort of attention. Attention is the faculty by which we steady the rope, so that one particular strand is kept continuously uppermost. The inattentive mind is one in which the rope keeps slipping round and refuses to be steadied in this manner; and in unquiet sleep, when the faculty of attention is dormant, the strands come quite irregularly and haphazard to the surface, and we have the phantasmagoria of dreams.

In the dog or the ape the rope is presumably incomparably simpler. But that it is of the nature of a rope we may, perhaps, not improbably surmise. Interest and the attention it commands steady the rope. Animals differ widely in their power of attention, as every one knows who has endeavoured to educate his pets. Darwin tells us that those who buy monkeys from the Zoological Gardens, to teach them to perform, will give a higher price if they are allowed a short time in which to select those in which the power of attention is most developed. And when animals dream, their consciousness-rope is slipping round unsteadily. That they do apparently dream is, so far, evidence of their possessing linked chains of memories.

In speaking of the faculty of attention in animals, it may be well to note that attention is of two kinds—perceptual or direct, and conceptual or indirect. In perceptual attention its motive is directly suggested by the object which stimulates this concentration of the faculties; a menacing dog, for example, stimulates my perceptual attention. In conceptual attention the motive is ulterior and indirect. The concentrated attention which a man devotes to the acquisition of Sanscrit does not arise directly out of the symbols over which he pores; it is of intellectual origin.

In the normal life of animals the attention is of the

perceptual order; it is a direct stimulation of the faculties through a perceptual presentation of sense or representation in memory which gives rise to an appetence or aversion. The importance of such a faculty is obvious. As M. Ribot well says, it is no less than a condition of life. The carnivorous animal that had not its attention roused on sight of prey would stand but a poor chance of survival; the prey that had not its attention roused by the approach of its natural enemy would stand but a poor chance of escape. The emperor moth that had not its attention roused by the scent of the virgin female would stand but a poor chance of propagating its species.

We are not, however, at present in a position further to discuss this matter. For there is a factor in the process which we shall have to consider more fully hereafter—the emotional factor. The hungry lion is in a very different position, so far as attention is concerned, from the satiated animal. The force and volume of the attention depends not merely, or even mainly, upon the intensity of the stimulus, but on the emotional state of the recipient organism.

Endeavour to divert the attention of any animal which is intent upon some action connected with the main business of its life—nutrition, self-defence, or the propagation of the species—the force of attention will at once be obvious.

In the training of animals (and young children) artificial associations, pleasurable or painful, have to be established in connection with certain actions. Abnormal appetences and aversions have to be introduced into the mental constitution. In this process much depends on the plasticity of the constitution. In the absence of such plasticity it is impossible to establish new associations.

We have seen that words are arbitrary* symbols, which we associate with objects, or qualities, or actions. Can animals, we may ask, form such arbitrary associations?

* I use the word "arbitrary" in the sense that they form no part of the normal construct such as would be formed by the animal.

There can be little question that they can. Many of the higher animals understand perfectly some of *our* words. The word "cat" or "rats" will suggest a construct to the dog on which he may take very vigorous action. How far they are able to communicate with each other is a somewhat doubtful matter. But the signs by which such communication is effected are probably far less arbitrary. And, in any case, the communication would seem to refer only to the here and the now. A dog may be able to suggest to his companion the fact that he has descried a worriable cat; but can a dog tell his neighbour of the delightful worry he enjoyed the day before yesterday?

I imagine that what a dog can suggest to his neighbour is what we symbolize by the simple expression "Come." But I am fully aware that other observers will interpret the facts in a different way. Here is an anecdote that is communicated to me by Mr. Robert Hall Warren, of Bristol. "My grandfather," he says, "a merchant of this city, or, as Thomas Poole, of Stowey, would have preferred calling him, 'a tradesman,' had two dogs, one a small one and another larger, who, being fierce, rejoiced in the appropriate name of Boxer. On one of his business journeys into Cornwall he took the smaller dog with him, and for some reason left it at an inn in Devonshire, promising to call for him on his return from Cornwall. When he did so, the landlord apologized for the absence of the dog, and said that, some time after my grandfather left, the little dog fought with the landlord's dog, and came off much the worse for the fight. He then disappeared, and some time afterwards returned with another and larger dog, who set upon his enemy, and, I think, killed him. Then the two dogs walked off, and were no more seen. From the description given, my grandfather had no doubt that the larger dog was Boxer, and, on returning home, found that the little dog had come back, and that both dogs had gone away, and, after a time, had returned home, where he found them." Now, some will say that the little dog told Boxer all about it; but I am inclined to believe that the facts may be explained by the communication "Come."

Dogs can also communicate their wishes to us. The action of begging in dogs is a mode of communication with us. Mr. Romanes tells of a dog that was found opposite a rabbit-hutch begging for rabbits. When I was at the Diocesan College near Capetown, a retriever, Scamp, used to come in and sit with the lecturers at supper. He despised bread, but used to get an occasional bone, which he was not, however, allowed to eat in the hall. He took it to the door, and stood there till it was opened for him. On one occasion he heard without the excited barking of the other dogs. He trotted round the hall, picked up a piece of bread which one of the boys had dropped, and stood with it in his mouth at the door. When it was opened, he dropped the bread, and raced off into the darkness to join in the fun. In a similar way, but with less marked intelligence, I have seen a dog begging before a door which he wished opened. My cat has been taught to touch the handle of the door with his paw when he wishes to leave the room. Mr. Arthur Lee, of Bristol, tells me that a favourite cat has a habit of knocking for admittance by raising the door-mat and letting it fall. This is an action similar to those communicated by several observers to *Nature*, where cats have learnt either to knock for admittance or to ring the bell—an action which, as my friend, Mr. J. Clifton Ward, informed me, was also performed by a dog of his. I think, therefore, that it is unquestionable that the higher animals are able to associate arbitrary signs with certain objects and actions, and to build these signs into the constructs that they form. Sir John Lubbock has tried some experiments with his intelligent black poodle Van, with the object of ascertaining how far the dog could be taught to communicate his wishes by means of printed cards. "I took," he says,[*] "two pieces of cardboard, about ten inches by three, and on one

---

[*] "The Senses of Animals," p. 277.

of them printed in large letters the word 'FOOD,' leaving the other blank. I then placed the two cards over two saucers, and in the one under the 'Food' card put a little bread-and-milk, which Van, after having his attention called to the card, was allowed to eat. This was repeated over and over again till he had had enough. In about ten days he began to distinguish between the two cards. I then put them on the floor, and made him bring them to me, which he did readily enough. When he brought the plain card, I simply threw it back; while, when he brought the 'Food' card, I gave him a piece of bread, and in about a month he had pretty well learned to realize the difference. I then had some other cards printed with the words 'Out,' 'Tea,' 'Bone,' 'Water,' and a certain number also with words to which I did not intend him to attach any significance, such as 'Nought,' 'Plain,' 'Ball,' etc. Van soon learned that bringing a card was a request, and soon learned to distinguish between the plain and printed cards; it took him longer to realize the difference between words, but he gradually got to recognize several, such as 'Food,' 'Out,' 'Bone,' 'Tea,' etc. If he was asked whether he would like to go out for a walk, he would joyfully fish up the 'Out' card, choosing it from several others, and bring it to me or run with it in evident triumph to the door.

"A definite numerical statement always seems to me clearer and more satisfactory than a mere general assertion. I will, therefore, give the actual particulars of certain days. Twelve cards were put on the floor, one marked 'Food' and one 'Tea.' The others had more or less similar words. I may again add that every time a card was brought, another similarly marked was put in its place. Van was not pressed to bring cards, but simply left to do as he pleased.*

* As I understand the observations here tabulated, the twelve cards lay always within Van's reach and sight. An ordinary untrained dog would have taken no notice of them. But Van, when he wanted food or tea, went and fetched the appropriate card, and got what he wanted in exchange. In twelve days he only made two mistakes, bringing "Nought" once and "Door" once.

"Day 1. Van brought 'Food' 4 times, 'Tea' 2 times.
"   2.   "   "   6   "
"   3.   "   "   8   "   " 2 "
"   4.   "   "   7   "   " 3 "
"   5.   "   "   6   "   " 4 "
"   6.   "   "   6   "   " 3 "   'Nought' once.
"   7.   "   "   8   "   " 2 "
"   8.   "   "   5   "   " 3 "
"   9.   "   "   4   "   " 2 "
" 10.   "   "   10   "   " 4 "   'Door' once.
" 11.   "   "   10   "   " 3 "
" 12.   "   "   6   "   " 3 "
                                 ———   ———
                                  80   31

"Thus, out of 113 times, he brought 'Food' 80 times, 'Tea' 31 times, and [one out of] the other 10 cards only twice. Moreover, the last time he was wrong he brought a card—namely, 'Door'—in which three letters out of four were the same as in 'Food.'"

These experiments and observations are of great interest. But, of course, no stress whatever must be laid on the fact that *words* chanced to be printed on the cards instead of any other arrangements of lines. I draw attention to this because I have heard Sir John Lubbock's interesting experiments quoted, in conversation, as evidence that the dog understands the meaning of words, not only spoken, but written! What they show is that Van is able, under human guidance, to associate certain arbitrary symbols with certain objects of appetence; and, desiring the object, will bring its symbol. It would have been better, I think, because less misleading to the general public, had Sir John Lubbock selected other arbitrary symbols than the printed words we employ. Then no one could have run away with the foolish notion that the dog *understands* the meaning of these words. No doubt if they had been written in Greek or Hebrew, some people would have been interested, but not surprised, to learn that a dog can be taught to understand with perfect ease these languages!

The next question is—Have the higher animals the power of analyzing their constructs and forming isolates or abstract ideas of qualities apart from the constructs of which these qualities are elements? Can we say, with Mr. Romanes,* "All the higher animals have general ideas of 'good-for-eating' and 'not-good-for-eating,' *quite apart from any particular objects of which either of these qualities happens to be characteristic*"? Or with Leroy,† that a fox "will see snares when there are none; his imagination, distorted by fear, will produce deceptive shapes, to which he will attach *an abstract notion of danger*"?

Now, this is a most difficult question to answer. But it seems to me that, if we take the term "abstract idea" in the sense in which I have used the word "isolate," we must answer it firmly, but not dogmatically (this is the last subject in the world on which to dogmatize), in the negative. Fully admitting, nay, contending, that this is a matter in which it is exceedingly difficult to obtain anything like satisfactory evidence, I fail to see that we have any grounds for the assertion that the higher animals have abstract ideas of "good-for-eating" or "not-good-for-eating," quite apart from any particular objects of which either of these qualities happens to be characteristic. ‡

The particular example is well chosen, since the idea of food is a dominant one in the mind of the brute. There can be no question that the quality of eatability is built in by the dog into a great number of his constructs. But I question whether this quality can be isolated by the dog, and can exist in his mind divorced from the eatables which suggest it. If it can, then the dog is capable of forming a concept as I have defined the term. I can quite understand

---

* "Mental Evolution in Man," p. 27.
† "Intelligence of Animals," p. 121.
‡ Mr. Romanes also says ("Mental Evolution in Animals," p. 235), "This abstract idea of ownership is well developed in many if not in most dogs." By an abstract idea of ownership I understand a conception of ownership which, to modify Mr. Romanes's phrase, is quite apart from any objects or persons of which such ownership happens to be characteristic. Even if we believe that a dog can regard this or that man as his owner, or this or that object as his master's property, still even this seems to me a very different thing from his possessing an abstract idea of ownership.

that a hungry dog, prowling around for food, has, suggested by his hunger, vague representations in memory of things good to eat, in which the element of eatability is predominant and comparatively distinct, while the rest is vague and indistinct. And that this is a concept in Mr. Sully's use of the term, I admit. But it appears to me that there is a very great difference between a perceptual construct with eatability predominant and the rest vague, and a conceptual isolate or abstract idea of eatability quite apart from any object or objects of which this quality is characteristic. And to mark the difference, I venture to call the prominent quality a *predominant* as opposed to the *isolate* when the quality is floated off from the object. *No doubt it is out of this perceptual prominence of one characteristic and vagueness of its accompaniments that conceptual isolation of this one characteristic has grown, as I believe, through the naming of predominants.* But I should draw the line between the one and the other somewhere distinctly above the level of intelligence that is attained by any dumb animal. I am not prepared either to affirm or deny that this line should be drawn exactly between brute intelligence and human intelligence and reason, though I strongly incline to the view that it should. I am not sure that every savage and yokel is capable of isolation, that he raises the predominant to the level of the isolate, or abstract idea. I am not sure that these simple folk submit the phenomena of nature around them, and of their own mental states to analysis. But they have in language the instrument which can enable them to do so, even if individually some of them have not the faculty for using language for this purpose. That is, however, a different question. But I do not at present see satisfactory evidence of the fact that animals form isolates, and I think that the probability is that they are unable to do so. I am, therefore, prepared to say, with John Locke, that this abstraction "is an excellency which the faculties of brutes do by no means attain to."

I am anxious, however, not to exaggerate my divergence,

more apparent, I believe, than real, from so able a student of animal psychology as Mr. Romanes. Let me, therefore, repeat that it is the power of analysis—the power of isolating qualities of objects, the power of forming " abstract ideas quite apart from the particular objects of which the particular qualities happen to be characteristic," as I understand these words—that I am unable to attribute to the brute. Animals can and do, I think, form predominants; they have not the power of isolation.

Furthermore, it seems to me that this capacity of analysis, isolation, and abstraction constitutes in the possessor a new mental departure, which we may describe as constituting, not merely a specific, but a generic difference from lower mental activities. I am not prepared, however, to say that there is a difference in kind between the mind of man and the mind of the dog. This would imply a difference in origin or a difference in the essential nature of its being. There is a great and marked difference in kind between the material processes which we call physiological and the mental processes we call psychical. They belong to wholly different orders of being. I see no reason for believing that mental processes in man differ thus in kind from mental processes in animals. But I do think that we have, in the introduction of the analytic faculty, so definite and marked a new departure that we should emphasize it by saying that the faculty of perception, in its various specific grades, differs generically from the faculty of conception. And believing, as I do, that conception is beyond the power of my favourite and clever dog, I am forced to believe that his mind differs generically from my own.

Passing now to the other vertebrates, the probabilities are that their perceptual processes are essentially similar to those of the higher animals; but, in so far as these creatures differ more and more widely from ourselves, we may, perhaps, fairly infer that their constructs are more and more different from ours. Still, the thrush that listens attentively on the lawn and hops around a particular spot must have a vague construct of the worm he hopes to have a more particular acquaintance with ere long. The cobra that I watched on the basal slopes of Table Mountain, and that raised his head and expanded his hood when I pitched a pebble on to the granite slope over which he was gliding, must have had a vague percept suggested thereby. The trout that leaps at your fly so soon as it touches the water must have a vague percept of an eatable insect which suggests his action. The carp* that come to the sound of a bell must have, suggested by that sound, vague percepts of edible crumbs. And no one who has watched as a lad the fish swimming curiously round his bait can doubt that they are by examination defining their percepts, and drawing unsatisfactory inferences of a perceptual nature.

And here let us notice that the whole set of phenomena which have been described in previous chapters under the heads of recognition-marks, of warning coloration, and of mimicry, involve close and accurate powers of perception. Recognition-marks are developed for the special purpose of enabling the organisms concerned rapidly and accurately to form particular perceptual constructs. Of what use would warning coloration be if it did not serve to suggest to the percipient the disagreeable qualities with which it is associated? The very essence of the principle of mimicry is that misleading associations are suggested. Here a false construct, untrue to fact, that is to say, one that verification would prove to be false, is formed; just as a well-executed imitation orange, in china or in soap, may lead a child to form a false construct, one that is proved to be incorrect so soon as the suggestions of sight are submitted to verification by touch, smell, and taste.

No one who has carefully watched the habits of birds can have failed to notice how they submit a doubtful object to examination. Probably the avoidance of insects protected by warning colours is not perfectly instinctive. I

---

* Doubt has recently been thrown on this fact. Mr. Bateson has shown that some fishes do not hear well, and has suggested that the carp may be attracted by seeing people come to the edge of the pond.

have seen young birds, after some apparent hesitation, peck once or twice doubtfully at such insects. A young baboon with whom I experimented at the Cape seemed to have an undefined aversion to certain caterpillars, which he could not be induced to taste, though he smelt at them. Scorpions he darted at, twisted off the sting, and ate with greedy relish.

If nudibranchs and other marine invertebrates be protectively coloured, there must be corresponding perceptual powers in the fishes that are thus led to avoid them; for there seems to be definite avoidance, and not merely indifference. This, however, might be made the subject of further experiment, not only with fishes, but with other animals. I tried some chickens with currant-moth caterpillars, to each of which I tied with thread a large looper. Some of them would have nothing to do with the unwonted combination. But one persistently pecked at the looper, and tried to detach it from its fellow-prisoner. Though, on the whole, there was some tendency for aversion to the currant-moth caterpillar to overmaster the appetence for the looper, I was not altogether satisfied with the result of the experiment. But I think that if the protectively coloured larva had been regarded with mere indifference (*i.e.* neither aversion nor appetence), the appetence for the loopers should have made the chickens seize them at once.

To return to fishes. It is probably difficult or impossible for us to imagine what their constructs are like; but that they, too, proceed to define them by examination seems to be a legitimate inference from some of their actions. Mr. Bateson says, " The rockling searches [for food] by setting its filamentous pelvic fins at right angles to the body, and then swimming about, feeling with them. If the fins touch a piece of fish or other soft body, the rockling turns its head round and snaps it up with great quickness. It will even turn round and examine uneatable substances, as glass, etc., which come in contact with its fins, and which presumably seem to it to require explanation."* And, speaking of the sole, the same observer says,† "In searching for food the sole creeps about on the bottom by means of the fringe of fin-rays with which its body is edged, and, thus slowly moving, it raises its head upwards and sideways, and gently pats the ground at intervals, feeling the objects in its path with the peculiar viliform papillæ which cover the lower (left) side of its head and face. In this way it will examine the whole surface of the floor of the tank, stopping and going back to investigate pieces of stick, string, or other objects which it feels below its cheek."

If we admit the fact that carp come to be fed at the sound of a bell, we have evidence that some fishes can associate an arbitrary sound with the advent of things good to eat. But it is, perhaps, better at present to regard the fact as one requiring verification.

That some birds can associate arbitrary signs with their percepts will be admitted by all who have watched their habits. And from its peculiar and almost unique power of articulation, the parrot shows us that not only may the words suggest a construct, but that the sight of the construct may suggest the word that it has heard associated with the object by man. Mr. Romanes gives evidence which satisfies him that a parrot which had associated the word "bow-wow" with a particular dog, uttered this sound when another dog entered the room. The word was here suggested at sight, not of the same object, but of an object which the bird recognized as similar. A somewhat similar case is furnished by one of my own correspondents (Miss Mabel Westlake). "We left London," she says, "in December, 1888, and brought our grey parrot with us; but left behind with a friend our favourite cat, a dark tortoise-shell with a white breast, the forehead clearly marked with a division down the middle to the tip of the nose. This led to our calling her 'Demi.' For a week or two after

---

* Journal of Marine Biological Association, New Series, vol. i. No. 2, p. 214. I should not myself have used the word "explanation."

† Ibid. vol. i. No. 3, p. 240.

our arrival in Bristol, a black-and-white cat belonging to the people formerly living here frequented the house. The parrot seemed delighted to see this cat, which was larger than our old cat, and called it Dem, as she had been accustomed to do in London. From that time until the commencement of January (1890), which was over a year, the parrot had not seen a cat that we are aware of, nor had we heard her call it for a long time. About six weeks ago, as I was coming along Kingsdown Parade, a large black kitten followed me home. We took it in and fed it. The next day it came into the room where the parrot was, and she immediately said 'Puss! puss! puss! Hullo, dear!' and during the day called it by the same name, 'Dem! Dem! Dem!' that she had called our cat in London."

We may here notice that, in most of the tricks which animals are taught to perform, the action is suggested by a form of words (or the tone and manner in which they are uttered). Mr. John G. Naish, J.P., of Ilfracombe,* has taught his cockatoo the following trick (I quote Mr. Naish's own words): "I give him a shilling, which he puts into the slit of a money-box. This is 'enlisting.' After that, I say to him, 'Will you die for the queen, like a loyal soldier?' Then he lies on his back, with his paws together, for as long as I hold up my finger. 'Now live for your master!' He takes hold of my finger and resumes his erect posture. Last year I took him into the street near my house, and collected on our 'Hospital Saturday.' He worked for more than an hour before he became impatient. And then he would do no more, but flung the coins over his head or at the giver in the funniest way. He went to sleep for a long time after that performance; and when he awoke and I took him, he covered my face with kisses, as if he was glad to find his bad dream was over." The weariness and failure to perform the trick when tired, and the long sleep which succeeded, are interesting points.

* I have to thank this gentleman for a most interesting account of the intelligence of his favourite bird.

What I wish especially to notice is, however, that the actions are suggested by certain forms of words; but that there is no evidence that the form of words is in any sense understood. When the onlooker sees a bird lie on its back when asked if it will die for the queen, and get up again when told to live for its master, he is apt to think that, since *he* understands the form of words, the bird must understand them too. But I am convinced that Mr. Naisb's intelligent cockatoo could have been taught with equal ease to lie down at the command "Abracadabra," and to stand up again at "Hocus pocus." Tricks taught to animals involve the performing animal and the human onlooker. The form of words introduced is *for the sake of the latter*, not for the sake of the former.

So much has been written concerning the intelligence of the parrot, and so much has been said concerning its imitative power of speech, that I must say somewhat on this head. I have received from Miss Mildred Sturge, of Clifton, an interesting account of an African West Coast parrot which was possessed by Miss Tregelles, of Falmouth. This parrot used the phrases it had learnt appropriately in time and place. "At dinner, when he saw the vegetable-dishes, he generally said, 'Polly wants potato;' at tea he would say, 'Polly wants cake,' or 'Polly's sop,' or 'Polly's toast.' Our grandmother's house was not far from the station, and almost before people could hear it, Polly would announce, 'Grandmamma, the train is coming,' and presently the train would quietly go by. Besides repeating much poetry, Polly made new editions by putting lines together from different authors; but the remarkable thing was that he always got the right rhyme. One of his favourite mixtures was, 'Sing a song of sixpence' and 'I love little pussy.' One day my mother overheard—

> "'Four and twenty blackbirds,
> When they die,
> Go to that world above,
> Baked in a pie.'"

Now, we must not underrate nor overrate the evidence

afforded by parrot-talk. The rhyme-association is interesting; but since we cannot suppose that the poetry is more to the parrot than a linked series of sounds, there does not seem much evidence of intelligence here, though the evidence of memory is important. The correct association of words and phrases with appropriate objects and actions is of great interest. But the fact that they are words and phrases does not give them a higher value than that of imitative actions in the dog or other animal. What parrot-talk does give us evidence of is (1) remarkable powers of memory; (2) an almost unique power of articulation; (3) a great faculty of imitation; (4) and some intelligence in the association of certain linked sounds which we call phrases with certain objects or actions. The teaching of phrases to the parrot is certainly not more remarkable than the teaching of clever tricks to many birds. But the fact that word-sounds are articulated throws a glamour over these special tricks, and leads some people to speak of the parrot's using language, instead of saying that the parrot can imitate some of the sounds made by man, and can associate these sounds with certain objects.

Coming now to the invertebrates, much has been written concerning the psychology and intelligence of ants and bees. What shall we say concerning their constructs? For reasons already given, I think we may suppose that they are analogous to ours; but it can scarcely be that they in any way closely resemble ours. Their sense-organs are constructed on a different plan from ours; they have probably senses of which we are wholly ignorant. Is it conceivable, by any one who has grasped the principle of construction, that with these differently organized senses and these other senses than ours, the world they construct can much resemble the world we construct? Remember how largely our perceptual world is the product of our geometrical senses—of our delicate and accurate sense of touch, and of our binocular vision, with its delicate and accurate muscular adjustments. Remember how largely these muscular adjustments enter into our perceptual world as constructed in vision. And then remember, on the other hand, that the bee is encased in a hard skin (the chitinous exoskeleton), and that its tactile sensations are mainly excited by means of touch-hairs seated thereon. Remember its compound eye with mosaic vision, coarser by far than our retinal vision, and its ocelli of problematical value, and the complete absence of muscular adjustment in either the one or the other. Can we conceive that, with organs so different, anything like a similar perceptual world can be elaborated in the insect mind? I for one cannot. Admitting, therefore, that their perceptions may be fairly surmised to be analogous, that their world is the result of construction, I do not see how we can for one moment suppose that the perceptual world they construct can in any accurate sense be said to resemble ours. For all that, the processes of discrimination, localization, outward projection; the formation of vague constructs, their definition through experience, and the association of reconstructs or representations;—all these processes are presumably similar in kind to those of which we have evidence in ourselves.

In considering such organisms as ants and bees, however, we must be careful to avoid the error of supposing that, because they happen to have no backbones, they are necessarily low in the scale of life and intelligence. The tree of life has many branches, and, according to the theory of evolution, these divergent branches have been growing up side by side. There is no reason whatever why the bee and the ant, in their branch of life, should not have attained as high a development of structure and intelligence as the elephant or the dog in their branch of life. I do not say that they have. As it is difficult to compare their structure, in complexity and efficiency, with that of vertebrates, so is it difficult to compare their intelligence. The mere matter of size may have necessitated the condensation of intelligence into instinct in a far higher degree than was required in the big-brained mammals. Still, their intelligence, though of a different order and on a different plane, may well be as high. And Darwin has said that the so-called brain of the ant may perhaps be regarded as the most wonderful piece of matter in the world.

That ants have some power of communication seems to be proved by the interesting experiments of Sir John Lubbock. He found that they could carry information to the nest of the presence of larvæ, and that the greater the number of larvæ to be fetched, the greater the number of ants brought out to fetch them in a given time. On one occasion Sir John Lubbock put an ant to some larvæ. "She examined them carefully, and went home without taking one. At this time no other ants were out of the nest. In less than a minute she came out again with eight friends, and the little group made straight for the heap of larvæ. When they had gone two-thirds of the way, I imprisoned the marked ant; the others hesitated a few minutes, and then, with curious quickness, returned home." This is only one observation out of many; and it shows (1) that since the marked ant took no larva home, she must have given information which led the others to come out—unless we can suppose that the smell of the larvæ she had examined still hung about her; and (2) that the communication was not detailed, and probably was no more than "Come," for, when the leader of the party was removed, the rest knew not * where to go—very possibly knew not why they had been summoned.

Passing now to creatures of lower organization, it is exceedingly difficult so to divest ourselves of our own special mental garments as to imagine what their simple and rudimentary constructs are like. Perhaps we may fairly surmise that, as visual olfactory and auditory organs develop, and differentiate from a common basis of more simple sensation, the process of outward projection has its rudimentary inception. The earthworm, which finds its way

* Professor Max Müller suggests to me that perhaps the ants were frightened.

to favourite food-stuffs buried in the earth in which it lives, would seem to possess the power of outward projection in a dim and possibly not very definite form. Through their marginal bodies—simple auditory or visual organs—the medusæ may have a rudimentary form of this capacity. In any case, they seem to have the power of localization. Mr. Romanes says,[*] "A medusa being an umbrella-shaped animal, in which the whole of the surface of the handle and the whole of the concave surface of the umbrella is sensitive to all kinds of stimulation, if any point in the last-named surface is gently touched with a camel-hair brush or other soft (or hard) object, the handle or manubrium is (in the case of many species) immediately moved over to that point, in order to examine or brush away the foreign body." And the same author thus describes [†] the process of discrimination in the sea-anemone: "I have observed that if a sea-anemone is placed in an aquarium tank, and allowed to fasten upon one side of the tank near the surface of the water, and if a jet of sea-water is made to play continuously and forcibly upon the anemone from above, the result, of course, is that the animal becomes surrounded by a turmoil of water and air-bubbles. Yet, after a short time, it becomes so accustomed to this turmoil that it will expand its tentacles in search of food, just as it does when placed in calm water. If now one of the expanded tentacles is gently touched with a solid body, all the others close around that body in just the same way as they would were they expanded in calm water. That is to say, the tentacles are able to discriminate between the stimulus which is supplied by the turmoil of the water, and that which is supplied by their contact with the solid body, and they respond to the latter stimulus notwithstanding that it is of incomparably less intensity than the former."

Here, in discrimination, we reach the lowest stage of mental activity. It is exceedingly difficult, however, to determine how far such simple responses to stimuli are

[*] "Mental Evolution in Animals," p. 82.   [†] Ibid. p. 48.

merely organic, and how far there enters a psychological element.

I ought not, perhaps, to pass over in perfect silence the subject of protozoan psychology. M. Binet has published a little book on "The Psychic Life of Micro-Organisms," in the preface of which he says, "We could, if it were necessary, take every single one of the psychical faculties which M. Romanes reserves for animals more or less advanced on the zoological scale, and show that the *greater part* of these faculties belonged equally to micro-organisms." He says that "there is not a single infusory that cannot be frightened, and that does not manifest its fear by a rapid flight through the liquid of the preparation," and he speaks of infusoria fleeing "in all directions like a flock of frightened sheep." He attributes memory to *Folliculina*, and instinct "of great precision" to *Difflugia*. He regards some of these animalculæ as "endowed with memory and volition," and he describes the following stages:—

"1. The perception of the external object.

"2. The choice made between a number of objects.

"3. The perception of their position in space.

"4. Movements calculated either to approach the body and seize it or to flee from it."

But when we have got thus far, we are brought up by the following sentence: "We are not in a position to determine whether these various acts are accompanied by consciousness, or whether they follow as simple physiological processes." Since, therefore, the fear, memory, instinct, perception, and choice, spoken of by M. Binet, may be merely physiological processes (though, of course, they *may be* accompanied by some dim unimaginable form of consciousness), it seems scarcely necessary to say more about them here.

I have now said all that is necessary, and all that I think justified by the modest scope of this work, concerning the process of construction in animals, and the nature of the constructs we may presume that they form. The process I hold to be similar in kind throughout the animal kingdom wherever we may presume that it occurs at all. But the products of the process seem to me to be presumably widely different. If we steadily bear in mind the fact that the world of man is a joint product of an external existence and the human mind, and then ask whether it is conceivable that the joint products of this external existence and the dog-mind, the bird-mind, the fish-mind, the bee-mind, or the worm-mind are exactly or even closely similar, we must, it seems to me, answer the question with an emphatic negative.

We will now consider the nature of the inferences of animals. It will be remembered that a distinction was drawn between perceptual inferences and inferences involving a conceptual element. As I use the words, perceptual inferences are a matter, at most, of intelligence; but conceptual inferences involve the higher faculty of reason.

It will be necessary here to say somewhat more than I have already said concerning inference. When I see an orange, that object is mentally constructed at the bidding of certain sight-sensations. All that is actually received is the stimulus of the retinal elements; the rest is suggested and supplied by the activity of the mind. It is sometimes said that this complementary part of the perception is inferred. So, too, when I hear a howl in the street which suggests the construct dog, it may be said that I infer the presence of the dog. And again, when the dog is perceived to be in pain, it may be said that this is an inference. Now, although the use of the word "inference" to denote the complementary part of a percept seems a little contrary to ordinary usage, still there are some advantages in so—with due qualification—employing it. But since, as it seems to me, the characteristic of the inference, if so we style it, in the formation of constructs by immediate association is its unconscious nature (*i.e.* unconscious as a process) we may perhaps best meet the case by speaking of these as unconscious inferences. When the inference is not immediate and unconscious, but involves a more individual conscious act of the mind in the perceptual sphere, we may speak of it as intelligent; and when the inference can only be reached by analysis and the use of concepts, we may call it rational.

Defining, therefore, "inference" as the passing of the mind from something immediately given to something not given but suggested through association and experience, we have thus three stages of inference: (1) unconscious inference on immediate construction (perceptual); (2) intelligent inference, dealing with constructs and reconstructs (perceptual); and (3) rational inference, implying analysis and isolation (conceptual).

Concerning unconscious inferences in animals, I need add nothing to that which I have already said concerning the process of construction. It is concerning the intelligent inferences * of animals that I have now to speak.

I do not propose here to bring forward a number of new observations on the highly intelligent actions which animals are capable of performing. Mr. Romanes has given us a most valuable collection of anecdotes on the subject in his volume on "Animal Intelligence." It is more to my purpose to discuss some of the more remarkable of these, and endeavour to get at the back of them, so as to estimate what are the mental processes involved. In doing so, the principle I adopt is to assume that the inferences are perceptual, unless there seem to be well-observed facts which necessitate the analysis of the

---

* These fall under the "practical intelligence" of Mr. Mivart. All their intelligent activities, in his view, are performed by the exercise of merely sensitive faculties, through their "consentience." I agree to so large an extent with Mr. Mivart in his estimate of animal intelligence, and in his psychological treatment, that I the more regret our wide divergence when we come to the philosophy of the subject. I am with him in believing that conception and perception, in the sense he uses the words, are beyond the reach of the brute. But I see no reason to suppose that these higher faculties differ *in kind* from the lower faculties possessed by animals. They differ generically, but not in kind. I believe that, through the aid of language, the higher faculties have been developed and evolved from the lower faculties. Here, therefore, I have to part company from Mr. Mivart.

phenomena, the formation of isolates, and therefore the employment of reason (*as I have above defined it*). In doing this, I shall *seem* to differ very widely from Mr. Romanes and other interpreters of animal habits and intelligence. But I believe that the divergence is less wide than it seems. I believe that it is largely, but I fear not entirely, a question of the terms we employ.

Why, then, rediscuss the question under these new terms? Because I believe that such rediscussion may place the matter in a fresh and, perhaps, clearer light. The question of the relation of animal intelligence to human reason is one upon which there is a good deal of disagreement, and one that has been discussed and re-discussed. I seek to put it in a somewhat new light. I have endeavoured to define carefully and accurately the terms I use, and the sense in which I use them. I have coined for my own purposes unfamiliar terms such as "construct," "isolate," and "predominant," that I might thereby be enabled to avoid the use of terms which, from the different senses in which they are employed by different writers, have become invested with a certain ambiguity. I trust, therefore, that even those with whom I seem most to disagree will allow that my aim has not been mere disputation, but scientific accuracy and precision in a difficult subject where these qualities are of essential importance.

I take first some observations communicated by Mr. H. L. Jenkins to Mr. Romanes, since, though they raise a point which we have already shortly considered, they form a transition from unconscious to perceptual inferences. Speaking of the intelligence of the elephant, Mr. Jenkins says,[*] "What I particularly wish to observe is that there are good grounds for supposing that elephants possess abstract ideas; for instance, I think it is impossible to doubt that they acquire, through their own experience, notions of hardness and weight." He then details observations which show that elephants at first hand up things

[*] Romanes, "Animal Intelligence," p. 401.

of all kinds to their mahouts with considerable force, but that after a time the soft articles are handed up rapidly and forcibly as before, but that hard and heavy things are handed up gently. "I have purposely," he says, "given elephants things to lift which they could never have seen before, and they were all handled in such a manner as to convince me that they recognized such qualities as hardness, sharpness, and weight."

Now, the question I wish here to ask is—Do the observations of Mr. Jenkins, the nature of which I have indicated, afford good or sufficient reasons for supposing that these animals possess abstract ideas? And I reply— That depends upon what is meant by abstract ideas. If it is implied that the abstract ideas are *isolates;* that is, qualities considered quite apart from the objects of which they are characteristic, I think not. But if Mr. Jenkins means that elephants, in a practical way, "recognize such qualities as hardness, sharpness, and weight" as *predominant* elements in the constructs they form, I am quite ready to agree with him. I much question, however, whether there is any conscious inference in the matter. The elephant sees a new object, and unconsciously and instinctively builds the element hardness or weight into the construct that he forms. And he shows his great intelligence by dealing in an appropriate manner with the object thus recognized. But I do not think any reasoning is required; that is to say, any process involving an analysis of the phenomena with subsequent synthesis, any introduction of the conceptual element.

Let us consider next an observation which shows a very high degree of perceptual intelligence on the part of the dog. Several observers have described dogs, which had occasion to swim across a stream, entering the water at such a point as to allow for the force of the current. And both Dr. Rae and Mr. Fothergill communicated to Mr. Romanes instances * of the dog's observing whether the tide was ebbing or flowing, and acting accordingly. Now,

---

* "Animal Intelligence," p. 465.

I believe that the dog performs this action through intelligence, and that man explains it by reason. The dog has presumably had frequent experience of the effect of the stream in carrying him with it. He has been carried beyond the landing-place, and had bother with the mud; but when he has entered the stream higher up, he has nearly, if not quite, reached the landing-stage. His keen perceptions come to his aid, and he adjusts his action nicely to effect his purpose.

On the bank sits a young student watching him. He sees in the dog's action a problem, which he runs over rapidly in his mind. Velocity of stream, two miles an hour. Width, one-eighth of a mile. Dog takes ten minutes to swim one-eighth of a mile. Distance flowed by the stream in ten minutes, one-third of a mile. Clever dog that! He allows just about the right distance. A little short, though! Has rather a struggle at the end.

The dog intelligently performs the feat; the lad reasons it out.

I do not know whether I am making my point sufficiently clear. A wanton boy is constantly throwing stones at birds and all sorts of objects. He does not know much about the force of gravitation or the nature of the curve his stone marks out; but he allows pretty accurately for the fall of the stone during its passage through the air. He acquires a catapult; and, being an intelligent lad, he perceives that he must aim a little above the object he wishes to hit. This is a perceptual inference. Reason may subsequently step in and explain the matter, or very possibly, being human, sparks of reason fly around his intelligent action.

Am I using the word "reason" in an unnatural and forced sense? I think not. My use is in accord with the normal use of the word by educated people. Two men are working in the employ of a mechanical engineer. Listen to their employer as he describes them. "A most intelligent fellow is A; he does everything by rule of thumb; but he's wonderfully quick at perceiving the bearing of a new bit of work; he sees the right thing to do, though he cannot tell you why it should be done. Now, B is a very different man; he is slow, but he reasons everything out. A knows the right thing to do; and B can tell you why it must be done. A has the keenest intelligence, but B the clearest reasoning faculty. If I have occasion to question them about any mechanical contrivance, A says, 'Let me see it work;' but B says, 'Let me think it out.'"

In other words, A, the intelligent man, deals with phenomena as wholes, and his perceptual inferences are rapid and exact; while B, the reasoner, analyzes the phenomena, and draws conceptual inferences about them.

Let us take next Dr. Rae's* most interesting description of the cunning of Arctic foxes. These clever animals, he tells us, soon learn to avoid the ordinary steel and wooden traps. The Hudson Bay trappers, therefore, set gun-traps. The bait is laid on the snow, and connected with the trigger of the gun by a string fifteen or twenty feet long, five or six inches of slack being left to allow for contraction from moisture. The fox, on taking up the bait, discharges the gun and is shot. But, after one or more foxes have been shot, the cunning beasts often adopt one of two devices. Either they gnaw through the string, and then take the bait; or they tunnel in the snow at right angles to the line of fire, and pull the bait *downwards*, thus discharging the gun, but remaining uninjured. This is regarded by Dr. Rae as a wonderful instance of "abstract reasoning."

Here, again, it is the "abstract reasoning" that I question. Do the clever foxes resemble the intelligent workman A, or the abstract reasoner B? I believe that their actions are the result of perceptual inferences. They adopt their cunning devices *after one or more foxes have been shot*. Their keen perceptions (let me repeat that the perceptions of wild animals are extraordinarily keen) lead them to see that this food, quiet as it seems, has to be taken with caution.

* "Animal Intelligence," p. 430; and *Nature*, vol. xix. p. 409.

With regard to the devices adopted, I think we need further information. Do Arctic foxes tunnel in the snow for any other purposes? What is the proportion of those who adopt this device to those who gnaw through the string? Have careful and reliable observers watched the foxes? or are their actions, as described by Dr. Rae, inferences, on the part of the trappers, from the state of matters they found when they came round to examine their traps? Without fuller information on these points, it is undesirable to discuss the case further. Even if we had full details, however, we should be as little able to get at the process of perceptual inference in the case of the fox as we are in the case of the intelligent workman, who sees the right thing to do, but cannot tell you how he reached the conclusion.

No one can watch the actions of a clever dog without seeing how practical he is. He is carrying your stick in his mouth, and comes to a stile. A young puppy will go blundering with the stick against the stile, and, perhaps, go back home, or get through the bars and leave the stick behind. But practical experience has taught the clever dog better. He lays down the stick, takes it by one end, and draws it backwards through the opening at one side of the stile. A friend tells me of a dog which was carrying a basket of eggs. He came to a stile which he was accustomed to leap, poked his head through the stile, deposited the basket, ran back a few yards, took the stile at a bound, picked up the basket, and continued on his course. "Intelligent fellow!" I exclaim. "Yes," says my friend, "he *knew the eggs would break* if he attempted to leap with the basket!" This is just the little gratuitous, unwarrantable, human touch which is so often filled in, no doubt in perfect good faith, by the narrators of anecdotes. Against such interpolations we must be always on our guard. It is so difficult not to introduce a little dose of reason.

Mr. Romanes obtained from the Zoological Gardens at Regent's Park a very intelligent capuchin monkey, on which his sister made a series of most interesting and valuable observations. This monkey on one occasion got hold of a hearth-brush, and soon found the way to unscrew the handle. After long trial, he succeeded in screwing it in again, and throughout his efforts always turned the handle the right way for screwing. Having once succeeded, he unscrewed it and screwed it in again several times in succession, each time with greater ease. A month afterwards he unscrewed the knob of the fender and the bell-handle beside the mantelpiece. Commenting on these actions, Mr. Romanes speaks * of "the keen satisfaction which this monkey displayed when he had succeeded in making any little discovery, such as that of the mechanical principle of the screw."

I once watched, near the little village of Ceres, in South Africa, a dung-beetle trundling his dung-ball over an uneven surface of sand. The ball chanced to roll into a sand hollow, from which the beetle in vain attempted to push it out. The sides were, however, too steep. Leaving the ball, he butted down the sand at one side of the hollow, so as to produce an inclined plane of much less angle, up which he then without difficulty pushed his unsavoury sphere.

Now, it seems to me that, if we say, with Mr. Romanes, that the brown capuchin discovered the *principle of the screw*, we must also say that the dung-beetle that I observed in South Africa was acquainted with the *principle of the inclined plane*. Such an expression, I contend, involves an unsatisfactory misuse of terms. A mechanical principle is a concept,† and as such, in my opinion, beyond the reach of the brute—monkey or beetle. That of which the monkey is capable is the perceptual recognition of the fact that certain actions performed in certain ways produce certain results. Why they do so he neither knows nor cares to know. What the brown capuchin discovered was

* "Animal Intelligence," p. 497.

† Mr. Romanes regards it as, in the case of the capuchin, a *recept*. But when he speaks of a generic idea of causation, and generic ideas of principles, and of qualities as recepts, I find it exceedingly difficult to follow him. They seem to me to be concepts supposed to be formed in the absence of language.

not the principle of the screw, but that the action of screwing produced the results he desired—a very different matter. My friend, Mr. S. H. Swayne, tells me that the elephant at the Clifton Zoo, having taking a tennis-racket from a boy who had been plaguing him, broke it by leaning it against a step and deliberately stepping on it in the middle, where it was unsupported. A most intelligent action. And it would have been a capital piece of exercise for the lad's reasoning power, had he been required to analyze the matter, to show why the elephant's action had the desired effect, and set forth the principle involved. I do not think the elephant himself possesses the faculty requisite for such a piece of reasoning. He is content with the practical success of his actions; principles are beyond him.

I will now give two instances of intelligence in vertebrates which exemplify phases of inference somewhat different from those which we have so far considered. Mr. Watson, in his "Reasoning Power of Animals,"* tells of an elephant which was suffering from eye-trouble, and nearly blind. A Dr. Webb operated on one eye, the animal being made to lie down for the purpose. The pain was intense, and the great beast uttered a terrific roar. But the effect was satisfactory, for the sight was partially restored. On the following day the elephant lay down of himself, and submitted quietly to a similar operation on the other eye. No doubt the elephant's action here was, in part, the result of its wonderful docility and training. But there was also probably the inference that, since Dr. Webb had already given him relief, he would do so again. The anticipation of relief outmastered the anticipation of immediate discomfort or pain. I do not think, however, that any one is likely to contend that any rational analysis of the phenomena is necessarily involved in the elephant's behaviour.

The other instance I will quote was communicated by Mr. George Bidie to *Nature*.† He there gives an account

* Page 54.   † Vol. xx. p. 96.

of a favourite cat which, during his absence, was much plagued by two boys. About a week before his return the cat had kittens, which she hid from her tormentors behind the book-shelves in the library. But when he returned she took them one by one from this retreat, and carried them to the corner of his dressing-room where previous litters had been deposited and nursed. Here abnormal circumstances and the reign of anarchy and persecution forced her to adopt a hiding-place where she might bring forth her young; but the return of normal conditions, sovereignty, and order led her to take up her old quarters under the protection of her master. Now, look at the description I have given in explanation of her conduct. See how it bristles with conceptual terms: "abnormal," with its correlative "normal;" "anarchy and persecution," "protection" and "order." All this, I believe, is mine, and not the cat's. For her there was a practical perception, in the one case of plaguing boys, in the other case of protecting master; and her action was the direct outcome of these perceptions through the employment of her intelligence.

Some stress has been laid on the occasional use of tools by animals. Mr. Peal* observed a young elephant select a bamboo stake, and utilize it for detaching a huge elephant-leech which had fixed itself beneath the animal's fore leg near the body. "Leech-scrapers are," he says, "used by every elephant daily." He also saw an elephant select and trim a shoot from the jungle, and use it as a switch for flapping off flies. How far, we may ask, do such actions imply "a conscious knowledge of the relation between the means employed and the ends attained"?† That, again, depends upon how much or how little is implied in this phrase.

A boy picks up a stone and throws it at a bird; he comes home and unlocks the garden-gate with a key; he enters his room, and removes the large "Liddell and

---

* *Nature*, vol. xxi. p. 34.

† Romanes, "Animal Intelligence," p. 17: Definition of *reason*.

Scott" which he uses as a convenient object to keep the lid of his play-box shut; he opens the box, and cuts himself a slice of cake with his pocket-knife. Then he goes to his tutor, who is teaching him about means and ends, and their relation to each other. He is told that the throwing of the stone was the means by which the death of the bird, or the end, was to be accomplished; that the use of the knife was the means by which the end in view, the severance of a piece of cake, was to be effected, and so on. He is led to see that the employment of a great many different things, differing in all sorts of ways—stones, keys, lexicons, and knives—may be classified together as means; and that a great many various effects, the death of a bird or the cutting a bit of cake, may be regarded as ends. He is told that when he thinks of the means and the ends together, as means and end, he will be thinking of their relationship. And it is explained to him that means and ends and their relationships are concepts, and involve the exercise of his reasoning powers.

Weary and sick to death of concepts and relationships and reason, at length he escapes to the garden. Picking up a light stick, he sweeps off the heads of some peculiarly aggravating poppies, and determines to think no more of means and ends, continuing to use the stick meanwhile as a most appropriate means to the end of decapitating the poppies. By all which I mean to imply that there is a great difference between selecting and using a tool for an appropriate purpose, and possessing a conscious knowledge of the relation between the means employed and the ends attained. I do not think that any conception of means, or end, or relationship is possible to the brute. But I believe that the elephant can perceive that this stick will serve to remove that leech. And if this is what Mr. Romanes means by its possessing a conscious knowledge of the relation between the means employed and the ends attained, then I am, so far, at one with him in the interpretation of the facts, though I disagree with his mode of expressing them.

I do not propose to consider particular instances of intelligent inferences as displayed by the invertebrates. Bees in the manipulation of their comb, ants in the economy of their nest, spiders in the construction of their web and the use they make of their silken ropes, show powers of intelligent adaptation which cannot fail to excite our wonder and admiration. But apart from the fact that insect psychology is more largely conjectural than that of the more intelligent mammals, a consideration of these actions would only lead me to reiterate the opinion above frequently expressed. In a word, I regard the bees in their cells, the ants in their nests, the spiders in their webs, as workers of keen perceptions and a high order of practical intelligence. But I do not, as at present advised, believe that they reason upon the phenomena they deal with so cleverly. Intelligent they are; but not rational.

Once more, let me repeat that the sense in which I use the words "rational" and "reason" must be clearly understood and steadily borne in mind. Mr. Romanes uses them in a different sense. "Reason," he says,* "is the faculty which is concerned in the intentional adaptation of means to ends. It therefore implies the conscious knowledge of the relation between means employed and ends attained, and may be exercised in adaptation to circumstances novel alike to the experience of the individual and to that of the species. In other words, it implies the power of perceiving analogies or ratios, and is in this sense equivalent to the term 'ratiocination,' or the faculty of deducing inferences from a perceived equivalency of relations. This latter is the only sense of the word that is strictly legitimate."

It is not my intention to criticize this use of the term "reason." Whether animals are capable of a conscious knowledge of the relation between means employed and ends attained, depends, as we have already seen, upon how much is implied by the word "knowledge"—whether the knowledge is perceptual or conceptual. My only care is

* "Mental Evolution in Animals," p. 318.

to indicate what seem to me the advantages of the usage (legitimate or illegitimate) I adopt.

I repeat, then, that the introduction of the process of analysis appears to me to constitute a new departure in psychological evolution; that the process differs generically from the process of perceptual construction on which it is grafted. And I hold that, this being so, we should mark the departure in every way that we can. I mark it by a restriction of the word "intelligence" to the inferences formed in the field of perception; and the use of the word "reason" when conceptual analysis supervenes. Whether I am justified in so doing, whether my usage is legitimate or not, I must leave others to decide. But, adopting this usage, I see no grounds for believing that the conduct of animals, wonderfully intelligent as it is, is, in any instances known to me, rational.

I say that the introduction of the process of analysis appears to me to constitute a new departure. This, however, must not be construed to involve any breach of continuity.

I do not believe that there is or has been any such breach of continuity. Take a somewhat analogous case. I regard the introduction of aerial respiration in animal life as a new departure. Organisms which had hitherto been water-breathers became air-breathers. But I do not imagine that there was any breach of continuity in respiration. The tadpole begins life as a water-breather only; the frog into which he develops is an air-breather; but there is no breach of continuity between the one state and the other. So, too, the little child dwells in the perceptual sphere; the man into whom he develops is capable of conceptual thought; but there is no breach of continuity in the mental life of the child. It is true that, with all our talk on the subject, we cannot say exactly when in this continuous mental life the new departure is made. But this is no proof whatever that there is no new departure. In a sigmoidal curve there is a new departure where the convex passes into the concave. We may find it difficult to mark the exact point of change. But that does not invalidate the fact that the change does actually take place.

If I be asked how, in the course of mental evolution, the new departure was rendered possible, I reply—Through language. The first step was, I imagine, *the naming of predominants*. If Noiré and Professor Max Müller be correct in their views, language took its origin in the association of an uttered sound with certain human activities. The action thus named was, so to speak, floated off by its sign. By diacritical marks attached to the word, the agent, the action, and the object of the action were distinguished, and thus came to be differentiated the one from the other. Inseparable in fact, they came henceforth to be separable in thought. Here was analysis in the germ. The action or activity was isolated, and henceforth stood forth as an element in abstract thought. All the busy world around was interpreted in terms of activities. The host of heaven and all the powers of earth were named according to their predominant activities. The moon became the measurer, the sun the shining one, the wind the one who bloweth, the fire the purifier, and so forth. Our verbs and nouns, then, being named predominants (agents, actions, or objects), adjectives and adverbs were subsequently introduced to qualify these by naming a quality less predominant, or to indicate the how, the when, and the where.

When once the different activities and different qualities came to be named or symbolized, they were, as I say, floated off from the agents or objects, and through isolation entered the conceptual sphere. *The named predominant became an isolate.* Body and mind became separable in thought; the self was differentiated from the not-self; the mind was turned inwards upon itself through the isolation of its varying phases; and the consciousness of the brute became the self-consciousness of man.

Language, and the analytical faculty it renders possible, differentiates man from the brute. "If a brute," says Mr. Mivart,[*] "could think 'is,' brute and man would be

---

[*] "Lessons from Nature," pp. 226, 227.

brothers. 'Is' as the copula of a judgment implies the mental separation and recombination of two terms that only exist united in nature, and can, therefore, never have impressed the sense except as one thing. And 'is,' considered as a substantive verb, as in the example, 'This man is,' contains in itself the application of the copula of judgment to the most elementary of all abstractions—'thing' or 'something.' Yet if a being has the power of thinking 'thing' or 'something,' it has the power of transcending space and time by dividing or decomposing the phenomenally one. Here is the point where instinct [intelligence] ends and reason begins." I regard this as one of the truest and most pregnant sentences that Mr. Mivart has written.

And when once the Logos had entered into the mind of man, and made him man, it slowly but surely permeated his whole mental being. Hence language is not only involved in our concepts, but also in our percepts, in so far as they are ours. Professor Max Müller goes so far as to question whether an unnamed percept is possible. And adult intellectual man is so permeated by the Logos that I am not prepared to disagree with him when he says that he has no unnamed perceptions. Nevertheless, the actions of the speechless child and our dumb companions show that they (children and animals) are capable of forming mental products of the perceptual order. But here, once more, we must not forget that it is in terms of these adult human percepts that we interpret the percepts of children and animals; that in doing so we cannot divest ourselves of the garment of our conceptual thought, that we cannot banish the Logos, and that, therefore, these percepts other than ours cannot be identical with ours, though they are of the same order, saving their conceptual element. We may put the matter thus—

(1) $x \times$ dog-mind
(2) $x \times$ cat-mind
(3) $x \times$ infant-mind
$\Big\} = \Big\{$ Percepts to be interpreted in terms of (4), being analogous thereto but not identical therewith.

(4) $x \times$ adult human mind = the percepts of psychologists, named or namable.

If the views that I have thus very briefly sketched (for I have no right to offer an opinion on a question of linguistic science) be correct, language has made analysis, isolation, and conceptual thought possible. But there may have been a transitory stage when the word-signs stood for predominants, not yet for isolates. Granting the possibility or probability of this, I am prepared to follow Professor Max Müller in his contention that language and thought, from the close of that stage onward, are practically inseparable, and have advanced hand-in-hand. It is true that I can now think out a chemical or physical problem without the use of words—the stages of the experimental work being visualized, just as a chess-player may think out a game in pictures of the successive moves. But, historically, I believe the power to do this has been acquired through language; and if I am able temporarily to isolate and analyze without language, thought being at times a little ahead of naming, yet the fact remains that language is absolutely necessary to make such advances good, if not for me, at any rate for man.

And here I would make one more suggestion. Professor Max Müller, as the result of analysis of the Aryan language, finds a comparatively small number of roots which he says are in all cases symbolic of concepts. Yes, for us now they symbolize concepts. But in their inception may they not have been symbolic of predominants? Have we not in them the signs for predominants not yet converted for the primitive utterers into isolates? May not these have been the stepping-stones from the perceptual predominants of animal man, to the conceptual isolates of rational man? Or, to modify the analogy, may they not have been the embryonic wings by which the human race were floated off from the things of sense into the free but tenuous air of abstract thought?

Lastly, before taking leave of the subject of this chapter, I am most anxious that it should not be thought that, in contending that intelligence is not reason, I wish in any way to disparage intelligence. Nine-tenths at least of the actions of average men are intelligent and not rational. Do we not all of us know hundreds of practical men who are in the highest degree intelligent, but in whom the rational, analytic faculty is but little developed? Is it any injustice to the brutes to contend that their inferences are of the same order as those of these excellent practical folk? In any case, no such injustice is intended; and if I deny them self-consciousness and reason, I grant to the higher animals perceptions of marvellous acuteness and intelligent inferences of wonderful accuracy and precision —intelligent inferences in some cases, no doubt, more perfect even than those of man, who is often distracted by many thoughts.

CHAPTER X.

THE FEELINGS OF ANIMALS: THEIR APPETENCES AND EMOTIONS.

There is one aspect of the mental processes of men and animals that we have so far left unnoticed—the aspect of feeling, the aspect of pleasure and pain. Quite distinct from, and yet intimately associated with, our perception of a beautiful scene, is the pleasure we derive therefrom; and quite distinct from, and yet inseparably bound up with, our perception of a discordant clang, is the painful effect that it produces.

We have, however, no separate organs for the appreciation of pleasure and pain. These feelings arise out of, and are bound up with, our sensations, our perceptions, and especially with the conscious exercise of our bodily activities. There may be, at any rate in some cases, separate nerves for the appreciation of the pleasurable and the painful; but even if this be so, these shades of feeling are so closely associated with our other activities, mental and bodily, that we may for the present regard them simply as the accompaniments of these activities.

The question has been raised and much discussed whether all our activities are accompanied by some shade or colouring of feeling, pleasurable on the one hand, or painful on the other; or whether some of these activities may not be indifferent in this respect, affording us neither pleasure nor pain. Put in this way, I think we may say that there may be activities which are thus indifferent. But if it be asked whether, in addition to the pleasurable and painful feelings, there is a third class of *feelings*, which we may call indifferent or neutral, I am inclined to answer it in the negative. I hold that every feeling, as such, must belong either to the painful or pleasurable class, and that if the pleasurable and painful, so to speak, exactly balance each other, then feeling, as such, does not emerge into consciousness at all. For, as Lotze says, "We apply the name 'feelings' exclusively to states of pleasure and pain, in contrast with sensations as [the elements of] indifferent perceptions of a certain content."

The broadest division of the feelings is, therefore, into pleasurable on the one hand, and painful on the other.

Another general question with regard to the feelings is—With what condition or state of the bodily organization are they associated? In answer to this question we may say (1) that any very violent and abnormal stimulus produces pain; (2) that the conditions of pleasure are to be sought within the limits of the healthy and normal exercise of the bodily functions and mental activities; (3) that within these limits the changes of activity consequent upon the rhythmic flow of normal organic processes bring with them, in the aggregate, pleasure, the delight of healthy life; (4) that within these limits, again, we experience pleasure or pain, enjoyment or weariness, ease or discomfort, happiness or unhappiness, with the continued rise and fall of our life-tide. For, as Spinoza says, "We live in perpetual mutation, and are called happy or unhappy according as we change for the better or the worse." So long as our activities remain at a dead level, there is indifference—neither pleasure nor pain. A rise of the tide of activity brings pleasure, a fall the reverse. Lastly, we may say (5) that beyond the limits of healthy and normal exercise there is, on the one hand, excessive exercise which, carried far enough, may give rise, first to fatigue, and then to acute pain; and, on the other hand, deficient exercise, which may produce that dull and numb form of pain which we call discomfort, or a sense of craving or want.

Pleasures and pains may thus be either massive or acute, diffused or locally concentrated. On the whole, we may say, with Mr. Grant Allen,* that "the acute pains, as a class, arise from the action of surrounding destructive agencies; the massive pains, as a class, from excessive function or insufficient nutriment." But since massive pains, when pushed to an extreme, merge into the acute class, "the two classes are rather indefinite in their limits, being simply a convenient working distinction, not a natural division." "Massive pleasure can seldom or never attain the intensity of massive pain, because the organism can be brought down to almost any point of innutrition or exhaustion; but its efficient working cannot be raised very high above the average. Similarly, any special organ or plexus of nerves can undergo any amount of violent disruption or wasting away, giving rise to very acute pains; but organs are very seldom so highly nurtured and so long deprived of their appropriate stimulant as to give rise to very acute pleasure." The amount of pleasure varies, according to Mr. Grant Allen, whose discussion of the subject is, perhaps, the best and clearest we have, directly as the number of nerve-fibres involved, and inversely as the natural frequency of their excitation. No doubt the principles above sketched out are somewhat vague and general; but we are scarcely justified in formulating any that are more precise and exact.

Accepting now the theory of evolution, we may say, furthermore, that during the long process of the moulding of life to its environment, there has been a constant tendency to associate pleasure with such actions as contribute towards the preservation and conservation of the individual and the race, and to associate pain with such actions as tend to the destruction or detriment of the individual or the race. For there can be little doubt that pleasure and pain are the primary incentives to action. Without the association of pleasure with conservative action, and pain with detrimental action, it is difficult to conceive how the evolution of conscious creatures would be possible. Conservative action, if it is to be persisted

---

* "Physiological Æsthetics:" chapter on "Pleasure and Pain."

in by a conscious creature, must be associated directly or indirectly with pleasurable feelings; nay, more, if it is to be persistently persevered in, its non-performance must be associated with that dull form of pain which we call a craving or want. Only under such conditions could activities which tend to the survival of the individual and the race be fostered and furthered.

It must be remembered, however, that such association is founded on experience, and has no necessary validity beyond experience. That quinine, though unpleasant to the taste, is, under certain circumstances, beneficial to the individual, and that acetate of lead, though sweet-tasted, is harmful, cannot be fairly urged in opposition to this principle, since the effects of these drugs form no part of the normal experience of the individual and the race. Nor can it be fairly objected that animals transported to new countries often eat harmful and poisonous plants presumably because they are nice; for these plants form part of an unwonted environment. Nor, again, is the fact that the association of pleasure with conservative action and pain with harmful action is not always perfect, in any sense fatal to the general principle. For the establishment of the association is still in progress; and with the increase in the complexity of life its accurate establishment is more and more difficult. No one is likely to contend that what appears to be a general principle must also be an invariable rule. The general principle is that under the joint influence of pleasure (attractive) and pain (repellent) the needle of animal life sets towards the pole of beneficial action. That the needle does not always point true only illustrates the fact that life-activities are still imperfect.

Let us notice that it is under the *joint* action of pleasure and pain that the needle sets. We must not think only of the positive aspect, and neglect the negative. What we know as wants, cravings, appetites, desires, and dissatisfactions, are dull and continuous pains,* which tend to

---

* All of these, at any rate, satisfy Mr. Herbert Spencer's definition. Pleasure he describes as a feeling which we seek to bring into consciousness

drive us to actions by which they shall be annulled, and the performance of which shall give us the pleasures of gratification. Dr. Martineau regards a felt want as a mainspring of our energy. "Life," he says,* "is a cluster of wants, physical, intellectual, affectional, moral, each of which may have, and all of which may miss, the fitting object. Is the object withheld or lost? There is pain: is it restored or gained? There is pleasure: does it abide or remain constant? There is content. The two first are cases of disturbed equilibrium, and are so far dynamic that they will not rest till they reach the third, which is their posture of stability and their true end." To this I would only add that the content which follows on the keen pleasure of satisfaction is evanescent, and ere long lapses into indifference, on which in due time follows the dull pain resulting from the recurrent pressure of the want or desire.

It is clear that, in introducing these wants and desires, we are entering the sphere of the emotions, and it is sometimes said that the emotions have their basis in pleasure and pain. If by this it is meant that the emotions often exhibit more or less prominently one or other of these two aspects of feeling, we may agree with the statement. It will be well, however, to lead up to our consideration of the emotions by taking a general review of the manner in which the organism responds to external stimuli.

A dog is lying dreamily on the lawn in the sunshine. Suddenly he raises his head, pricks his ears, scents the air, looks fixedly at the hedge, and utters a low growl. Place your hand upon his shoulder, and you will find that his muscles are all a-tremble. He can restrain himself no longer, and darts through the hedge. You follow him, look over the hedge, and see that it is his old enemy, the butcher's cur. They are moving slowly past each other, head down, teeth bared, back roughened. You whistle

and retain there; pain, as a feeling which we seek to get out of consciousness and keep out.

* "Types of Ethical Theory," vol. ii. p. 350.

softly. Such a whistle would generally bring him bounding to your feet. But now it is apparently unheard. The two dogs have a short scuffle, and the cur slinks off. Your dog races after him; but after a few minutes returns, jumps up at you playfully, and then lies down again on the grass. But every now and then, for ten minutes or so, he raises his head and growls softly.

Let us briefly analyze the dog's actions, reading into them, conjecturally, the accompaniments in consciousness. As he lies on the lawn, he receives a sense-stimulus, auditory or olfactory, which gives rise to the construction of the percept dog (perhaps particularized through olfactory discrimination). About the formation of constructs or percepts, however, we have already said enough; we have now to consider their effects. The head is raised, the ears pricked, and so on. The dog is on the alert. His attention is roused. What are the physiological effects? Certain motor-activities or tendencies to activity. These are of two kinds—first, in connection with the sense-organs, the muscles of which are brought into play in such a way as to bring the organs to bear upon the exciting object; secondly, in connection with many other muscles, which are innervated, so as to be ready to act rapidly and forcibly. The first motor-effect, that on the muscles of the sense-organs, is a very characteristic physical concomitant of the psychological state which we term "attention;" the second effect, the incipient innervation of muscles likely to be called into play, is equally characteristic of the psychological state we call alertness.

Meanwhile an emotional state is rising in the mind of the dog. We may call it, conjecturally, anger and combativeness. But what we name it does not much signify for our present purpose. It has a growing tendency to work itself out in a series of definitely directed actions. And this reaches its point of culmination when the dog rushes through the hedge and stands with bared teeth before his antagonist. A whole set of appropriate muscles are now strongly innervated. There is probably a double innervation—an innervation prompting to activity and an innervation inhibiting or restraining from activity. The attention is so concentrated that he heeds not, probably hears not, his master's whistle. He is keenly on the alert. Then he sees his chance; the inhibition or restraint is withdrawn, and he flies at his opponent. The emotional tendency works itself out in action. Even after he has resumed his place on the lawn, memories of the emotional state return, and lead him to lift his head, slightly bare his teeth, and growl.

Now, with regard to the emotional state here indicated, we may notice, first, that it is initiated by a percept; secondly, that associations of pleasure or pain are by no means the most important or predominant characteristics; thirdly, that the motor-tendencies seem to be essential, the emotional state being the psychological aspect of these motor-tendencies; and, fourthly, that we should perhaps be justified in speaking of a presentative emotion when the percept which gives rise to the emotion is presentative; and a representative emotion where the originating percept is represented in memory. And with regard to the attention which was incidentally introduced, we may notice that it, too, has motor-concomitants, and that it is directly associated with the emotional state. If no emotional state is aroused by a percept, attention is not specially directed to the object. The concentration of the attention is directly proportional to the intensity of the emotion evoked.

Emotions, then, would seem from this illustration to be certain psychological states which accompany activities or tendencies to activity. They are evoked by appropriate objects perceived or remembered. Where the tendency is towards the object, as in the sexual emotions, we may speak of it as an *appetence;* where it is away from the object, as in the emotion of fear, we may speak of it as an *aversion.* Appetences are normally pleasurable; aversions, painful.

It is clear that the organism must be in a condition fitting it to carry out its various activities. And this condition is more or less variable. In the terms of our previous analogy (Chapter II.) the tissues are "explosive." After a series of explosions have taken place in a tissue, its store of explosive material becomes exhausted, and a powerful stimulus is required to liberate further energy in the exhausted tissue. A period of rest is required to enable the plasmogen to generate a fresh store of explosive material. As this store increases to its maximum pitch, the tissue becomes more and more ready to respond at the slightest touch. Responsiveness to external stimuli is spoken of as *sensitiveness;* emotional responsiveness is called *sensibility.* What we have before spoken of as a want or craving is a state of heightened sensibility, which often gives rise to a painful state of general uneasiness. It may also give rise to perceptual representations in memory, as may be seen in the dreams experienced during a state of extreme sexual sensibility. If we seek a basis for the emotional states, therefore, we shall find it in sensibility rather than in pleasure and pain.

The motor-accompaniments of the emotional states have long been known under the title of the "expression" of the emotions. The term is too deeply rooted to be altered; but we may notice that what is called the expression of an emotion is really *its partial fulfilment in action.* Some psychologists, dissatisfied with the term "expression of the emotions," as seeming to imply that the emotion is one thing and its expression another, go so far as to say that the motor-accompaniments are the objective aspect of what, under its subjective aspect, *is* the emotion. It is quite possible, however, to experience an emotion without any motor-accompaniments at all. Nevertheless, there is, I believe, in such cases an unfulfilled tendency to action.

A most important feature in general physiology and psychology is *the postponement or suppression of action.* The physiological faculty on which it is based is inhibition. I do not propose to discuss the somewhat conflicting views on the physiological mechanism of inhibition. It is, however, a fact of far-reaching importance which no one is likely to deny. In its higher ranges it is the objective basis and aspect of self-restraint.

A stimulus gives rise to sensation and perception; the perception gives origin to an emotional state; and the emotional state is fulfilled in appropriate motor-activities. The process is a continuous one, and, in the absence of inhibition, would in all cases inevitably fulfil itself. But through the faculty of inhibition, the final state of activity may be postponed or suppressed. We may place side by side the physiological series and the accompanying psychological series thus—

$$\left.\begin{array}{l}\text{Stimulus of}\\ \text{sense-organ}\end{array}\right\} \rightarrow \text{nervous processes in brain} \rightarrow \left\{\begin{array}{l}\text{Stimulus of}\\ \text{motor-organs.}\end{array}\right.$$

$$\left.\begin{array}{l}\text{Consciousness of}\\ \text{sense-stimulus}\end{array}\right\} \leftarrow \text{perception, emotion} \rightarrow \left\{\begin{array}{l}\text{Consciousness of}\\ \text{activity.}\end{array}\right.$$

The arrows pointing away from perception and emotion are intended to indicate the fact that the consciousness of sense-stimulus on the one hand, and of activity on the other hand, are accompaniments of the nervous processes in the brain, and are referred outwards to the sense-organ or the motor-organ, as the case may be. It must be remembered that the two series, physiological and psychological, belong to distinct phenomenal orders. If one speaks of emotion being fulfilled in activity, and thus seems to jump from the psychological to the physiological series, one does so merely to avoid the appearance of pedantry.

Now, by the postponement or suppression of action, the process is either arrested in its middle phase, the motor-organs not being innervated at all, or, as I believe to be more probable, the motor-organs are doubly innervated, a stimulus to activity being counteracted by an inhibitory stimulus, the two neutralizing each other either in the motor-organ or the efferent nerves which convey the stimuli. In any case, there is no consciousness * of activity. And the mind occupies itself more and more completely with the central processes, perception, and emotion, and also, in

---

* Such consciousness of activity is probably associated with the innervation of afferent, not efferent, nerves.

human beings, conceptual thoughts and emotions. Nevertheless, at any rate *so long as we confine ourselves to the perceptual sphere*, these processes have their normal fulfilments in action, and, if they become sufficiently intense, actually do so fulfil themselves.

Now, since the emotions with which we are now dealing (we may call them emotions in the perceptual sphere) are stages in the fulfilment of activities (though the activities themselves may be suppressed), it is clear that there may be as many emotional states as there are modes of activity. Hence, no doubt, the extreme difficulty of anything like a satisfactory classification of these emotions, especially when the activities are regarded as a merely extraneous expression.

Moreover, when certain emotions reach a high pitch of intensity, they may defeat their own object, and give rise, not to definite well-executed motor-activities, but to helpless contradictory actions, affections of glandular and other organs, and a general condition of collapse. The emotion of fear, for example, will lead to motor-activities tending to remove a man from the source of danger; but when it reaches the degree of dread, or its culmination terror, the effects are markedly different. The countenance pales, the lips tremble, the pupils of the eyes become dilated, and there is an uncomfortable sensation about the roots of the hair. The bowels are often strongly affected, the heart palpitates, respiration labours, the secretions of the glands are deranged, the mouth becomes dry, and a cold sweat bursts from the skin. The muscles cease to obey the will, and the limbs will scarcely support the weight of the body. Here we have all the effects of a prolonged struggle to escape. Just as such a prolonged struggle will at length produce these motor and other effects accompanied by the emotion of terror; so, if the emotion of terror be produced directly, these motor and other effects are seen to accompany it.

Mr. Charles Richardson, the well-known engineer of the Severn Tunnel, has recorded several instances of railway servants and others being so affected by the approach of a train or engine that they have been unable to save themselves by getting out of the way, though there was ample time to do so. This may have been through the effect of terror. But one man, who was nearly killed in this way, only just saving himself in time, informed me that he experienced no feeling of terror; he was unable to explain why, but he couldn't help watching the train as it darted towards him. In this case it seems to have been a sort of hypertrophy of attention. His attention was so rivetted that he was unable to make, or rather he felt no desire to make, the appropriate movements. He said, "I had to shake myself, and only did so just in time. For in another moment the express would have been on me. When it had passed, I came over all a cold sweat, and felt as helpless as a baby. I was frightened enough *then*." Cases of so-called fascination in animals may be due in some cases to terror, but more often, perhaps, to a hypertrophy of attention, such as is seen in the hypnotic state. Speaking of the effects of artificial light on fish, Mr. Bateson says,* "Bass, pollack, mullet, and bream generally get quickly away at first, but if they can be induced to look steadily at the light with both eyes, they generally sink to the bottom of the tank, and on touching the bottom commonly swim away. . . . In the case of mullet, effects apparently of a mesmeric character sometimes occur, for a mullet which has sunk to the bottom as described will sometimes lie there quite still for a considerable time. At other times it will slowly rise in the water until it floats with its dorsal fin out of the water, as though paralyzed. . . . When the light is first shown, turbot generally take no notice of it, but after about a quarter of an hour I have three times seen a turbot swim up, and lie looking into the lamp steadily. It seemed to be seized with an irresistible impulse like that of a moth to a candle, and throws itself open-mouthed at the lamp." As a boy I used frequently to "mesmerize" chickens by making them look at a chalk

* Journal of Marine Biological Association, New Series, vol. i. No. 2, pp. 216, 217.

mark. They would then lie for some time perfectly motionless. Some such effect has, perhaps, led to the instinct displayed by some animals of "shamming dead."

Returning now to the emotions as displayed in man, we may take one more example in anger. This is an emotion that arises from the idea of evil having been inflicted or threatened. "Under moderate anger," says Darwin, "the action of the heart is a little increased, the colour heightened, and the eyes become bright. The respiration is likewise a little hurried; and as all the muscles serving for this purpose act in association, the wings of the nostrils are sometimes raised to allow of a free draught of air; and this is a highly characteristic sign of indignation. The mouth is commonly compressed, and there is almost always a frown on the brow. Instead of the frantic gestures of extreme rage, an indignant man unconsciously throws himself into an attitude ready for attacking or striking his enemy, whom he will, perhaps, scan from head to foot in defiance. He carries his head erect, with his chest well expanded, and the feet planted firmly on the ground. With Europeans the fists are generally clenched." "Under rage the action of the heart is much accelerated, or, it may be, much disturbed. The face reddens, or it becomes purple from the impeded return of the blood, or may turn deadly pale. The respiration is laboured, the chest heaves, and the dilated nostrils quiver. The whole body often trembles. The voice is affected. The teeth are clenched or ground together, and the muscular system is commonly stimulated to violent, almost frantic, action. But the gestures of a man in this state usually differ from the purposeless writhings and struggles of one suffering from an agony of pain; for they represent more or less plainly the act of striking or fighting with an enemy."

These examples will serve to remind the reader of the nature of those complex aggregates of organized feelings which we call emotions, and will also show the close connection of these emotions with the associated bodily movements and activities which constitute their normal fulfilment. So close is this connection, that the assumption of the appropriate attitude will conjure up a faint revival of the associated emotion. Let any one stand with squared shoulders, clenched fists, and set muscles, and he will find the respiration affected, and perhaps also the heart-beat, and will experience a faint revival of the emotion of anger. Very different will be his feelings as he reseats himself, abandons his limbs to a posture of leisurely repose, and allows a pleasant smile to steal over his features.

The next point to notice about these emotions is that they are to a large extent instinctive, and are evidenced in the infant at so early a period that individual acquisition is out of the question. In any case, the basis of sensibility is innate. As Mr. Sully says,* " There are instinctive capacities of emotion of different kinds, answering to such well-marked classes of feeling as fear, anger, and love. These emotions arise uniformly when the appropriate circumstances occur, and for the most part very early in life. Thus there is an instinctive disposition in the child to feel in the particular way known as anger or resentment when he is annoyed or injured."

In this, as in other cases of instinctive action, of which we shall have more to say in the next chapter, it is, of course, impossible to say for certain how far the activities observed are associated with psychological states. The activities are undoubtedly instinctive. And their performance by an adult would be accompanied by an emotional state. It is, therefore, probable that in the very young child they have their emotional concomitants. Still, we must remember that oft-repeated actions tend to become automatic, that the accompanying consciousness sinks into evanescence, and that it is, therefore, *possible* that the emotional state may not have that vividness which the activities seem to bespeak.

There only remains, before passing on to consider the feelings and emotions of animals, to indicate what Mr. Sully terms † " the three orders of emotion." The first

* " Outlines of Psychology," p. 481. † Ibid. p. 494.

order comprises the individual and personal emotions—those which are self-interested and have sole reference to the individual who feels, enjoys, or suffers. They take origin in percepts, either in presentations of sense or representations in memory. The second order introduces the sympathetic emotions. They are evoked on sight of the sufferings or emotional states of others. If we see a woman insulted, we are filled with indignation; and this emotion has a sympathetic origin. The third order comprises the complex feelings known as *sentiments*. They have reference to certain qualities of objects or activities of individuals which inspire admiration or disapprobation. They are abstract in their nature, and belong to the conceptual sphere. Such are love of truth, beauty, virtue, liberty, justice. To become operative on conduct, however, they need, at any rate in the case of most people, to be particularized and individualized, or brought within the perceptual sphere, ere they arouse anything that is emotional in much more than in name. As Dr. McCosh has well said, "No man ever had his heart kindled by the abstract idea of loveliness, or sublimity, or moral excellence, or any other abstraction. That which calls forth our admiration is a lovely scene; that which raises wonder or awe is a grand scene; that which calls forth love is not loveliness in the abstract, but a lovely and loving person; that which evokes moral approbation is not virtue in the abstract, but a virtuous agent performing a virtuous act. The contemplation of the beautiful and the good cannot evoke deep or lively emotion. He who would create admiration for goodness must exhibit a good being performing a good action."

Turning now to the lower animals, the first question that suggests itself is—What are their capacities for pleasure and pain? A very difficult question to answer. We cannot, I think, hope to know how much or how little the invertebrates feel—to what degree they are psychologically sensitive. Even among the higher vertebrates we are very apt, I imagine, to over-estimate the intensity of their feelings. Among human-folk it is not he who halloas loudest that is necessarily most hurt. And it is only through the expression of their feelings in cries and gestures that we can conjecture the feelings of animals. There are grounds for supposing that savages are far less keenly sensitive than civilized people. And we have some reason for believing and hoping that our dumb companions are less sensitive to pain than we are. Mr. G. A. Rowell, for example, in his "Essay on the Beneficent Distribution of the Sense of Pain," tells us that "a post-horse came down on the road with such violence that the skin and sinews of both the fore fetlock joints were so cut that, on his getting up again, the bones came through the skin, and the two feet turned up at the back of the legs, the horse walking upon the ends of its leg-bones. The horse was put into a field close by, and the next morning it was found quietly feeding about the field, with the feet and skin forced some distance up the leg-bones, and, where it had been walking about, the holes made in the ground by the leg-bones were three or four inches deep." Mr. Lamont gives a somewhat similar observation in the case of the reindeer. "On one occasion," he says, "we broke one of the fore feet of an old fat stag from an unseen ambush; his companions ran away, and the wounded deer, after making some attempts to follow them, which the softness of the ground and his own corpulence prevented him doing, looked about him a little, and then, seeing nothing, actually began to graze on his three remaining legs, as if nothing had happened of sufficient consequence to keep him from his dinner." Colonel Sir Charles W. Wilson, in his work "From Korti to Khartoum," gives similar instances with regard to camels. "The most curious thing," he says,* "was that they showed no alarm, and did not seem to mind being hit. One heard a heavy thud, and, looking round, saw a stream of blood oozing out of the wound, but the camel went on chewing his cud as if nothing at all had happened, not

* Page 70.

even giving a slight wince to show he was in pain." And, again,* "I heard the rush of the shot through the air, and then a heavy thud behind me. I thought at first it had gone into the field-hospital; but, on looking round, found it had carried away the lower jaw of one of the artillery camels, and then buried itself in the ground. The poor brute walked on as if nothing had happened, and carried its load to the end of the day."

With regard to this question, then, of the susceptibility of animals to pleasure and pain, no definite answer can be given. That they feel more or less acutely we may be sure; how keenly they feel we cannot tell; but it is better to over-estimate than to under-estimate their sensitiveness. In any case, whether their pain be acute or dull, whether their pleasures be intense or the reverse, we should do all in our power to increase the pleasures and diminish the pains of the dumb creatures who so meekly and willingly minister to our wants.

That the bodily feelings and wants occupy a large relative space in the conscious life of brutes can scarcely be questioned. On the one hand are the dull pains resulting from the organic wants and appetences, and driving the animal to their gratification; the keen pleasure that accompanies this gratification, when intelligence is so far developed that it can be foreseen, being a pull in the same direction. And on the other hand are the pleasures of the normal and healthy exercise of the sense-organs and bodily activities giving rise to the pleasures of existence, the joys of active and vigorous life. In the main, these bodily feelings, or sense-feelings, as they are sometimes called, seem to cluster round three chief centres—food, sex, and the free exercise of the bodily activities, including in some cases what seems to be play. Give a wild creature liberty and the opportunity of gratifying its appetites; allow its bodily functions the alternating rhythm of healthy and vigorous exercise and restorative repose; and its life is happy and joyous. It is not troubled by the pressure

* Page 104.

of unfulfilled ideals. The very struggle for existence, keen as it often is, by calling into play the full exercise of the activities, ministers to the health and happiness of brutes as well as men. Sir W. R. Grove has preached[*] the advantages of antagonism. Speaking of the rabbit, he says, "To keep itself healthy, it must exert itself for its food; this, and perhaps avoiding its enemies, gives it exercise and care, brings all its organs into use, and thus it acquires its most perfect form of life. An estate in Somersetshire, which I once took temporarily, was on the slope of the Mendip Hills. The rabbits on one part of it, that on the hillside, were in perfect condition, not too fat nor too thin, sleek, active, and vigorous, and yielding to their antagonists, myself and family, excellent food. Those in the valley, where the pasturage was rich and luxuriant, were all diseased, most of them unfit for human food, and many lying dead on the fields. They had not to struggle for life; their short life was miserable and their death early; they wanted the sweet uses of adversity—that is, of antagonism." Without endorsing the view that these rabbits were unhealthy *only* because they had too much food and comfort—for the food, though abundant, may have been in some way noxious, and the damp situation may have been prejudicial—we may still believe that a struggle for life is better for animals (and men) than unlimited ease and plenty.

Under the influence, then, of these bodily pleasures and wants, the activities of animals are drawn out and guided. As Darwin says, in his autobiography,[†] "An animal may be led to pursue that course of action which is most beneficial to the species by suffering, such as pain, hunger, thirst, and fear; or by pleasure, as in eating and drinking, and in the propagation of the species; or by both means combined, as in the search for food. But pain or suffering of any kind, if long continued, causes depression, and lessens the power of action, yet it is adapted to make a creature guard itself against any great or sudden evil.

[*] *Nature*, vol. xxxvii. p. 619.   [†] Vol. i. p. 310, under date 1876.

Pleasurable sensations, on the other hand, may be long continued without any depressing effect; on the contrary, they stimulate the whole system to increased action. Hence it has come to pass that most or all sentient beings have been developed in such a manner, through natural selection, that pleasurable sensations serve as their habitual guides. We see this in the pleasure from exertion, even occasionally of great exertion, of the body or mind—in the pleasure of our daily meals, and especially in the pleasure derived from sociability, and from loving our families. The sum of such pleasures as these, which are habitual or frequently recurrent, give, as I can hardly doubt, to most sentient beings an excess of happiness over misery, although they occasionally suffer much. Such suffering is quite compatible with belief in natural selection; which is not perfect in its action, but tends only to render each species as successful as possible in the battle for life with other species, in wonderfully complex and changing circumstances."

Passing now from the bodily feelings and wants to the emotions, there can be no question that the simpler emotions, of which I have taken fear and anger as typical, are shared with us by the dumb brutes. And the interesting observations of Mr. Douglas Spalding showed beyond doubt that they are instinctive—their manifestation being prior to, and not the outcome of, individual experience. Writing in *Macmillan's Magazine*, he says, "A young turkey, which I had adopted when chirping within the uncracked shell, was, on the morning of the tenth day of its life, eating a comfortable breakfast from my hand, when the young hawk in a cupboard just beside us gave a shrill 'Chip! chip! chip!' Like an arrow, the poor turkey shot to the other side of the room, and stood there, motionless and dumb with fear, until the hawk gave a second cry, when it darted out at the open door right to the extreme end of the passage, and there, silent and crouched in a corner, remained for ten minutes. Several times during the course of that day it again heard these alarming sounds, and in every instance with similar manifestations of fear." And as an example of combined fear and anger, Mr. Spalding says, "One day last month, after fondling my dog, I put my hand into a basket containing four blind kittens three days old. The smell my hand had carried with it sent them puffing and spitting in a most comical fashion."

A remarkable instance of inherited antipathy in the dog was communicated by Dr. Huggins to Mr. Darwin. He possessed an English mastiff, Kepler, which was brought when six weeks old from the stable in which he was born. The first time Dr. Huggins took him out he started back in alarm at the first butcher's shop he had ever seen, and throughout his life he manifested the strongest and strangest antipathy to butchers and all that pertained to them. On inquiry, Dr. Huggins ascertained that in the father, in the grandfather, and in two half-brothers of Kepler the same curious antipathy was innate. Of these, Paris, a half-brother, on one occasion, at Hastings, sprang at a gentleman who came into the hotel at which his master was staying. The owner caught the dog, and apologized, saying he had never known him to behave thus before except when a butcher came into the house. The gentleman at once said that was his business.

That many animals display affection towards their offspring and their mates, towards man and towards other companions, is a matter of familiar observation. Often the attachments are strange, as of cats and horses, or contrary to instinctive tendencies, as between cats and dogs. Sometimes they are capricious, as when Mr. Romanes's wounded widgeon conceived a strong, persistent, and unremitting attachment to a peacock;* or even insane, as where a pigeon became the victim of an infatuation for a ginger-beer bottle. Strong attachment to man is often exhibited. Every one knows the story which Mr. Darwin tells † of the little monkey who bravely rushed at the dreaded baboon which had attacked his keeper. A friend

---

* "Mental Evolution in Animals," p. 318.
† "Descent of Man," pt. i. chap. iii.

of my own (the Rev. George H. R. Fisk, of Capetown) tells me the following story (which may be added to the many similar cases reported of dogs) concerning a favourite cat he had as a boy. It happened that the children of the house, my friend among the number, were confined to their room by measles. Their mother remained with the children by day and night until they were convalescent. She then came down and resumed her usual daily life, but was shocked at the appearance of the cat, which was little more than skin and bones, and would not touch food or milk. The cat seemed to know that Mrs. Fisk could help her, and gave her no peace till she had taken her upstairs to the convalescent patients. To Mrs. Fisk's surprise, the cat snarled and beat the young master with her paws. Why the cat chose this peculiar method of venting her feelings it is difficult to say. But immediately afterwards she went down into the kitchen, ate the meat and drank the milk which she had before refused to touch. Early next morning she mewed outside the young master's room; and, having gained admittance, sat at the foot of the bed until he woke, and then licked his face and hair.

This leads us on to the class of sympathetic emotions. For the sympathetic emotions are those which centre, not round the self, but round some other self in whose welfare an interest is, in some way and for some reason, aroused. Not long ago, at the Hamburg Zoological Gardens, I saw two baboons fighting savagely. One at last retreated vanquished, with his arm somewhat deeply gashed. He climbed to a corner of the cage and sat down, moodily licking his wound. Thither followed him a little capuchin, and, though his bigger friend took mighty little notice of his overtures, seemed anxious to comfort him, nestling against him, and laying his head against his side. So far as one could judge, it was not curiosity, but sympathy, that prompted his action.

The following example of sympathetic action on the part of a dog towards a stranger-dog is communicated to me by Mrs. Mann, a friend of mine at the Cape. Carlo was a favourite black retriever, and a highly intelligent animal. "One day," says Mrs. Mann, "a miserable-looking white dog came into our yard. Carlo went up to him, looking displeased, dog-fashion, and ready to fly at the intruder. It was clear, however, that some communication passed between them, for Carlo's wrath seemed disarmed, and he trotted into the kitchen, coming out again with a chop-bone (one with a good deal of meat on it) which the cook had given him. On looking into the yard, the miserable cur was seen enjoying the bone, Carlo sitting straight up watching him with a look of satisfaction." *

That dogs feel sympathy with man will scarcely be questioned by any one who has known the companionship of these four-footed friends. At times they seem instinctively to grasp our moods, to be silent with us when we are busy, to lay their shaggy heads on our knees when we are worried or sad, and to be quickened to fresh life when we are gay and glad—so keen are their perceptions. Their life with man has implanted in them some of the needs of social beings; and as they are ever ready to sympathize with us, so do they rejoice in our sympathy. To be deprived of that sympathy, to be neglected, to have no attention bestowed on them, is to some dogs a punishment more bitter than direct reproof. Mr. Romanes quotes † an account given him by Mrs. E. Picton of a Skye terrier who had the greatest aversion to being washed, snarling and biting during the operation. Threats, beating, and starvation were all of no avail; but the animal was reduced to submission by persistent neglect on the part of his mistress. At the end of a week or ten days he looked wretched and forlorn, and yielded himself quite quietly and

---

* Miss Nellie Maclagan describes how her Newfoundland similarly took a roll to a hungry pauper-friend (*Nature*, vol. xxviii. p. 150). Mr. Duncan Stewart gives (*Nature*, vol. xxviii. p. 31) the case of a cat who used frequently to provide her blind mother with food. Sir Harry Lumsden states that during the cold autumn of 1878 some tame partridges in Aberdeenshire brought two wild coveys to be fed near the doorstep of the house. And a case has been communicated to me by Miss Agnes Tauner, of Clifton, of a thrush that pulled up worms on the lawn for a lame companion.

† "Animal Intelligence," p. 440.

patiently to one of the roughest ablutions it had ever been his lot to experience.

So far I have been content to credit animals with very general and simple forms of emotion—anger, fear, antipathy, affection, and some form of sympathy. If, on the perusal of familiar anecdotes, we also credit them with jealousy, envy, emulation, pride, resentment, cruelty, deceitfulness, and other more complex emotional states, we must remember that every one of these, as we know them, is essentially human. It is necessary to insist on the need of caution and the danger of anthropomorphism. This is, perhaps, even more necessary in the case of the emotions than in that of the perceptions, which we have before considered. Even among men, different individuals and different races probably vary far more in their emotions than in their perceptions. The emotions of civilized man have assumed their present form in the midst of complex social surroundings. They one and all bear ineffaceably stamped upon them the human image and superscription. In terms of these complex human emotions we have to decipher the simpler emotional states of the lower animals. We call them by the same names; we think of them as like unto those that we experience. And we can do no otherwise, if we are to consider them at all. But let us not lose sight of the fact that all we can ever hope to see in the mirror of the animal mind is a distorted image of our own mental and emotional features. And since the mirrors are of varying and unknown curvature, we can never hope to be in a position accurately to estimate the amount of distortion.

Remembering this, it is always well to look narrowly at every anecdote of animal intelligence and emotion, and endeavour *to distinguish observed fact from observer's inference*. If we take the great number of stories illustrative of revenge, consciousness of guilt, an idea of caste, deceitfulness, cruelty, and so forth, in the higher mammalia, we shall find but few that do not admit of a different interpretation from that given by the narrator. A cat's treatment of a mouse is adduced by a number of witnesses as illustrative of cruelty; but others see in this conduct, not cruelty, but practice and training in an important branch of the business of cat-life. That is to say, the act, though objectively cruel from the human standpoint, is not on this view performed from a motive of cruelty. Some time ago I ventured to stroke the nose of a little lion-cub which had tottered, kitten-like, to the bars of its cage. "I wish," I said shortly afterwards to a distinguished animal painter, "you could have caught the look of conscious dignity (I speak anthropomorphically) with which the lioness turned and seemed to say, 'How dare you meddle with my child!'" "I have seen such a look and attitude," said Mr. Nettleship; "but I attributed it, not to pride, but to fear." Mr. Romanes quotes,* as typically illustrative of an "idea of caste," the case of Mr. St. John's retriever, which struck up an acquaintance with a rat-catcher and his cur, but at once cut his humble friends, and denied all acquaintanceship with them, on sight of his master. I, on the other hand, should regard this case as parallel with that which I have noted a hundred times. My dogs would go out with the nurse and children when I was busy or absent; but if I appeared within sight, they raced to me. The stronger affection prevailed. A dog is described † as "showing a deliberate design of deceiving," because he hobbled about the room as if lame and suffering from pain in his foot. I would suggest that there was no pretence, no "deliberate design of deceit," in this case, but a direct association of ideas between a hobbling gait and more sympathy and attention than usual. I am not denying objective deceitfulness to the dog any more than I deny objective cruelty to the cat. My only question is whether the *motive* is deceit. We must not forget that the deceitful intent is a piece, not of the observed fact, but of the observer's inference. Mr. Romanes, for example, tells ‡ of a black retriever who was asleep, or apparently asleep, in the kitchen of a certain dignitary of the Church. The

* "Animal Intelligence," p. 442. † Ibid. p. 444. ‡ Ibid. p. 451.

cook, who had just trussed a turkey for roasting, was suddenly called away. During her temporary absence, "the dog carried off the turkey to the garden, deposited it in a hollow tree, and at once returned to resume his place by the fire, where he pretended to be asleep as before." Unfortunately, a perfidious gardener had watched him, and brought back the turkey, so that the retriever did not enjoy the feast he had reserved for a quiet and undisturbed moment. Assuming that the gardener and cook were accurate in their statement of fact, the deceitful intent is an inference on their part, or that of the dignitary of the Church, or Mr. Romanes. I do not deny its correctness from the objective standpoint. Deceitfulness is apparently exhibited by children at a very tender age. But for us civilized adults deceit and its converse, truthfulness in action, mean something a good deal more definite than for dogs and infants.

Animals are often described as harbouring feelings of revenge and vindictiveness. To test this in the elephant, Captain Shipp gave an elephant a sandwich of cayenne pepper. "He then waited," says Mr. Romanes,[*] "for six weeks before again visiting the animal, when he went into the stable, and began to fondle the elephant as he had previously been accustomed to do. For a time no resentment was shown, so that the captain began to think that the experiment had failed; but at last, watching an opportunity, the elephant filled his trunk with dirty water, and drenched the captain from head to foot." Here the facts are that an injury was received, and that the retaliation followed after an interval of six weeks. The inference seems to be that the elephant harboured feelings of revenge or vindictiveness during this period. It may have been so. It may be, however, that the elephant never once pictured the captain during the six weeks; but, on seeing him again, remembered the injury, and, as we say, paid him out. But what we understand by revenge and vindictiveness is the keeping of an injury before the mind for the express

[*] "Animal Intelligence," p. 387.

purpose of ultimately avenging it. And this the elephant, to say the least of it, may not have done.

In Miss Romanes's interesting observations on the Cebus monkey, she says,* "He bit me in several places to-day when I was taking him away from my mother's bed after his morning's game there. I took no notice; but he seemed ashamed of himself afterwards, hiding his face in his arms, and sitting quiet for a time." But, in a footnote, we read, "On subsequent observation, I find this quietness was not due to shame at having bitten me; for whether he succeeds in biting any person or not, he always sits quiet and dull-looking after a fit of passion, being, I think, fatigued." I quote this to illustrate the difference which I am endeavouring to insist upon between observed fact and observer's inference.

Mr. Romanes comments† on the remarkable change which has been produced in the domestic dog as compared with wild dogs, with reference to the enduring of pain. "A wolf or a fox will sustain the severest kinds of physical suffering without giving utterance to a sound, while a dog will scream when any one accidentally treads upon its toes. This contrast," says Mr. Romanes, "is strikingly analogous to that which obtains between savage and civilized man: the North American Indian, and even the Hindoo, will endure without a moan an amount of physical pain—or, at least, bodily injury—which would produce vehement expressions of suffering from a European. And, doubtless, the explanation is in both cases the same; namely, that refinement of life engenders refinement of nervous organization, which renders nervous lesions more intolerable." I cannot accept this as the most probable explanation. In the first place, the human beings referred to have different ideals in the matter of conduct under pain and suffering. The American Indian and the Hindoo have a stoic ideal, which does not influence the average European. On the other hand, the dog, from his association with man, has learnt more and more to

* "Animal Intelligence," p. 486.   † Ibid. p. 141.

give expression to his feelings in barks, whines, and yelpings. To howl at every little pain would do a wolf no good, but rather advertise him to his enemies; to howl when his toes are trodden on makes most men look where they are stepping, and probably pet the sufferer for his pains. In the one case, to howl is disadvantageous; in the other, it is advantageous. I do not, however, put forward my own explanation as necessarily more correct than that given by Mr. Romanes (though I regard it myself as more probable). My object is to show that it is possible for two observers to regard the same activities of animals, and read into them different psychological accompaniments. Throughout the sections of Mr. Romanes's work which deal with the emotions, I feel myself forced at almost every turn to question the validity of his inferences.

From all that I have said in the last chapter, it will be gathered that I am not prepared to credit our dumb companions with a single *sentiment*. A sense of beauty, a sense of the ludicrous, a sense of justice, and a sense of right and wrong,—these abstract emotions or sentiments, *as such*, are certainly impossible to the brute, if, as I have contended, he is incapable of isolation and analysis. But, as we have already seen, even with us these emotions have to be particularized and brought within the perceptual sphere ere they are strongly operative on conduct. We are not roused to indignation by an abstract sense of injustice, but by the particular performance of an unjust deed. Even so, however, the emotional state aroused carries with it in us some of the spirit of the conceptual sphere from which it has descended. The analogous emotions in animals cannot possess, if I am right, any tincture of this conceptual spirit. And since we cannot divest ourselves of our conceptual spirituality, we cannot justly estimate what these emotional states, in dog or ape, are like. Remembering this, let us see what can be said in favour of a perceptual sense of injustice, guilt, the ludicrous, and the beautiful. In evidence of a sense of justice, we have the oft-quoted case of the turnspit-dog reported by Arago the astronomer.* This dog refused, with bared teeth, to enter out of his turn the drum by the revolution of which the spit was rotated. M. Arago, for whom the pullet on the spit was being dressed, requested that the dog's companion, after turning the spit for a short time, should be released. Whereupon the dog who had before been so refractory seemed satisfied that his turn for drudgery had come, and, entering the wheel of his own accord, began without hesitation to turn it as usual. Many will be prepared to maintain that dogs resent unjust chastisement. A gentleman I met near Rio de Janeiro possessed a dog whose sensitiveness was such that, after a reproof, he would leave the house, and sometimes not return for several days. His owner assured me of his belief that in such cases the reproof had always been undeserved; and he told me of one definite instance in which the reproof—never more than verbal—had been for a theft which was afterwards found to have been committed by his garden-boy. On this occasion the dog was away for three days, and returned in a wretched and miserable condition. What shall we say of such cases? Seeing how complex is what we call a sense of justice, I am not prepared to credit the dog therewith; and I am disposed to regard such actions as I have just described as the result of a breach of normal association. Dogs, like men, are creatures of habit; and breaches of normal association—occurrences contrary to expectation—give rise to uneasiness, dissatisfaction, and consequent resentment.

Conversely, many of the cases where dogs and other animals are said to know when they have done wrong, and to suffer the pricks of conscience, may probably be satisfactorily explained by association. When my friend, coming down into his drawing-room, sees Tim's "guilty" look, he suspects that the dog has, contrary to rule, been taking a nap on one of the chairs; and his suspicions are not a little strengthened by the unnatural warmth of the easiest armchair. "Ah! Tim always knows when he has done

* "Animal Intelligence," p. 443.

wrong," says my friend. But not improbably the association in Tim's mind is a direct one between a nap on that chair and his master's displeasure. What Tim knows is, perhaps, not that he *has* done wrong, but that he will "catch it." It is the expectation of a reproof, or something more, that gives rise to his look of conscious guilt. In the same way, the look of "conscious rectitude" we often see in some dogs may be due to the anticipation of a word of commendation. And, in general, I fancy that the association in an animal's mind is between the performance of a given act and the occurrence of certain consequences. When this association becomes definite it must, I imagine, draw after it a dislike of such actions as have been accompanied by evil consequences, and a delight in such actions as have been accompanied by pleasant consequences. And eventually this dislike or delight is transferred from his own actions to the similar actions of others. Thus dogs punish their puppies for acts of uncleanliness, while cats are even more particular in this respect. A correspondent in *Nature* * gives a case of a cat chastising by a violent blow with her paw her kitten, who was about to enjoy a herring which had been set down before the fire to keep hot. So, too, according to Mr. Darwin,† "when the baboons in Abyssinia plunder a garden, they silently follow their leader, and, if an imprudent young animal makes a noise, he receives a slap from the others to teach him silence and obedience." And Mr. Schaub communicated to Professor Nipher ‡ a case of a black-and-tan terrier bitch, whose pup had stolen a stocking from his bedroom, and who followed the young offender, took the stocking from him, and returned it to the owner. Her action gave evidence, he says, of displeasure at the action of the pup. And Mr. Schaub contrived to have the offence committed on many successive mornings, the same performance being repeated each time.

* Mr. Alexander Mackennal, vol. xxi. p. 397.
† "Descent of Man," pt. i. chap. iii., quoted from Brehm's "Thierleben."
‡ *Nature*, vol. xxviii. p. 32.

In this connection I will give two anecdotes of Carlo, communicated to me by Mrs. Mann. "Once I came upon Carlo sitting in the dining-room doorway, Dulceline, the cat, angrily watching him from the stairs, and also evidently having an eye on a leg of mutton half dragged off the dish on the dining-table. Carlo had clearly caught the thief in the act. He was on guard; and he seemed much relieved when higher powers came on the scene. Honesty seemed part of Carlo's nature. In this matter we never had to give him any lessons. Nor could he bear to see dishonesty in others. One Sunday, one of the little girls saw Carlo coming along looking so anxiously at her that she knew he wanted her to come. She therefore followed him, and Carlo took her to the store-room, the door of which her sister had left open. In the doorway Carlo stopped, and looked first up at his mistress and then into the store-room, as much as to say, 'What can we think of this?' And truly there was a certain little black-and-tan terrier, whose principles were by no means of a high order, regaling himself with some cold meat that he had dragged on to the floor. Toby knew he was in the wrong, and tried to flee. But Carlo stopped him as he endeavoured to fly past. And when Toby was thereupon duly slapped, Carlo sat straight up, with a face of conscious rectitude."

These anecdotes, communicated to me by a lady of culture and intelligence, illustrate how, in describing the actions of animals, phraseology only, in strictness, applicable to the psychology of man, is unwittingly and almost unavoidably employed. Toby's "*principles* were not of a high order," yet he "*knew he was in the wrong,*" while Carlo watched him receive his punishment, and "sat straight up, with a face of *conscious rectitude.*"

Coming now to a sense of humour or a sense of the ludicrous, Darwin himself said,* "Dogs show what may fairly be called a sense of humour, as distinguished from mere play; if a bit of stick or other such object be thrown

* "Descent of Man," quoted by Romanes, p. 445.

to one, he will often carry it away for a short distance; and then, squatting down with it on the ground close before him, will wait until his master comes close to take it away. The dog will seize it and rush away in triumph, repeating the same manœuvre, and evidently enjoying the practical joke." Mr. Romanes had a dog who used to perform certain self-taught tricks, "which clearly had the object of exciting laughter. For instance, while lying on his side and violently grinning, he would hold one leg in his mouth. Under such circumstances, nothing pleased him so much as having his joke duly appreciated, while, if no notice was taken of him, he would become sulky." To these I may add an observation of my own. I used sometimes, when staying at Lancaster with a friend, to take his dog Sambo, a highly intelligent retriever, to the seashore. His chief delight there was to bury small crabs in the sand, and then stand watching till a leg or a claw appeared above the surface, upon which he would race backwards and forwards, giving short barks of keen enjoyment. This I saw him do on many occasions. He always waited till a helpless leg appeared, and then bounded away as if he could not contain the canine laughter that was in him. Who shall say, however, what was passing through the mind of the dog in any of these three cases? The motive of Mr. Darwin's dog may have been to prolong the game, though I expect there was something more than this. Mr. Romanes's dog exemplified, perhaps, the sense of satisfaction at being noticed. Sambo's performance is now, as it was years ago, beyond me. But a sense of humour, involving a delicate appreciation of the minor incongruities of life, is, I imagine, too subtle an emotion for even Sambo.

I pass now to the sense of beauty, and I shall consider this at greater length, because of its bearing on sexual selection and the origin of floral beauty.

The interesting experiments of Sir John Lubbock already alluded to seem to establish the fact that bees have certain colour-preferences. Blue and pink are the most attractive colours; yellow and red are in less favour. No doubt these preferences have arisen in association with the flowers from which the bees obtain their nectar. They have a practical basis of biological value. But there seems no doubt that certain colours are now for them more attractive than others. Bees and other insects are, undoubtedly, attracted by flowers; these flowers excite in us an æsthetic pleasure; the bees are, therefore, supposed to be attracted to the flowers through their possession of an æsthetic sense. Now, this does not necessarily follow. It is the nectar, not the beauty of the flower, that attracts the bee. So long as the flower is sufficiently *conspicuous* to be rapidly distinguished by the insect, the conditions of the case are met so far as insect psychology is concerned. The fact remains, however, that the flowers thus conspicuous to the insect are fraught with beauty *for us*.

In the case of sexual selection among birds, again, I believe that the gorgeous plumage has its basis of origin in that pre-eminent vitality which Mr. Tylor and Mr. Wallace have insisted on. But, as before indicated, this will not serve to explain its special character for each several species of birds. Here, again, conspicuousness and recognition are unquestionably factors. But that the bright plumage of male birds awakens emotional states in the hens, that it probably also arouses sexual appetence, seems to be shown by the manner in which the finery is displayed by the male before the female. I think it is probable, also, that pleasure, becoming thus associated with bright colours in the mate, is also aroused by bright colours in other associations. Thus the gardener bower-bird, described by Dr. Beccari,* collects in front of its bower flowers and fruits of bright and varied colours. It removes everything unsightly, and strews the ground with moss, among which it places the bright objects from among which the cock bird is said to select daily gifts for his mate's acceptance! Dr. Gould states that certain humming-birds decorate their nests "with the utmost taste," weaving into their structure beautiful pieces of flat lichen. If by crediting birds with

* *Nature*, vol. xl. p. 327.

a sense of beauty we mean that in them pleasurable emotions may be aroused on sight of objects which we regard as beautiful, I am not prepared to deny them such a sense of beauty, nay, I fully believe that such pleasurable feelings are aroused in them. When, however, it is said that the gorgeous plumage of male birds has been produced by the æsthetic choice of their mates, I am not so ready to agree. A consciously æsthetic motive has not, I believe, been a determining cause. The mate selected has been that which has excited the strongest sexual appetence; his beauty has probably not, as such, been distinctly present to consciousness. Here, then, we have again the question which arose in conection with floral beauty—How is it that the sight of the mates selected by hen birds excites in us, in so many cases, an æsthetic pleasure?

It is clear that this is a matter rather of human than of animal or comparative psychology. As such, except for purposes of illustration, it does not fall within the scope of this work. I can, therefore, say but a few words on the subject. The view that I think erroneous is that either floral beauty or the beauty of secondary sexual characters has been produced on æsthetic grounds, that is to say, for the sake of the beauty they are seen by man to possess. It is, therefore, to the point to draw attention to the fact that many of the objects and scenes which excite in us this æsthetic sense have certainly not been produced for the sake of their beauty. Their beauty is an adjunct, a by-product of rarest excellence, but none the less a by-product.

Nothing can be more beautiful in its way than a well-grown beech or lime tree; and yet it cannot be held to have been produced for its beauty's sake. The leaves of many trees, shrubs, and plants are scarcely less beautiful than the flowers. But *they* cannot have been produced by the æsthetic choice of insects. From the depth of a mine there may be brought up a specimen of ruby copper ore, or malachite, or a nest of quartz crystals, or an agate, or a piece of veined serpentine, which shall be at once pronounced a delight to the eye. But for the eye it was not evolved. The grandeur of Alpine scenery, the charm of a winding river, the pleasing undulations of a flowing landscape,—no one can say that these were evolved for the sake of their beauty. The fact of their being beautiful is, therefore, no proof that the blue gentian, or the red admiral, or the robin redbreast were evolved for the sake of, or by means of, the beauty that they possess. Again, one leading feature in the beauty of flowers is their symmetry. The beauty is, so to speak, kaleidoscopic beauty. It is not so much the single veined or marbled petal that is so lovely, as the group of similar petals symmetrically arranged. But this symmetry can hardly be said to have been selected for its æsthetic value; it is rather part of the natural symmetry of the plant. Even with butterflies and birds and beasts the symmetrical element is an important one in their beauty.*

I must not attempt to analzye our sense of beauty or endeavour to trace its origin. It appears to involve a pleasurable stimulation of the sense-organs concerned,

---

* Another example of beauty which can hardly be said to have been evolved for beauty's sake is to be seen in birds' eggs. Mr. Henry Seebohm regards the bright colours of some birds' eggs as a difficulty in the way of the current interpretation of organic nature. "Few eggs," he says (*Nature*, vol. xxxv. p. 237), "are more gorgeously coloured [than those of the guillemot], and no eggs exhibit such a variety of colour. [They are sometimes of a bluish green, marbled or blotched with full brown or black; sometimes white streaked with brown; sometimes pale green or almost white with only the ghosts of blotches and streaks; and sometimes the reddish brown extends so as to form the ground-tint which is blotched with deeper brown.] It is impossible to suppose that protective selection can have produced colours so conspicuous on the white ledges of chalk cliffs; and sexual selection must have been equally powerless. It would be too ludicrous a suggestion to suppose that a cock guillemot fell in love with a plain-coloured hen because he remembered that last season she laid a gay-coloured egg."

If we connect colour with metabolic changes, its occurrence in association with the products of the highly vascular oviduct will not be surprising. Some *guidance* is, however, on the principles advocated in Chapter VI., required to maintain a standard of coloration. In many cases such guidance is found in protective selection, as in the plover's eggs in our frontispiece. In the guillemot's egg such protective selection seems to be absent, and, as Mr. Seebohm himself says, "no eggs exhibit such a variety of colour."

In our present connection, however, the point to be noticed is that many eggs are undoubtedly beautiful. But they cannot have been in any way selected for the sake of their beauty.

together with perceptions of symmetry, of diversity and contrast, and of proportion, with a basis of unity. It is rich in suggestions and associations. It is heightened by sympathy. A beautiful scene is doubly enjoyable if a congenial companion is by our side.

"The whole effect of a beautiful object, so far as we can explain it," says Mr. Sully,* "is an harmonious confluence of these delights of sense, intellect, and emotion, in a new combination. Thus a beautiful natural object, as a noble tree, delights us by its gradations of light and colour, the combination of variety with symmetry in its contour or form, the adaptation of part to part, or the whole to its surroundings; and, finally, by its effect on the imagination, its suggestions of heroic persistence, of triumph over the adverse forces of wind and storm. Similarly, a beautiful painting delights the eye by supplying a rich variety of light and shade, of colour, and of outline; gratifies the intellect by exhibiting a certain plan of composition, the setting forth of a scene or incident with just the fulness of detail for agreeable apprehension; and, lastly, touches the many-stringed instrument of emotion by an harmonious impression, the several parts or objects being fitted to strengthen and deepen the dominant emotional effect, whether this be grave or pathetic on the one hand, or light and gay on the other. The effect of beauty, then, appears to depend on a simultaneous presentment in a single object of a well-harmonized mass of pleasurable material or pleasurable stimulus for sense, intellect, and emotion."

This, too, is what I understand by an æsthetic sense of beauty; and if a hen bird has her sexual appetence evoked by the bright display of her mate, the emotional state she experiences is something very different from what we know as a sense of beauty. The adjective "æsthetic" should in any case, I think, be resolutely excluded in any discussion of sexual selection.

Æsthetics, like conceptual thought, accompany the sup-

* "Outlines of Psychology," p. 537.

pression or postponement of action. As we have already seen, the normal and primitive series is (1) sense-stimulus; (2) certain nerve-processes in the brain which are associated with perception and emotion; and (3) certain resulting activities. By the suppression of action the mind comes to occupy itself more and more completely with the central processes. Perception blossoms forth into conceptual thought; emotion blossoms forth into æsthetics.

"'Throughout the whole range of sensations, perceptions, and emotions which we do not class as *æsthetic*,'* says Mr. Herbert Spencer, 'the states of consciousness serve simply as aids and stimuli to guidance and action. They are transitory, or, if they persist in consciousness some time, they do not monopolize the attention; that which monopolizes the attention is something ulterior, to the effecting of which they are instrumental. But in the states of mind we class as æsthetic the opposite attitude is maintained towards the sensations, perceptions, and emotions. These are no longer links in the chain of states which prompt and guide conduct. Instead of being allowed to disappear with merely passing recognition, they are kept in consciousness and dwelt upon, their natures being such that their continued presence in consciousness is agreeable.' The action which is the normal consequent on sensation is here postponed or suppressed; and thus we are enabled to make knowledge or beauty an end to be sought for its own sake; and thus, too, we are able to make progress, otherwise impossible, in science and in art. Sensations and perceptions are the roots from which spring the sturdy trunk of action, the expanded leaves of knowledge, and the fair blossoms of art. The leaves and the flowers are the terminal products along certain lines of development; but the function of the leaves is to minister to the growth of the wood, and the function of the flowers is to minister to the continuance and well-being of the race. So, too, in human affairs. Knowledge and art are justified by their influence on conduct; truth and beauty must ever

* I should add, "or as *conceptual thought*."

guide us towards right living; and æsthetics are true or false according as they lead towards a higher or a lower standard of moral life." *

To sum up, then, concerning this difficult subject, the following are the propositions on which I would lay stress: (1) What we term an æsthetic sense of beauty involves a number of complex perceptual, conceptual, and emotional elements. (2) The fact that a natural object excites in us this pleasurable emotion does not carry with it the implication that the object was evolved for the sake of its beauty. (3) Even if we grant, as we fairly may, that brightly coloured flowers, in association with nectar, have been objects of appetence to insects; and that brilliant plumage, in association with sexual vigour, has been a factor in the preferential mating of birds;—this is a very different thing from saying that, either in the selection of flowers by insects, or in the selection of their mates by birds, a consciously æsthetic motive has been a determining cause. (4) In fine, though animals may be incidentally attracted by beautiful objects, they have no æsthetic sense of beauty. A sense of beauty is an abstract emotion. Æsthetics involve ideals; and to ideals, if what has been urged in these pages be valid, no brute can aspire.

What applies thus to æsthetics applies also to ethics. Few, however, will be found to contend that animals can be moral or immoral, or have any moral ideas properly so called. Mr. Romanes does indeed state, in the table he prefixes to his works on Mental Evolution, that the anthropoid apes and dogs are capable of "indefinite morality." He leaves this to be explained, however, in a future work. In the published instalment of "Mental Evolution in Man" he seems to contend,† or, at least, admit, "that the fundamental concepts of morality are of later origin than the names by which they have been baptized." But he says nothing of indefinite morality, which still remains for con-

* This paragraph is quoted from the author's "Springs of Conduct," p. 263.
† Page 347.

sideration in another work. In the mean while we may, I think, confidently assume that ethics, like conceptual thought and æsthetics, are beyond the reach of the brute. Morality is essentially a matter of ideals, and these belong to the conceptual sphere.

I have now said enough* to indicate what I mean by advocating the exercise of extreme caution in our inferences concerning the emotional states of animals. We must remember, first, how liable to error are our inferences in these matters; we must remember, next, how complex and essentially human are our own emotions. I do not for one moment deny that in animals are to be found the perceptual germs of even the higher emotional states. Nevertheless, if we employ, in our interpretation of the actions of animals, such terms as "consciousness of guilt," "sense of right and wrong," "idea of justice," "deceitfulness," "revenge," "vindictiveness," "shame," and the rest, we must not forget that these terms stand for human products, that they are saturated with conceptual thought, and that they must be to a large extent emptied of their meaning before they can become applicable to the emotional consciousness of brutes.

* I have said nothing about the emotions of invertebrates, because I have nothing special to say. They have, no doubt, emotions analogous to fear, anger, and so on. But it is difficult to interpret their actions. The "angry" wasp is, perhaps, a good deal more frightened than furious. Sir John Lubbock's interesting experiments seem to show that ants have what is termed the instinct of play. But this admirable observer has rendered it probable that sympathy and affection in ants and bees have been somewhat exaggerated.

CHAPTER XI.

ANIMAL ACTIVITIES: HABIT AND INSTINCT.

So soon as one of the higher animals comes into the world a number of simple vital activities are already in progress or are at once initiated. Some of these are what are termed "automatic actions," or actions which take their origin within the organ which manifests the activity; such are the heart-beat and the rhythmical contractions of the intestines by which the food is pushed onwards through the alimentary canal. Some are reflex, or responsive, actions, taking origin from a stimulus coming from without; such are the contraction of the pupil of the eye under bright light, the pouring forth of the secretions on the presence of food in the alimentary canal, taking the breast, sneezing, and so forth. Some are partly automatic and partly reflex; such is the rhythm of respiration.

In addition to these vital activities, there is a vast body of more complex activities, for the performance of which the animal brings with it innate capacities. Some of these, which we term "instinctive," are performed at once and without any individual training, as when a chicken steps out into the world, runs about, and picks up food without learning or practice. Others, which we term "habitual," are more or less rapidly learnt, and are then performed without forethought or attention. The store of innate capacity is often very large; and a multitude of activities are ere long performed with ease and certainty so soon as the animal has learnt to use the organization it thus inherits. And lastly, built upon this as a basis, by recombining of old activities in new modes, and by special application of the activities to special circumstances, we have the activities which we term "intelligent;" and here again the activities are sometimes divided into two classes, answering respectively to the reflex and the automatic, but on a higher plane, according as they are responsive to stimuli coming more or less directly from without, or spontaneous and taking their origin from within. But it is probably rather the remoteness and indirectness of the responsive element than its absence that characterizes these spontaneous activities.

Another classification of activities is into voluntary and involuntary. Voluntary actions are consciously performed for the attainment of some more or less definite end or object. Involuntary actions, though they may be accompanied by consciousness, and though they may be apparently purposive, are performed without intention. Notwithstanding the conscious element, they may, perhaps, be regarded as rather physiological than psychological. The simple vital activities belong to this class. But some are much more complex. If, when I am watching the cobra at the Zoo, it suddenly strikes at the glass near my face, I involuntarily start back. The action is apparently purposive, that is to say, an observer of the action would perceive that it was performed for a definite end, the removal from danger; it is also accompanied by consciousness; but it is unintentional, no representation of the end to be gained or the action to be performed being at the moment of action framed by the mind. On the other hand, if I perform a voluntary act, such as selecting and lighting a cigar, there is first a desire or motive directed to a certain end in view, involving an ill-defined representation of the means by which that end may be achieved; and this is followed by the fulfilment of the desire through the application of the means to the performance of the act.

In the carrying out of voluntary activities, then, both perception and emotional appetence are involved. There are construction and reconstruction, memory and anticipation, and interwoven therewith the motive elements of appetence or aversion. It is emotion that gives force and power to the motive. And this must be regarded as the dynamic element in voluntary activity, while intelligence is the directive element. Feeling is the horse in the carriage of life, and Intelligence the coachman.

Let us here note that, in speaking of the activities of animals and the motives by which they are prompted, we are forced, if we would avoid pedantry, to leap backwards and forwards across the chasm which separates the mental from the physical. Motives, as we know them, are mental phenomena; the activities, as we see them, are physical phenomena. The two sets of phenomena belong to distinct phenomenal categories. In ordinary speech, when we pass and repass from motives to actions, and from actions to the feelings they may give rise to, we are apt to be forgetful of the depth of the chasm we so lightly leap. And this is no doubt because the chasm, though so infinitely deep, is so infinitely narrow. There are, however, no physical analogies by which we can explain the connection between the physical and the mental, between body and mind. The so-called connection is, in reality, as I believe, identity. Viewed from without, we have a series of physical and physiological phenomena; felt from within, we have a series of mental and psychological phenomena. It is the same series viewed from different aspects. This is no explanation; it is merely a way, and, as I believe, the correct way, of stating the facts. Why certain physiological phenomena should have a totally different aspect to the organism in which they occur from that which they offer to one who watches them from without, is a question which I hold to be insoluble. All we have to remember, however, is that, in passing from the mental to the physical, we are changing our point of view. The series may be set down thus—

External aspect: Physical stimulus $\rightarrow$ interneural processes $\rightarrow$ activities.
Inner aspect: Accompanying consciousness $\leftarrow$ mental states $\rightarrow$ accompanying consciousness.

The physical stimulus and the resulting activities are

occurrences in the external world, and more or less lie open to our view. But the intervening physical and physiological neural processes are hidden from us. As occurring in ourselves, however, the mental states which are the inner aspects of these neural processes stand out clearly in the light of consciousness. When, therefore, we are watching the life-activities of others, we naturally fill in between the physical stimulus and the activities, not the neural processes of which we are so ignorant, but mental states analogous to those of which we are conscious under similar conditions. Thus we leap from the physical to the mental, and back again to the physical, as represented by the diagonal lines in the above scheme. And there can be no objection to our doing so if we bear in mind that we are thus changing our point of view.

The human organism, then—for at present we may regard the matter from man's own position—is a wonderfully delicate piece of organization, with mental (inner) and physical (outer) aspects. It is in a condition of the most delicate equipoise. Under the influence of a perception associated with an appetence, or of a conception accompanied by a desire, it is thrown into a state of unstable equilibrium; the performance of the action which leads to the fulfilment or satisfaction of the appetence or the desire restores the stability of the system. The instability is caused by the conjoint action of an attraction towards some state represented as desirable, and a repulsion from the existing state which is relatively undesirable. In some cases the attraction, and in others the repulsion, is predominant. When we are in an uncomfortable position, the discomfort is predominant, and we seek relief by changing our attitude. When the bright sunshine tempts us to go out for a walk, the attraction is predominant. But if the uncomfortable attitude is enforced and prolonged, we have a mental representation of the relief we long for; and this is attractive. And if we have work which keeps us indoors, the irksome restraint brings with it an aversion to our present lot.

Inseparably associated with the appetence or aversion there is a representation of the activity which constitutes the fulfilment of the emotion. On the physiological side this is probably an incipient excitation of the muscles or other organs concerned in the requisite actions. The miser's fingers itch to clutch the gold, the possession of which he desires. Our muscles twitch as we long to join in the race or the active contention of a game of football. Our horse grows restive as the hunt goes by. Our dog can scarce restrain himself from racing after the rabbits in the park. Under the influence of emotion, then, the body is prepared for activity, the organs and muscles are beginning to be innervated, and, if the appetence or desire be sufficiently strong, the appropriate actions are initiated, and the organism tends to pass from the state of unstable equilibrium arising out of a pressing need to the stable condition of satisfied appetence. The function of the will in this process we shall have briefly to consider presently.

Let us here notice, with regard to the activities, what we have before seen with regard to the process of perceptual construction. We there noticed that, at the bidding of a relatively simple suggestion, a complex object may be constructed by the mind. This presupposes a highly complex mental organization ready to be set in motion by the appropriate stimulus. The organization has been established by association and through evolution in the individual and his ancestors. It is the same with the activities. They, too, are the outcomes of associations and experiences established and registered during generations of ancestral predecessors. At the bidding of the appropriate stimulus arousing impulse or appetence, a train of activities of great intricacy may be set agoing with remarkable accuracy and precision. It is true that a certain amount of individual education is required to draw out and establish the latent powers of the body, as also of the mind; but *the ability is inborn*, and only requires to be cultivated. Every one of us inherits an organization rendering him capable of performing a vast amount of mental construction and a great number of bodily activities. All he has to do is to learn how to use it and to make himself master of the powers that are given him.

At first, the acquisition of this mastery over the innate powers, even in the performance of comparatively simple muscular adjustments, may require a good deal of attention and practice. But, as time goes on, the frequent repetition of the ordinary activities of everyday life leads to their easier and easier performance. In simple responsive actions the appropriate activity follows readily on the appropriate stimulus. And, ere long, many acts which at first required intelligent attention are performed easily and without consciousness of effort or definite intention. A close association between certain oft-recurring stimuli and the appropriate response in activity is thus established, and the action follows on the stimulus without hesitation or trouble. With fuller experience and further practice in the ordinary avocations of life, the responsive activities link themselves more and more closely in association, become more and more complex, are combined in series and classes of activity of greater length and accuracy, and thus become organized into *habits*. Under this head fall those activities which we learn with difficulty in childhood, and perform with ease in after-life. At first voluntary and intentional, they have become, or are becoming, through frequency and uniformity of performance, more or less involuntary and unintentional.

"The work of the world is," we are told, "for the most part done by people of whom nobody ever hears. The political machine and the social machine are under the ostensible control of personages who are well to the front; but these brilliant beings would be sorely perplexed, and the machinery would soon come to a standstill, but for certain experienced, unambitious, and unobtrusive members of society." So is it also in the economy of animal life. The work of life is—to paraphrase Mr. Norris's words—for the most part done by habits of which nobody ever thinks. The bodily organization is ostensibly under the control of intellect and reason; but these brilliant qualities would be sorely perplexed, and the machinery would soon come to a standstill, but for certain unobtrusive, habitual activities which are already as well trained in the routine work of life as are the permanent clerks in the routine work of a Government office.

The importance of the establishment of these habitual activities is immense. As the muscular and other responses of ordinary everyday life become habitual, the mind is, so to speak, set free from any special care with regard to their regulation and co-ordination, and can be concentrated on the end to be attained by such activities. The cat that is creeping stealthily upon the bird has all her attention rivetted on the object of her appetence, and has not to trouble herself about the movements of her body and limbs. When the swallows are wheeling over our heads in the summer air, their sweeping curves and graceful evolutions are not the outcome of careful planning, but are just the normal exercise of activities which from long practice have become habitual. To swim, to skate, to cycle, to row, to play the piano or the violin,—all these require our full attention at first. But with practice they become habitual, and during their performance the attention may be devoted to quite other matters. This is a great gain. Without it complex trains of activities could not be performed with ease by man or beast.

When once habits have been firmly established, their normal performance is accompanied by a sense of satisfaction. But if their performance is prevented or thwarted, there arises a sense of want or dissatisfaction. The pining of a caged wild animal for liberty is a craving for the free performance of its habitual activities. In an animal born into captivity the craving is probably less intense, though, for reasons which will presently become evident, it is presumably by no means absent. Animals are, to a very large extent, creatures of habit. Much of the pleasure of their existence lies in the performance of habitual activities. Our zoological gardens, interesting as they are to us, are probably centres of an amount of misery and discomfort, from unfulfilled promptings of habit and instinct, which we can hardly realize.

From habitual activities we may pass by easy steps to those which are instinctive. Both habits and instincts, or, to use a more convenient and satisfactory mode of expression for our present purpose, both habitual and instinctive activities, are based upon innate capacity. But whereas habitual activities always require some learning and practice, and very often some intelligence, on the part of the individual, instinctive activities are performed without instruction or training, through the exercise of no intelligent adaptation on the part of the performer, and either at once and without practice (perfect instincts) or by self-suggested trial and practice (incomplete instincts).*

There is some little difficulty in distinguishing between instinctive activities and reflex actions. Mr. Herbert Spencer defines or describes instinct as compound reflex action. Mr. Romanes defines instinct as reflex action into which there is imported the element of consciousness. But, on the one hand, many instincts involve something more than compound reflex action, since there is an organized sequence of activities; and, on the other hand, the difficulty (which Mr. Romanes admits) or impossibility (as I contend) of applying the criterion of consciousness renders unsatisfactory the introduction of the mental element as distinctive. I would say, therefore, that (1) reflex actions are those comparatively isolated activities which are of the nature of organic or physiological responses to more or less definite stimuli, and which involve rather the several organs of the organism than the activities of the organism as a whole; and that (2) instinctive activities are those organized trains or sequences of co-ordinated activities which are performed by the individual

---

* I use the term "incomplete," and not "imperfect," because Mr. Romanes, in his admirable discussion of the subject, applies the term "imperfect instinct" to cases where the instinct is not perfectly adapted to the end in view (see "Mental Evolution in Animals," p. 167).

in common with all the members of the same more or less restricted group, in adaptation to certain circumstances, oft-recurring or essential to the continuance of the species.

These instinctive activities may, as I have said, be performed at once and without practice (perfect instincts) or by self-suggested trial and practice (incomplete instincts). Most young mammals require some little practice in the use of their limbs before they are able to walk or run. But young pigs run about instinctively so soon as they are born. Thunberg, the South African traveller, relates, on the testimony of an experienced hunter, the case of a female hippopotamus which was shot the moment she had given birth to a calf. "The Hottentots," he said, "who imagined that after this they could catch the calf alive, immediately rushed out of their hiding-place to lay hold of it; but, though there were several of them, the new-born calf got away from them, and at once made the best of its way to the river."

Even in cases where some practice is apparently necessary, the activities may be, and often are, perfectly instinctive. They cannot, however, be performed immediately on birth, because the nervous and muscular mechanism is not at that time sufficiently developed. They might, perhaps, with advantage be termed "deferred instincts." If time be given for this development, the activities are carried out at once and without practice. Throw a new-born puppy into the river, and, after some helpless floundering, he will be drowned. Throw his brother when fully grown into the river, and, though he may never have been in the water in his life, he will swim to shore. He has not to learn to swim; this is with him an instinctive activity. The dog inherits the power which the boy must with some little difficulty acquire. He probably has to pay no special attention to the muscular adjustments involved. The act is accompanied by consciousness, but not that directed consciousness we call "attention." When the boy has acquired the habit, he is scarcely conscious of the special muscular co-ordinations as he swims across the river; he is only conscious of a desire to pick the water-lilies near the further bank.

Birds, especially those which are called prœcoces, in contradistinction from the altrices, which are hatched in a helpless, callow condition, come into the world prepared at once to perform complex activities. Mr. Spalding writes,* "A chicken that had been made the subject of experiments on hearing [having been blindfolded at birth] was unhooded when nearly three days old. For six minutes it sat chirping and looking about it; at the end of that time it followed with its head and eyes the movements of a fly twelve inches distant; at ten minutes it made a peck at its own toes, and the next instant it made a vigorous dart at the fly, which had come within reach of its neck, and seized and swallowed it at the first stroke; for seven minutes more it sat calling and looking about it, when a hive-bee, coming sufficiently near, was seized at a dart, and thrown some distance much disabled. For twenty minutes it sat on the spot where its eyes had been unveiled without attempting to walk a step. It was then placed on rough ground, within sight and call of a hen with a brood of its own age. After standing chirping for about a minute, it started off towards the hen, displaying as keen a perception of the qualities of the outer world as it was ever likely to possess in after-life. It never required to knock its head against a stone to discover that there was 'no road that way.' It leaped over the smaller obstacles that lay in its path, and ran round the larger, reaching the mother in as nearly straight a line as the nature of the ground would permit. This, let it be remembered, was the first time it had ever walked by sight."†

Mr. Spalding's experiments also proved that, even

---

* *Macmillan's Magazine*, February, 1873. Professor Eimer, in his "Organic Evolution" (English translation, p. 245), narrates similar experiences.

† Mr. W. Larden states, in *Nature* (vol. xlii.), that his brother extracted, from the oviduct of a Vivora de la Cruz snake in the West Indies, two young snakelets six inches long. Both, though thus from their mother's oviduct untimely ripped, threatened to strike, and made the burring noise with the tail, characteristic of the snake.

among the altrices, young birds do not require to be taught to fly, but fly instinctively so soon as the bodily organization is sufficiently developed to render this activity possible. He kept young swallows caged until they were fully fledged, and then allowed them to escape. They flew straight off at the first attempt. They exhibited the instinctive power of flight in a perfect but deferred form.

It is, however, among the higher invertebrates—especially among the insects, and of them pre-eminently in the social hymenoptera, ants and bees, that the most remarkable and complete instincts are seen. There is, however, a tendency to ascribe all the habits of ants and bees to instinct, often, as it seems to me, without sufficient evidence that they are performed without instruction, and through no imitation or intelligent adjustment. This is, perhaps, a survival of the old-fashioned view that all the mental activities of the lower animals are performed from instinct, whereas all the activities of human beings are to be regarded as rational or intelligent. In popular writings and lectures, for example, we frequently find some or all of the following activities of ant-life ascribed to instinct: recognition of members of the same nest; powers of communication; keeping aphides for the sake of their sweet secretion; collection of aphid eggs in October, hatching them out in the nest, and taking them in the spring to the daisies, on which they feed, for pasture; slave-making and slave-keeping, which, in some cases, is so ancient a habit that the enslavers are unable even to feed themselves; keeping insects as beasts of burden, *e.g.* a kind of plant-bug to carry leaves; keeping beetles, etc., as domestic pets; habits of personal cleanliness, one ant giving another a brush-up, and being brushed-up in return; habits of play and recreation; habits of burying the dead; the storage of grain and nipping the budding rootlet to prevent further germination; the habits described by Dr. Lincecum, and to a large extent confirmed by Dr. McCook,* that Texan

---

* Dr. McCook confirms the observation that the clearings are kept clean, that the ant-rice alone is permitted to grow on them, and that the produce of

ants go forth into the prairie to seek for the seeds of a kind of grass of which they are particularly fond, and that they take these seeds to a clearing which they have prepared, and then sow them for the purpose, six months afterwards, of reaping the grain which is the produce of their agriculture; the collection by other ants of grass to form a kind of soil on which there subsequently grows a species of fungus upon which they feed; the military organization of the ecitons of Central America; and so forth. Now, the description of the habits of ants forms one of the most interesting chapters in natural history. But to lump them together in this way, as illustrations of instinct, is a survival of an old-fashioned method of treatment. That they have to a very large extent *an innate basis* may be readily admitted. But at present we are hardly in a position to say how far they are instinctive, that is, performed by each individual straight off, and without imitation, instruction, or intelligence; how far habitual, that is, performed after some little training and practice; how far there is the intelligent element of special adaptation to special circumstances; how far they are the result of imitation; to what extent, if any, individual training and instruction are factors in the process.

To put the matter in another way. Suppose that an intelligent ant were to make observations on human activities as displayed in one of our great cities or in an agricultural district. Seeing so great an amount of routine work going on around him, might he not be in danger of regarding all this as evidence of blind instinct? Might he not find it difficult to obtain satisfactory evidence of the establishment of our habits, of the fact that this routine work has to some extent to be learnt? Might he not say (perhaps not wholly without truth), "I can see nothing whatever in the training of the children of these men to fit them for their life-activities. The training of their children

---

this crop is carefully harvested; but he thinks that the ant-rice sows itself, and is not actually planted by the ants (see Sir John Lubbock's "Scientific Lectures," 2nd edit., p. 112).

has no more apparent bearing upon the activities of their after-life than the feeding of our grubs has on the duties of ant-life. And although we must remember," he might continue, "that these large animals do not have the advantage which we possess of awaking suddenly, as by a new birth, to their full faculties, still, as they grow older, now one and now another of their instinctive activities are unfolded and manifested. They fall into the routine of life with little or no training as the period proper to the various instincts arrives. If learning thereof there be, it has at present escaped our observation. And such intelligence as their activities evince (and many of them do show remarkable adaptation to uniform conditions of life) would seem to be rather ancestral than of the present time; as is shown by the fact that many of the adaptations are directed rather to past conditions of life than to those which now hold good. In the presence of new emergencies to which their instincts have not fitted them, these poor men are often completely at a loss. We cannot but conclude, therefore, that, although shown under somewhat different and less favourable conditions, instinct occupies fully as large a space in the psychology of man as it does in that of the ant, while their intelligence is far less unerring and, therefore, markedly inferior to our own."

Of course, the views here attributed to the ant are very absurd. But are they much more absurd than the views of those who, on the evidence which we at present possess, attribute all the varied activities of ant-life to instinct? Take the case of the ecitons, or military ants, or the harvesting ants, or the ants that keep draught-bugs as beasts of burden: have we sufficient evidence to enable us to affirm that these activities are purely instinctive and not habitual? That they are to a large extent innate, few are likely to deny; but then our own habitual acts have a basis that is, to a very large extent, innate. The question is not whether they have an innate basis, but whether all the varied manœuvres of the military ants, for example, are displayed to the full without any learning or imitation, without teaching and without intelligence on the part of every individual in the army.*

That in some cases there is something very like a training or education of the ant when it emerges from the pupa condition is rendered probable by the observations of M. Forel. As Mr. Romanes says,† "The young ant does not appear to come into the world with a full instinctive knowledge of all its duties as a member of a social community. It is led about the nest and 'trained to a knowledge of domestic duties, especially in the case of larvæ.' Later on, the young ants are taught to distinguish between friends and foes. When an ants' nest is attacked by foreign ants, the young ones never join in the fight, but confine themselves to removing the pupæ; and that the knowledge of hereditary enemies is not wholly instinctive in ants is proved by the following experiment, which we owe to Forel. He put young ants belonging to three different species into a glass case with pupæ of six other species—all the species being naturally hostile to one another. The young ants did not quarrel, but worked together to tend the pupæ. When the latter hatched out, an artificial colony was formed of a number of naturally hostile species, all living together after the manner of the 'happy families' of the showmen."

I have said that the varied activities of ants, though they may not in all cases be truly instinctive, are nevertheless the outcome of certain innate capacities. It seems to me necessary to distinguish carefully between innate

* The experiments, both of Sir John Lubbock and Mr. Romanes, show that the homing instinct of bees is largely the result of individual observation. Taken to the seashore at no great distance from the hive, where the objects around them, however, were unfamiliar (since the seashore is not the place where flowers and nectar are to be found), the bees were nonplussed and lost their way. Similarly, the migration of birds " is now," according to Mr. Wallace, " well ascertained to be effected by means of vision, long flights being made on bright moonlight nights, when the birds fly very high, while on cloudy nights they fly low, and then often lose their way" ("Darwinism," p. 442). This, of course, does not explain the migratory instinct—the internal prompting to migrate—but it indicates that the carrying out of the migratory impulse is, in part at least, intelligent.

† "Animal Intelligence," p. 59.

capacity and instinct. Every animal comes into the world with an innate capacity to perform the activities which have been necessary for the maintenance of the normal existence of its ancestors. This is part of its inherited organization. Only when these activities are performed at the bidding of impulse, through no instruction and from no tendency to imitation, can they, strictly speaking, be termed instinctive. The more uniform the conditions of ancestral life, and the more highly developed the organism when it enters upon the scene of active existence, the more likely are the innate capacities to manifest themselves at once and without training as perfect instincts. Among birds, the præcoces, which reach a high state of development within the egg, and among insects, those which undergo complete metamorphosis, and emerge from the pupa or chrysalis condition fully formed and fully equipped for life, display the greatest tendency to exhibit activities which are truly and perfectly instinctive. But man, whose ancestors have lived and worked under such complex conditions, and who comes into the world in so helpless and immature a state, though his innate capacities are enormous, exhibits but few and rudimentary instincts.

One marked characteristic of many of the habits and instincts of the lower animals is the large amount of blind prevision (if one may be allowed the expression) which they display. By blind prevision I mean that preparation for the future which, if performed through intelligence or reason, we should term "foresight," but which, since it is performed prior to any individual experience of the results, is done, we must suppose, in blind obedience to the internal impulse. The sphex, a kind of wasp-like insect, forms a little mud chamber in which she lays her eggs. She goes forth, finds a spider, stings it in such a way that it is paralyzed but not killed, and places it in the chamber for her unborn young, which she will never see. The hen incubates her eggs, though she may never have seen a chicken in her life. The caterpillars of an African moth weave a collective cocoon as large as a melon. All unite to weave the enveloping husk; each forms its separate cocoon within the shell, and all these separate cocoons are arranged round branch-passages or corridors, by which the moths, when they emerge from the chrysalis condition, may escape. Another caterpillar, that of a butterfly (*Thekla*) feeds within the pomegranate, but with silken threads attaches the fruit to the branch of the tree, lest, when withered, it should fall before the metamorphosis is complete. An ichneumon fly, mentioned by Kirby and Spence, "deposits its eggs in the body of a larva hidden between the scales of a fir-cone, which it can never have seen, and yet knows where to seek;" and thus provision is made for young which it will never know. Instances of such blind prevision might be quoted by the score. It is idle to speculate as to the accompaniments of consciousness of such acts. If it be asked—May there not be associated with the performance of the instinctive activity of incubation an inherited memory of a generalized chick? we can only answer that we do not know, but that we guess not.*

There is, however, one association, in the case of these and other instincts, which we may fairly surmise to be frequent, though, for reasons to be specified hereafter, it is probably not invariable. Just as we saw to be the case with habits, so too with instinctive activities, their performance is not infrequently associated with pleasurable feeling, their non-performance with pain and discomfort and a sense of craving or want. The animal prevented from performing its instinctive activities is often apparently unquiet, uneasy, and distressed. Hence I said that the animals in our zoological gardens, even if born and reared in captivity, may exhibit a craving for freedom and a yearning to perform their instinctive activities. This craving may be regarded as a blind and vague impulse, prompting the animal to perform those activities which are for its own good and for the good of the race to which it belongs. The satisfaction of the craving, the gratification

---

* The American expression, "I guess," is often far truer to fact than its English equivalent, "I think."

of the blind impulse, is accompanied by a feeling of relief and ease. Thus where a motive emerges at all into consciousness, that from which we may presume that instinctive activities are performed is not any foreknowledge of their end and purpose, but the gratification of an immediate and pressing need, the satisfaction of a felt want.

We have, so far, been concerned merely with the various kinds of activity presented by men and animals, and with some of their characteristics. The organism, in virtue of its organization, has an inherited groundwork of innate capacity. Surrounding circumstances and commerce with the world draw out and develop the activities which the innate capacity renders possible. First, there are automatic and reflex actions, which are comparatively isolated activities in response to definite stimuli, external or internal. Secondly, there are those organized trains or sequences of co-ordinated activities which are performed by the individual in common with all the members of the same more or less restricted group, in adaptation to certain circumstances, oft-recurring or essential to the continuance of the species. These are the instinctive activities. But no hard-and-fast line can be drawn between them and reflex actions. The instinctive activities may be either perfect or relatively imperfect, according to the accuracy of their adaptation to the purpose for which the activity is performed; but in either case they are carried out without learning or practice. In some cases, however, they cannot be performed until the organization is more perfectly developed than it is at birth; but when the proper time arrives they are perfect, and require no practice; these may be termed "deferred instincts." Where some practice, but only a little, is required, the instinctive activities may be regarded as incomplete; and these pass into those activities which require at first a good deal of practice, learning, and attention, but eventually run off smoothly and without special attention, at times almost or quite unconsciously. These are habitual activities. Finally, we have those activities which are performed in special adaptation to special circumstances. These are intelligent activities.

All of these may be, and the last, the intelligent actions, invariably are, accompanied by consciousness. The habitual activities, and those which are incompletely instinctive, are also, we may presume, accompanied by consciousness during the process of their organization and establishment. It is possible, however, that some of the perfectly instinctive activities may be performed unconsciously. When we consider how perfectly organized such activities are, and when we also remember that perfectly organized habitual activities are frequently in us unconscious, we shall see cause for suspecting that instinctive activities may, at any rate in some cases, be unconscious. No doubt the conditions of consciousness are not well understood. But let us accept Mr. Romanes's suggestion, that a physiological concomitant is ganglionic delay. "Now what," he asks,* "does this greater consumption of time imply? It clearly implies," he answers, "that the nervous mechanism concerned has not been fully habituated to the performance of the response required, and therefore that, instead of the stimulus merely needing to touch the trigger of a ready-formed apparatus of response (however complex this may be), it has to give rise in the nerve-centre to a play of stimuli before the appropriate response is yielded. In the higher planes of conscious life this play of stimuli in the presence of difficult circumstances is known as indecision; but even in a simple act of consciousness—such as signalling a perception—more time is required by the cerebral hemispheres in supplying an appropriate response to a non-habitual experience, than is required by the lower nerve-centres for performing the most complicated of reflex actions by way of response to their habitual experience. In the latter case the routes of nervous discharge have been well worn by use; in the former case these routes have to be determined by a complex play of forces amid

* "Mental Evolution in Animals," pp. 73, 74.

the cells and fibres of the cerebral hemispheres. And this complex play of forces, which finds its physiological expression in a lengthening of the time of latency, finds also a psychological expression in the rise of consciousness." Now, since in many instinctive activities the stimulus "merely needs to touch the trigger of a ready-formed apparatus of response," I think that they *may* be unconscious. And Mr. Romanes thus himself supplies the reason for rejecting his own definition of instinct as "reflex action into which there is imported the element of consciousness." Of course, logically, Mr. Romanes can reply, "It is merely a question of where we draw the line; if the activity is unconscious, it is a reflex action; if conscious, it is an instinct." I think this unsatisfactory, (1) because the criterion of consciousness, from its purely inferential nature, is practically impossible of application with accuracy; (2) because the same series of activities may probably at one time be unconscious and at another time conscious; and (3) because many actions which are almost universally regarded as reflex actions may at times be accompanied by consciousness, and would then have, on Mr. Romanes's view, to be regarded as instincts.

Having made this initial criticism, I may now state that I regard Mr. Romanes's treatment of instinct as most admirable and masterly. Building upon the foundation laid by Charles Darwin, he has worked out the theory of instinct in a manner at once broad and yet minute, lucid and yet close, definite in doctrine and yet not blind to difficulties. If I say that it is a piece of work worthy of the great master whose devoted disciple Mr. Romanes has proved himself, I am according it the highest praise in my power. I have ventured in this volume to criticize some of Mr. Romanes's conclusions in the field of animal intelligence. And lest I should seem to undervalue his work, lest our few divergences should seem to hide our many parallelisms, I take this opportunity of testifying to my great and sincere admiration of the results of his careful and exact observations, his patient and thoughtful inferences, and his lucid and often luminous exposition.

I do not propose to go over the ground so exhaustively covered by Mr. Romanes in his discussion of instinct. I shall first endeavour shortly to set forth his conclusions, and then review the subject in the light of modern views of heredity.

Admitting that some instincts may have arisen from the growth, extension, and co-ordination of reflex actions, Mr. Romanes regards the majority of instincts as of two-fold origin—first, from the natural selection of fortuitous unintelligent activities which chanced to be profitable to the agent (primary instincts); and, secondly, from the inheritance of habitual activities intelligently acquired. These are the secondary instincts, comprising activities which have become instinctive through lapsed intelligence. In illustration of primary instincts, Mr. Romanes cites the instinct of incubation. "It is quite impossible," he says,[*] "that any animal can ever have kept its eggs warm with the intelligent purpose of hatching out their contents, so that we can only suppose that the incubating instinct began by warm-blooded animals showing that kind of attention to their eggs which we find to be frequently shown by cold-blooded animals. . . . Those individuals which most constantly cuddled or brooded over their eggs would, other things equal, have been most successful in rearing progeny; and so the incubating instinct would be developed without there ever having been any intelligence in the matter."

Many of the instincts which exhibit what I have termed above "blind prevision" must, it would seem, belong completely or in the main to this class. The instincts of female insects, which lead them to anticipate by blind prevision the wants of offspring they will never see; the instincts of the caterpillars, which lead them to make provision for the chrysalis or imago condition of which they can have no experience; the instinct of a copepod crustacean, which lays its eggs in a brittle-star, that they

[*] "Mental Evolution in Animals," p. 177.

may therein develop, probably in the brood-sac, and may even destroy the reproductive powers of the host for the future good of her own offspring—these and many others would seem to have no basis in individual experience.

In illustration of the second class of instincts, those due to lapsed intelligence, Mr. Romanes cites the case of birds living on oceanic islands, which at first show no fear of man, but which acquire in a few generations an instinctive dread of him—for the wildness or tameness may become truly instinctive. "If," says Dr. Rae,* "the eggs of a wild duck are placed with those of a tame one under a hen to be hatched, the ducklings from the former, on the very day they leave the egg, will immediately endeavour to hide themselves, or take to the water if there is any water, should any person approach, whilst the young from the tame duck's eggs will show little or no alarm, indicating in both cases a clear instance of instinct or 'inherited memory.'"

It must not be supposed that these two modes of origin are mutually exclusive, and that any particular instinct must belong either to the one class or the other. On the contrary, many instincts have, as it were, a double root—the principle of selection combining with that of lapsing intelligence in the formation of a joint result. Intelligence may thus give a new direction to a primary instinct, and, the intelligent modification being inherited, what is practically a new instinct may arise. Conversely, selection may tend to preserve those individuals which perform some intelligent action, and may, therefore, aid the lapsing of intelligence in establishing and stereotyping an instinct.

Referring the reader to Mr. Romanes's work for the examples and illustrations by which he enforces his views, we may now proceed to consider the subject in the light of recently developed theories of heredity.

We have seen that a school of biologists has arisen who deny the inheritance of acquired characters. But Mr.

---

* *Nature*, vol. xxviii. p. 271, quoted in "Mental Evolution in Animals," footnote, p. 196.

Romanes's secondary instincts depend upon the inheritance of habits intelligently acquired. By the school of Professor Weismann, therefore (if we may so call it without injustice to Mr. Francis Galton), secondary instincts, in so far as any individual acquisition is concerned, are denied. Opposed to this school are those who lay great stress on the inheritance of acquired characters. Some of them seem driven to the opposite extreme in the matter of instinct, and appear to hold that instincts are entirely (or let us say almost entirely) due to lapsed intelligence. Professor Eimer, of Tübingen, for example, says,[*] "I describe as automatic actions those which, originally performed consciously and voluntarily, in consequence of frequent practice, come to be performed unconsciously and involuntarily. . . . Such acquired automatic actions can be inherited. Instinct is inherited faculty, especially is inherited habit." In his discussion of the subject, Professor Eimer seems to make no express allusion to primary instincts. And he regards at any rate some of those which are classed by Mr. Romanes as primary, as due to lapsed intelligence. "Every bird," he says [†] "must, from the first time it hatches its eggs, draw the conclusion that young will also be produced from the eggs which it lays afterwards, and this experience must have been inherited as instinct." He says [‡] that the infant takes the breast and sucks "in accordance with its acquired and inherited faculties." He believes [§] that "the original progenitors of our cuckoo, when they began to lay their eggs in other nests, acted by reflection and with design." Regarding the mason-wasps and their allies, which sting larvæ in the ganglia which govern muscular action, and thus provide their young with paralyzed but living prey, he exclaims,[∥] "What a wonderful contrivance! What calculation on the part of the animal must have been necessary to discover it!" Of the storing instincts of bees he remarks,[¶] "Selection cannot here have had much influence, since the

---

[*] "Organic Evolution," pp. 223, 224.   [†] Ibid. p. 263.
[‡] Ibid. p. 303.   [§] Ibid. p. 258.   [∥] Ibid. p. 279.   [¶] Ibid. p. 276.

workers do not reproduce. In order to make these favourable conditions constant, insight and reflection on the part of the animals, and inheritance of these faculties, were necessary." And he concludes,* "Thus, according to the preceding considerations, automatic action may be described as habitual voluntary action; instinct, as inherited habitual voluntary action, or the capacity for such action."

Professor Eimer would not probably deny the co-operation of natural selection in the establishment of these instincts, but he throws it altogether into the background. Now, such a view seems to me wholly untenable. Many of the instincts of insects are performed only once in the course of each individual life. Can it be supposed that the weaving of a cocoon by the caterpillar is mainly a matter of lapsed intelligence? Even if we credit the hen bird with the amount of reflection supposed by Professor Eimer, can we grant to the ancestors of the ichneumon fly such far-reaching observation and intelligence as really to foresee (not by blind prevision, but through intelligent foresight) the future development of the eggs which she lays in a caterpillar? Are we to suppose that the instinctive action of the young cuckoo, which, *the day after it is hatched*, will eject all the other occupants of a hedge-accentor's nest,† can have had its origin in lapsed intelligence? If, because of their purposive character, we are to regard such instincts as of intelligent origin, may we not be told that through intelligent design the pike has beset its jaws, palate, and gill-arches with innumerable teeth, all backwardly directed for the purpose of holding its slippery prey; and the eagle has protected its eye with a bony ring of sclerotic plates, like the holder of an optician's watch-glass? If mimicry in form and colour is due to natural selection, why not mimicry in habits and activities? If *structures* of a wonderfully purposive character have been evolved with-

* "Organic Evolution," p. 298. The late G. H. Lewes held somewhat similar views.

† See Mr. John Hancock, Natural History Transactions, Northumberland, Durham, and Newcastle-on-Tyne, vol. viii. (1886); and *Nature*, vol. xxxiii. p. 519.

out the intelligent co-operation of the organisms which possess them, why not some of the highly purposive *activities*?

And here the disciple of the school of Professor Weismann will echo and extend the question, and will say, "Yes! why not *all* instinctive activities? You are ready to admit," he will continue, "that many instincts, wonderfully purposive in their nature, are of primary origin, that is due to natural selection; why, then, invoke any other mode of origin? If lapsed intelligence be excluded in these cases, why introduce it at all? Why not admit, what our theory of heredity demands, that * 'all instinct is entirely due to the operation of natural selection, and has its foundation, not upon inherited experiences, but upon the variations of the germ'?"

Professor Weismann's contention needs much more serious consideration than that of Professor Eimer. I think there is force in the *à priori* argument (as an *à priori* argument) that since very complex instincts are probably of primary origin, there is no *à priori* necessity for the introduction of the hypothesis of lapsed intelligence. Let me first illustrate this further.

A certain beetle (*Sitaris*) lays its eggs at the entrance of the galleries excavated by a kind of bee (*Anthophora*), each gallery leading to a cell. The young larvæ are hatched as active little insects, with six legs, two long antennæ, and four eyes, very different from the larvæ of other beetles. They emerge from the egg in the autumn, and remain in a sluggish condition till the spring. At that time (in April) the drones of the bee emerge from the pupæ, and as they pass out through the gallery the sitaris larvæ fasten upon them. There they remain till the nuptial flight of the anthophora, when the larva passes from the male to the female bee. Then again they await their chance. The moment the bee lays an egg, the sitaris larva springs upon it. "Even while the poor mother is carefully fastening up her cell, her mortal enemy is be-

* Weismann, "On Heredity," p. 91.

ginning to devour her offspring; for the egg of the anthophora serves not only as a raft, but as a repast. The honey, which is enough for either, would be too little for both; and the sitaris, therefore, at its first meal, relieves itself from its only rival. After eight days the egg is consumed, and on the empty shell the sitaris undergoes its first transformation, and makes its appearance in a very different form. . . . It changes into a white, fleshy grub, so organized as to float on the surface of the honey, with the mouth beneath and the spiracles above the surface. . . . In this state it remains until the honey is consumed;"* and, after some further metamorphoses, develops into a perfect beetle in August.

Now, it seems to me difficult to understand how, at any stage of this long series of highly adaptive, instinctive activities, lapsed intelligence can have been a factor. And therefore I say, if such a complex series† can have resulted

---

* M. Fabre, as interpreted by Sir John Lubbock, "Scientific Lectures," 2nd edit., p. 45.

† In further illustration of the fact that purposiveness and complex adaptation of activities is no criterion of present or past direction by intelligence, we may draw attention to the action of the leucocytes, or white blood-corpuscles. Metchnikoff found that in the water-flea (*Daphnia*), affected by spores of *Monospora bicuspidata*, a kind of yeast which passes from the intestinal canal into the body-cavity, the leucocytes attacked and devoured the conidia. If a conidium were too much for one cell, a plasmodium, or compound giant-cell, was formed to repel the invader. The same thing occurs in anthrax, the bacilli being attacked and devoured by the leucocytes. "If we summarize," says Mr. Bland Sutton ("General Pathology," pp. 127, 128), "the story of inflammation as we read it zoologically, it should be likened to a battle. The leucocytes are the defending army, their roads and lines of communication the blood-vessels. Every composite organism maintains a certain proportion of leucocytes as representing its standing army. When the body is invaded by bacilli, bacteria, micrococci, chemical or other irritants, information of the aggression is telegraphed by means of the vaso-motor nerves, and leucocytes rush to the attack; reinforcements and recruits are quickly formed to increase the standing army, sometimes twenty, thirty, or forty times the normal standard. In the conflict, cells die and often are eaten by their companions; frequently the slaughter is so great that the tissue becomes burdened by the dead bodies of the soldiers in the form of pus, the activity of the cell being testified by the fact that its protoplasm often contains bacilli, etc., in various stages of destruction. These dead cells, like the corpses of soldiers who fall in battle, later become hurtful to the organism they were in their lifetime anxious to protect from harm, for they

from natural selection and non-intelligent adaptation, I see no *à priori* reason why any instinct, no matter how complex, should not have had a like origin.

Let us, however, next consider whether Professor Weismann's theory of the origin of instincts necessarily altogether excludes intelligence as a co-operating factor. The essential point on which that theory is absolutely insistent is that what is handed on through inheritance is *an innate, and not an individually acquired, character*. Now, since intelligent actions are characteristically individual, and performed in special adaptation to special circumstances, it would seem, at first sight, that the intelligent modification of an instinct could not, on Professor Weismann's view, be handed on. Let us consider whether this must be so.

Speaking of ants and bees, Darwin pointed out that their instincts could not possibly have been acquired by inherited habit, since they are performed by neuter insects, that is, by undeveloped females incapable of laying eggs and continuing their race. For a habit to pass into an instinct by inheritance, it is obviously necessary that the organism which performs the habitual actions should be capable of producing offspring by which these actions might be inherited. But in this case the parental forms do not possess these instincts, while the neuter insects which do possess them are sterile.

And how does Mr. Darwin meet this difficulty? "It is lessened, or, as I believe, disappears," he says,[*] "when it is remembered that selection may be applied to the family

---

are fertile sources of septicæmia and pyæmia—the pestilence and scourge so much dreaded by operative surgeons." Now, if the leucocytes were separate organisms, whose habits were being described, some might suppose that they were actuated by intelligence, individual or inherited. But in this case the activities are purely physiological. The marshalling of the cells during the growth of tissue (*e.g.* the antler of a stag before described) is of like import. And Dr. Verworn has shown that when a (presumably weak) electric current is passed through a drop of water containing protozoa, they will, when the current is closed, flock towards the negative pole, and when the current is opened will travel towards the positive pole. The implication of all this is that vital phenomena may be intensely purposive, and yet afford no evidence or indication of the present or ancestral play of intelligence.

[*] "Origin of Species," p. 230.

as well as to the individual. Breeders of cattle wish the flesh and fat to be well marbled together; an animal thus characterized has been slaughtered, but the breeder has gone with confidence to the same stock, and has succeeded. Such faith may be placed in the power of selection, that a breed of cattle always yielding oxen with extraordinarily long horns could, it is probable, be formed by carefully watching which individual bulls and cows, when matched, produced oxen with the longest horns; and yet no one ox would ever have propagated his kind. . . . Hence we may conclude that slight modifications of structure or of instinct, correlated with the sterile condition of certain members of the community, have proved advantageous; consequently, the fertile males and females have flourished, and transmitted to their fertile offspring a tendency to produce sterile members with the same modifications. This process must have been repeated many times, until that prodigious amount of difference between the fertile and sterile females of the same species has been produced which we see in many social insects."

Now let us apply this illustration to the case of habits intelligently acquired. Instead of the possession of long horns, suppose the performance of some habitual action be observed in the oxen. Then, by carefully watching which individual bulls and cows, when matched, produced oxen which performed this intelligent habitual action, a breed of cattle always yielding oxen which possessed this habit might, on Darwin's principles, be produced. The intelligence of oxen might in this way be enhanced. Such faith may be placed in the power of selection that a breed of cattle always yielding oxen of marked intelligence could, it is possible, be formed by carefully watching which individual bulls and cows, when matched, produced the most intelligent oxen; and yet no ox would ever have propagated its kind. Regarding, then, a nest of ants or bees as a social community, mutually dependent on each other, and subject to natural selection, that community would best escape elimination in which the queen produced two sets of offspring—one set in which the procreative faculty was predominant to the partial exclusion of intelligence, and another in which intelligent activities were predominant to the exclusion of propagation.

It is possible that I have weakened my case by introducing such a difficult problem as the instincts of neuter insects. And I would beg the reader to remember that this is only incidental. What I wish to indicate is that among the many variations to which organisms are subject, there are variations in their intelligent activities; that these are of elimination value, those animals which conspicuously possess them escaping elimination in its several modes; that those survivors which thus escape elimination are likely to hand on, through inheritance, that intelligence which enabled them to survive; that if, thoughout a series of generations, such intelligence be applied to some definite end, nervous channels will tend to be definitely established, and the intelligent activity will more and more readily become habitual; that eventually, through the lapsing of intelligence, these habitual activities may become so fixed and stereotyped as to become instinctive; that intelligence has thus been a factor in the establishment of these instinctive activities; that throughout the sequence there is no inheritance of anything individually acquired, the intelligent variations being throughout of germinal origin; and that, therefore, in the origin of instincts, the co-operation of intelligence and the lapsing of intelligence are not excluded on the principles advocated by Professor Weismann.

What, then, is excluded? Any *individually acquired increment*, either in the intelligence displayed or the stereotyping process. The subject of instinct and of animal intelligence has not at present been considered at any great length by Professor Weismann, but, judging by the general tenor of his writings, I take it that what he demands is definite proof that such individually acquired increment *is actually inherited*.

As before indicated in the chapter on "Heredity," such proof it is, from the nature of the case, almost impossible to produce. Suppose that we find evidence of a gradually increasing application of intelligence to some important life-activity, or a more and more defined stereotyping of some incompletely habitual or instinctive action; how are we to prove that the increment in either case is due to the inheritance of individual acquisitions, not to the selection of favourable innate (that is to say, germinal) variations? Such a hopeless task may at once be abandoned.

Are we, then, to leave the question as insoluble? I think not. It is still open to us to consider whether there are any cases in which the inheritance of acquired modifications is a more probable hypothesis than the selection of favourable germinal variations. Now, the acquisition of an instinctive dread of man, and the loss of this instinctive timidity under domestication, seem to be of this kind. And yet I doubt whether the evidence on this head is *convincing*. For the loss of instinctive timidity, Professor Weismann may invoke the aid of panmixia. But if there is truth in what I have already urged on this head, panmixia will not adequately account for the facts. On the other hand, he may contend that the instinctive dread is not due to the inheritance of individually acquired experience, but to the selection of the wilder birds and animals through the persistent elimination of those which are tame. And in support of this view, he may quote Darwin himself, who says,* "It is surprising, considering the degree of persecution which they have occasionally suffered during the last one or two centuries, that the birds of the Falklands and Galapagos have not become wilder; it shows that the fear of man is not soon acquired." It is questionable, however, whether this persecution, admittedly occasional, can have much elimination value. There is, however, the element of imitation and instruction to be taken into account, and the difficulty of proving that the timidity is really instinctive. It has frequently been observed that birds become, after a while, quite fearless of

* See Appendix to Mr. Romanes's "Mental Evolution in Animals," p. 361.

trains. Here elimination is practically excluded; but it has to be proved that this fearlessness is truly instinctive. Professor Eimer says,* "In my garden every sparrow and every crow know me from afar because I persecute these birds. Once, in the presence of a friend, I shot a crow from the roof of my house, while the pigeons and starlings on the same roof, to the great astonishment of my friend, to whom I had predicted it, remained perfectly quiet. They had learned by frequent experience at what my gun was aimed, and knew that it did not threaten them." There is nothing in this interesting observation, however, to show that what the pigeons had learnt had, by *inherited* experience, become instinctive. And Professor Weismann will not, in all probability, be prepared to accept as a logical inference "that this instinct of fear, because it can be dispelled by experience, must be founded on inherited, acquired experience." †

Fully admitting, then, that this is a matter of relative probability, and that the observations and inferences in this matter are not by themselves convincing, I still think that the balance of probability is here on the side of some inheritance of experience. Take next such an instinctive habit as that which dogs display of turning round in a narrow circle ere they lie down. In its origin the instinct probably arose with the object of preparing a couch in the long grass. Now, is this habit of elimination value? Can we suppose that it arose through the elimination of those ancestral animals which failed to perform this habit? I find it difficult to accept this view, though it is just possible that the animals which did this thereby escaped the observation of their enemies. It is also possible that this originally was a merely purposeless habit, a strange trick of manner, which has been inherited, and rendered constant and fixed. Here again, however, I think the balance of probability is that the habit was intelligently acquired and inherited.

I have before drawn attention to the more or less in-

* "Organic Evolution," p. 227.   † Ibid. p. 228.

completely instinctive avoidance, by birds and lizards, of insects with warning coloration. That the avoidance is not perfectly instinctive is shown by the fact that young birds sometimes taste these caterpillars or insects. But a very small basis of experience, often a single case, is sufficient to establish the association. And in young chicks the avoidance of bees and wasps seems to be perfectly instinctive. The effects on the young birds, however, can hardly be of elimination value. Mr. Poulton offered unpalatable insects "to animals from which all other food was withheld. Under these circumstances, the insects were eaten, although often after many attempts, and evidently with the most intense disgust."* I have caused bees to sting young chickens; the result was extreme discomfort, but in no cases permanent injury or death. If, then, the instinct is not of elimination value, that is to say, not such as to save the possessors from elimination, how can it have been established by natural selection? And if not due to natural selection, to what can it be due, save inherited antipathy?

Natural selection is such a far-reaching and ubiquitous factor in organic evolution, that it is not likely that many cases can be found in which the play of elimination can be rigidly excluded. But there are not a few in which elimination does not appear to be the most important factor. Mr. G. L. Grant has recently observed that the sparrows near Auckland, New Zealand, have taken to burrowing holes in sand-cliffs, like the sand-martin. The cliff-swallow of the Eastern United States has almost ceased to build nests in the cliffs, like its progenitors, and now avails itself of the protection afforded by the eaves of houses. The surviving beavers in Europe are said to have abandoned the instinct of building huts and dams. The race being no longer sufficiently numerous to live in communities, the survivors live in deep burrows. In Russian Lapland, under the persecution of hunters, the reindeer are reported to be abandoning the tundras, or open lichen-covered tracts, for

* "Colours of Animals," p. 180.

the forests. The kea (*Nestor notabilis*), a brush-tongued parrot of New Zealand, which normally feeds on honey, fruits, and berries, has, since the introduction of sheep, taken to a carnivorous diet. It is said to have begun by pecking at the sheep-skins hung out to dry; subsequently it began to attack living sheep; and now it has learnt to tear its way down to the fat which surrounds the kidneys. This habit, far from being the result of elimination, is rapidly leading to the elimination of the bird that has so strangely adopted it.

Now, although in these cases elimination has, I think, been a quite subordinate factor, I do not adduce them as convincing evidence that acquired habits are hereditary. Instruction and imitation in each successive generation may well have come into play. There is no proof that they are even incompletely instinctive. But I think that these are the kinds of activities, renewed and careful observations and, if possible, experiments on which, may lead to more decisive results. It would probably not be difficult to ascertain how far the carnivorous habit of the kea has become hereditary, and how far it is performed in the absence of instruction and without the possibility of imitation.

I confess that when I look round upon the varied habits of birds and mammals, when I see the frigate bird robbing the fish-hawk of the prey that it has captured from the sea, the bald-headed chimpanzee adopting a diet of small birds, a *Semnopithecus* in the Mergui Archipelago eating crustacea and mollusca, and the koypu, a rodent, living on shell-fish; when I consider the divergence of habits in almost every group of organisms, the ground-pigeons, rock-pigeons, and wood-pigeons, seed-eating pigeons and fruit-eating pigeons; the carrion-eating, insect-eating, and fruit-eating crows; the aquatic and terrestrial kingfishers, some living on fish, some on insects, some on reptiles;* the divergent habits of the ring-ousel and the water-ousel; and the peculiar habits of blood-sucking bats;—when I see these

* Wallace's "Darwinism," p. 109.

and a thousand other modifications and divergences of habit, I question whether the theory that they have all arisen through the elimination of those forms which failed to possess them may not be pushed too far; I am inclined to believe that the inheritance of acquired modifications has been a co-operating factor. It is not enough to say that these habits are all useful to their several possessors. *It has to be shown that they are of elimination value*—that their possession or non-possession has made all the difference between survival and elimination.

On the whole, then, as the result of a careful consideration of the subject of instinctive and habitual activities, and in accordance with my general view of organic evolution as set forth in previous chapters, I am disposed to accept the inheritance of individually acquired modifications of habit as a working hypothesis. I do not think that absolutely convincing evidence thereof can at present be produced. But to the best of my judgment, the probabilities are in favour of the inheritance of modifications of existing activities, due to intelligence, instruction, and imitation; always provided that the exercise of these modified activities is sufficiently frequent and definite to give rise to habits in the individual.

I recognize three factors in the origin of instinctive activities—

1. Elimination through natural selection.
2. Selection through preferential mating.
3. The inheritance of individually acquired modifications.

Of these I consider the first quite incontrovertible; the second as highly probable; and the third as probable in a less degree. In all three, intelligence may or may not have been a factor. Some of the habits which have survived elimination under the first factor may have been originally intelligent, some of them from the first unintelligent. Some of the love-antics (so called), which, through their tendency to excite sexual appetence in the female, have been selected under the second factor, may have had a basis in intelligence; many of them probably have not. And though the great majority of individually acquired modifications of habits have owed their origin to intelligent direction, still it is conceivable that some of them have not. An animal may have been forced by circumstances to modify its habits, without any exercise of intelligence; and this modification, forced, through changed conditions, upon all the members of a species, may, through inheritance, have passed into the stereotyped condition of an instinct. Under each factor, then, we have two several categories.

1. Elimination .. { *a.* of unintelligent activities.
{ *b.* of intelligent activities
2. Selection .. { *a.* of unintelligent activities.
{ *b.* of intelligent activities.
3. Inheritance .. { *a.* of unintelligent activities.
{ *b.* of intelligent activities.

In all cases, however, where intelligence has been a co-operating factor, this intelligence has lapsed so soon as the activity became truly instinctive.

From the co-operation of the factors it is almost impossible to give examples which shall illustrate the exclusive action of any one. The following table must therefore be regarded as indicating the probable predominance of the factor indicated:—

1. { *a.* Caterpillars spinning cocoons.
{ *b.* Instincts of social hymenoptera.
2. { *a.* Drumming of snipe.
{ *b.* Procedure of Queensland bower-bird.
3. { *a.* Ants forming nests in trees in flooded parts of Siam.
{ *b.* Instinctive fear of man.

In speaking of the instinct of caterpillars spinning cocoons as unintelligent, I am regarding the final purpose of the activity. Intelligence may very possibly have come into play in modifying the details of procedure. In giving the drumming of snipe as an example of unintelligent activities furthered by selection, I am assuming that it has a sexual import, and that the activity correlated with a narrowing of the tail-feathers was not, in its inception, intelligently performed with the object of exciting sexual appetence in the hen. The case of the ants of Siam is given by Mr. Romanes,* on the authority of Lonbière, who says "that in one part of that kingdom, which lies open to great inundations, all the ants made their settlements upon trees; no ants' nests are to be seen anywhere else." Now, this modification of habits may have been the result of intelligence; or it may have been forced upon the ants by circumstances. The floods drove them on to the trees; the instinctive impulse to build a settlement was imperative; hence the settlement had to be formed on the trees, because the ground was flooded. The difficulty of ascertaining whether intelligence has or has not been a factor is simply part of the inherent difficulty of comparative psychology—a difficulty on which sufficient stress has already been laid in an earlier chapter.

The great majority of the instinctive activities of animals have arisen through a co-operation of the factors, and it is exceedingly difficult in any individual case to assign to the factors their several values.

And here we must once more notice that the separation off of the instinctive activities from the other activities of animals is merely a matter of convenience in classification. In the living organism the activities—automatic actions, reflex actions, incompletely and perfectly established instincts, habits, and intelligent activities—are unclassified and commingled. They are going on at the same time, shading the one into the other, untrammelled by the limits imposed by a scientific method of treatment.

Once more, too, we must notice that the activities of animals are essentially the outcome and fulfilment of emotional states. When the emotional sensibility is high, the resulting activities are varied and vigorous. As we have before seen, this high state of emotional sensibility is correlated with a highly charged and sensitive condition of the organic explosives elaborated by the plasmogen of the cells. After repose, and at certain periodic times, this state of exalted sensibility is apt to occur. It is exemplified

* "Mental Evolution in Animals," p. 244.

in the so-called instinct of play, which manifests itself in varied activities in the early morning, in early life, and in the returning warmth of spring—at such times, in fact, as the life-tide is in full flood.

But perhaps the activities which result from a highly wrought state of sensibility are best seen at the periodic return of sexual appetence or impulse in animals of various grades of life and intelligence. Many organisms, at certain periods of the year, and in presence of their mates, are thrown into a perfect frenzy of sexual appetence. The love-antics of birds have been so frequently described that I will merely quote from Darwin * Mr. Strange's account of the satin bower-bird: "At times the male will chase the female all over the aviary, then go to the bower, pick up a gay feather or a large leaf, utter a curious kind of note, set all his feathers erect, run round the bower, and become so excited that his eyes appear ready to start from his head; he continues opening first one wing, and then the other, uttering a low, whistling note, and, like the domestic cock, seems to be picking up something from the ground, until at last the female goes gently towards him." Instances might be quoted from almost all classes of the animal kingdom. Many fish display "love-antics," for example, the gay-suited, three-spine stickleback, whose excitement is apparently intense. Newts display similar activities. Even the lowly snail makes play with its love-darts (*spiculæ amoris*), practical tangible darts of glistening carbonate of lime. Mr. George W. Peckham has recently described † the extraordinary "love-dance" of a spider (*Saitis pulex*). "On May 24 we found a mature female, and placed her in one of the larger boxes; and the next day we put a male in with her. He saw her as she stood perfectly still, twelve inches away; the glance seemed to excite him, and he at once moved towards her; when some four inches from her he stood still, and then began the

---

* "Descent of Man," pt. ii. chap. xiii.
† George W. and Elizabeth G. Peckham, "Occasional Papers of the Natural History of Wisconsin," vol. i. (1889), p. 37.

most remarkable performances that an amorous male could offer to an admiring female. She eyed him eagerly, changing her position from time to time, so that he might be always in view. He, raising his whole body on one side by straightening out the legs, and lowering it on the other by folding the first two pairs of legs up and under, leaned so far over as to be in danger of losing his balance, which he only maintained by sidling rapidly towards the lowered side. The palpus, too, on this side was turned back to correspond to the direction of the legs nearest it. He moved in a semicircle for about two inches, and then instantly reversed the position of the legs, and circled in the opposite direction, gradually approaching nearer and nearer to the female. Now she dashes towards him, while he, raising his first pair of legs, extends them upward and forward as if to hold her off, but withal slowly retreats. Again and again he circles from side to side, she gazing towards him in a softer mood, evidently admiring the grace of his antics. This is repeated until we have counted a hundred and eleven circles made by the ardent little male. Now he approaches nearer and nearer, and when almost within reach whirls madly around and around her, she joining and whirling with him in a giddy maze. Again he falls back and resumes his semicircular motions, with his body tilted over; she, all excitement, lowers her head and raises her body, so that it is almost vertical; both draw nearer; she moves slowly under him, he crawling over her head, and the mating is accomplished."

It can scarcely be doubted that such antics, performed in presence of the female and suggested at sight of her, serve to excite in the mate sexual appetence. If so, it can, further, scarcely be doubted that there are degrees of such excitement, that certain antics excite sexual appetence in the female less fully or less rapidly than others; yet others, perhaps, not at all. If so, again, it can hardly be questioned that those antics which excite most fully or most rapidly sexual appetence in the female will be perpetuated through the selection of the male which performs them.

This is sexual selection through preferential mating. And, I think, the importance of these activities, their wide range, and their perfectly, or at any rate incompletely instinctive nature, justifies me in emphasizing this factor in the origin of instinctive activities. It has hitherto, I think, not received the attention it deserves in discussions of instinct.

A few more words may here be added to what has already been said on the influence of intelligence on instinct. The influence may be twofold—it may aid in making or in unmaking instincts. We have seen that instincts may be modified through intelligent adaptation. A little dose of judgment, as Huber phrased it, often comes into play. The cell-building instinct of bees is one which is remarkably stereotyped; and yet it may be modified in intelligent ways to meet special circumstances. When, for example, honey-bees were forced to build their comb on the curve, the cells on the convex side were made of a larger size than usual, while those on the concave side were smaller than usual. Huber constrained his bees to construct their combs from below upwards, and also horizontally, and thus to deviate from their normal mode of building. The nest-construction of birds, again, may be modified in accordance with special circumstances. And, perhaps, it is scarcely too much to say that, whenever intelligence comes on the scene, it may be employed in modifying instinctive activities and giving them special direction.

Now, suppose the modifications are of various kinds and in various directions, and that, associated with the instinctive activity, a tendency to modify it *indefinitely* be inherited. Under such circumstances intelligence would have a tendency to break up and render plastic a previously stereotyped instinct. For the instinctive character of the activities is maintained through the constancy and uniformity of their performance. But if the normal activities were thus caused to vary in different directions in different individuals, the offspring arising from the union of these differing individuals would not inherit the instinct in the same purity. The instincts would be imperfect, and there would be an inherited tendency to vary. And this, if continued, would tend to convert what had been a stereotyped instinct into innate capacity; that is, a general tendency to certain activities (mental or bodily), the exact form and direction of which is not fixed, until by training, from imitation or through the guidance of individual intelligence, it became habitual. Thus it may be that it has come about that man, with his enormous store of innate capacity, has so small a number of stereotyped instincts.

But while intelligence, displayed under its higher form of originality, may, in certain cases, lead to all-round variation, tending to undermine instinct and render it less stereotyped, intelligence, under its lower form of imitation, has the opposite tendency. For young animals are more likely to imitate the habits of their own species than the foreign habits of other species, and such imitation would therefore tend towards uniformity.

Imitation is probably a by no means unimportant factor in the development of habits and instincts. Mr. A. R. Wallace, in his "Contributions to the Theory of Natural Selection," contends that the nest-building habit in birds is, to a large extent, kept constant by imitation. The instinctive motive is there, but the stereotyped form is maintained through imitation of the structure of the nest in which the builders were themselves reared. Mr. Weir, however, writing to Mr. Darwin, in 1868, says in a letter, which Mr. Romanes quotes,* "The more I reflect on Mr. Wallace's theory, that birds learn to make their nests because they have themselves been reared in one, the less inclined do I feel to agree with him. . . . It is usual with canary-fanciers to take out the nest constructed by the parent birds, and to place a felt nest in its place, and, when the young are hatched and old enough to be handled, to place a second clean nest, also of felt, in the box, removing the other. This is done to prevent acari. But I never knew that canaries so reared failed to make a nest when

---

* "Mental Evolution in Animals," p. 226.

the breeding-time arrived. I have, on the other hand, marvelled to see how like a wild bird's the nests are constructed. It is customary to supply them with a small set of materials, such as moss and hair. They use the moss for the foundation, and line with the finer materials, just as a wild goldfinch would do, although, making it in a box, the hair alone would be sufficient for the purpose. I feel convinced nest-building is a true instinct." On the other hand, Mr. Charles Dixon, quoted* in Mr. Wallace's "Darwinism," speaking of chaffinches which were taken to New Zealand and turned out there, says, "The cup of the nest is small, loosely put together, apparently lined with feathers, and the walls of the structure are prolonged for about eighteen inches, and hang loosely down the side of the supporting branch. The whole structure bears some resemblance to the nests of the hang-birds (*Icteridæ*), with the exception that the cavity is at the top. Clearly these New Zealand chaffinches were at a loss for a design when fabricating their nest. They had no standard to work by, no nests of their own kind to copy, no older birds to give them any instruction, and the result is the abnormal structure I have just described."

There is more evidence in favour of the view that the song of birds is, in part at least, imitative. That it has an innate basis is certain; and that it may be truly instinctive is shown by Mr. Couch's observation of a goldfinch which had never heard the song of its own species, but which sang the goldfinch-song, though tentatively and imperfectly. On the other hand, imitation is undoubtedly a factor. The Hon. Daines Barrington says (1773), "I have educated nestling linnets under the three best singing larks—the skylark, woodlark, and titlark—every one of which, instead of the linnet's song, adhered entirely to that of their respective instructors. When the note of the titlark linnet was thoroughly fixed, I hung the bird in a room with two common linnets for a quarter of a year. They were in full song, but the titlark linnet adhered steadfastly to that of

* "Darwinism," p. 76, from *Nature*, vol. xxxi. p. 533.

the titlark." Mr. Wallace, who quotes this, adds,* "For young birds to acquire a new song correctly, they must be taken out of hearing of their parents very soon, for in the first three or four days they have already acquired some knowledge of the parent's notes, which they afterwards imitate." Dureau de la Malle, as quoted by Mr. Romanes,† describes how he taught a starling the "Marseillaise," and from this bird all the other starlings in a canton to which he took it are stated to have learned the air!

That dogs, monkeys, and other mammalia have powers of imitation needs no illustration. And when we remember that it is only the imitation of strange and unusual actions that arrests our attention, while the imitation of normal activities is likely to pass unnoticed, we may, I think, fairly surmise that imitation is by no means an unimportant factor in the acquisition and development of habits. And where the young animal is surrounded during the early plastic and imitative period of life by its own kith and kin, imitation will undoubtedly have a conservative tendency.

The education of young animals by their parents has also a conservative tendency. Mr. Spalding's observations show that the flight of birds is instinctive; but the parent birds normally aid the development of the instincts by instruction. Ants, as we have seen, are instructed in the business of ant-life. Dogs and cats train their young. And Darwin tells us, on the authority of Youatt,‡ that lambs turned out without their mothers are very liable to eat poisonous herbs.

We may say, then, with regard to the influence of intelligence on instinctive activities, that it may lead them to vary along certain definite lines of increased adaptation; that it may, in some cases, lead them to vary along divergent lines, and hence tend to render stereotyped instincts more plastic; and that, through imitation and instruction, it may tend to render instinctive habits more uniform in a community, and hence, if the habits are tending to vary

---

* "Contributions." etc., p. 222.
† "Mental Evolution in Animals," p. 222.   ‡ "On Sheep," p. 404.

under changed circumstances in a given direction, may tend to draw the habits of all the members of the community in that given direction.

And with regard to the more general question of the variation of habits and instincts, we may say that, in addition to those variations in the origin and direction of which intelligence is a factor, there are other variations which take their origin without the influence of intelligence under the stress of changing circumstances, and yet others which may arise as we say "fortuitously" or "by chance," that is, from some cause or causes whereof we are at present ignorant, and which do not appear to be evoked directly by the stress of environing circumstances.

Granting, however, the existence of these variations in whatsoever way arising, and granting the influence of natural selection, of sexual selection, and perhaps of the inheritance of individually acquired modifications, those variations which are for the good of the race or species in which they occur will have a tendency to be perpetuated, while those which are detrimental will be weeded out and will tend to disappear.

Passing on now to consider the characteristics of those activities which we term "intelligent," we may first notice what Mr. Charles Mercier, in "The Nervous System and the Mind," calls the four criteria of intelligence. Intelligence is manifested, he says, first, in the novelty of the adjustments to external circumstances; secondly, in the complexity; thirdly, in the precision; and fourthly, in dealing with the circumstances in such a way as to extract from them the maximum of benefit.

Now, I think it is clear that, when it is our object to distinguish intelligent from instinctive activities, the precision of the adjustment cannot be regarded as a criterion of intelligence. Many instinctive acts are wonderfully precise. The sphex is said to stab the spider it desires to paralyze with unerring aim in the central nerve-ganglion. Other species, which paralyze crickets and caterpillars, pierce them in three and nine places respectively, according to the number of the ganglia. And yet this seems to be a purely instinctive action. So, too, to take but one more example, there is surely no lack of precision in the cell-making instinct of bees. We may say, then, that, granting that an action is intelligent, the precision of the adjustment is a criterion of the level of intelligence; but that, since there may be instinctive actions of wonderful precision, this criterion is not distinctive of intelligence. Nay, more, there are many reflex actions of marvellous precision and accuracy of adjustment; and there can be no question of intelligence, individual or ancestral, in many of these.

Nor can we regard prevision (which is sometimes advanced as a criterion of intelligence) as specially distinctive of intelligent acts regarded objectively in the study of the activities of animals. For, as we have already seen, there are many instincts which display an astonishing amount of what I ventured to term "blind prevision"—instance the instinctive regard for the welfare of unborn offspring, and the instinctive preparation for an unknown future state in the case of insect larvæ.

Nor, again, is the complexity of the adjustment distinctive of intelligence as opposed to instinct. The case of the sitaris, before given, the larva of which attaches itself to a male bee, passes on to the female, springs upon the eggs she lays, eats first the egg and then the store of honey,— this case, I say, affords us a series of sufficiently marked complexity. This instinct, the paralyzing, but not killing outright, of prey by the sphex; the marvellous economy of wax in the cell-building of the honey-bee; the affixing to their body, by crabs, of seaweed (*Stenorhynchus*), of ascidians (an Australian *Dromia*), of sponge (*Dromia vulgaris*), of the cloaklet anemone (*Pagurus prideauxii*); and other cases too numerous for citation;—these show, too, that the circumstances may be dealt with in such a way as to extract from them the maximum of benefit, probably without intelligence. It would be quite impossible intelligently to improve upon the manner of dealing with the circumstances displayed in many instinctive activities, even those which we have reason to believe were evolved without the co-operation of intelligence.

There remain, therefore, the novelty of the adjustment and the individuality displayed in these adjustments. And here we seem to have the essential features of intelligent activities. The ability to perform acts in special adaptation to special circumstances, the power of exercising individual choice between contradictory promptings, and the individuality or originality manifested in dealing with the complex conditions of an ever-changing environment,— these seem to be the distinctive features of intelligence. On the other hand, in instinctive actions there seems to be no choice; the organism is impelled to their performance through impulse, as by a stern necessity; they are so far from novel that they are performed by every individual of the species, and have been so performed by their ancestors for generations; and, in performing the instinctive action, the animal seems to have no more individuality or originality than a piece of adequately wound clockwork.

It may be said that, in granting to animals a power of individual choice, we are attributing to them free-will; and surely (it may be added), after denying to them reason, we cannot, in justice and in logic, credit them with this, man's choicest gift. I shall not here enter into the free-will controversy. I shall be content with defining what I mean by saying that animals have a power of individual choice. Two weather-cocks are placed on adjoining church pinnacles, two clouds are floating across the sky, two empty bottles are drifting down a stream. None of these has any power of individual choice. They are completely at the mercy of external circumstances. On the other hand, two dogs are trotting down the road, and come to a point of divergence; one goes to the right hand, the other to the left hand. Here each exercises a power of individual choice as to which way he shall go. Or, again, my brother and I are out for a walk, and our father's dog is with us. After a while we part, each to proceed on his own way. Pincher stands irresolute. For a while the impulse to follow me and the impulse to follow my brother are equal. Then the former impulse prevails, and he bounds to my side. He has exercised a power of individual choice. If any one likes to call this yielding to the stronger motive an exercise of free-will, I, for one, shall not say him nay. What I wish specially to notice about it is that we have here a sign of individuality. There is no such individuality in inorganic clouds or empty bottles. Choice is a symbol of individuality; and individuality is a sign of intelligence.

But though I decline here to enter into the free-will controversy, I may fairly be asked where I place volition in the series between external stimulus and resulting activity; and what I regard as the concomitant physiological manifestation. I doubt whether I shall be able to say anything very satisfactory in answer to these questions. I shall have to content myself with little more than stating how the problem presents itself to my mind.

I believe that volition is intimately bound up and associated with inhibition. I go so far as to say that, without inhibition, volition properly so called has no existence. When the series follows the inevitable sequence—

Stimulus : perception : emotion : fulfilment in action

—the act is involuntary. And such it must ever have remained, had not inhibition been evolved, had not an alternative been introduced, thus—

Stimulus : perception : emotion $\Big\langle$ fulfilment in action. / inhibition of action.

At the point of divergence I would place volition. Volition is the faculty of the forked way. There are two possibilities —fulfilment in action or inhibition. I can write or I can cease writing; I can strike or I can forbear. And my poor little wounded terrier, whose gashed side I was sewing up, clumsily, perhaps, but with all the gentleness and tenderness I could command, could close his teeth on my hand or could restrain the action.

I have here, so to speak, reduced the matter to its

simplest expression. It is really more complex. For volition involves an antagonism of motives, one or more prompting to action, one or more prompting to restraint. The organism yields to the strongest prompting, acts or refrains from acting according as one motive or set of motives or the other motive or set of motives prevails; in other words, according as the stimuli to action or the inhibitory stimuli are the more powerful.

And then we must remember that the perceptual volition of animals becomes in us the conceptual volition of man. An animal can choose, and is probably conscious of choosing. This is its perceptual volition. Man not only chooses, and is conscious of choosing, but can *reflect upon his choice;* can see that, under different circumstances, his choice would have been different; can even fancy that, under the same circumstances (external and internal), his choice might have been different. This is conceptual volition. Just as Spinoza said that desire is appetence with consciousness of self; so may we say that the volition of contemplative man is the volition of the brute with consciousness of self. No animal has consciousness of self; that is to say, no animal can reflect on its own conscious states, and submit them to analysis with the formation of isolates. Self-consciousness involves a conception of self, persistent amid change, and isolable in thought from its states. It involves the isolation in thought of phenomena not isolable in experience. We can think about the self as distinct from its conscious states and the bodily organization; but they are no more separable in experience than the rose is separable from its colour or its scent. Such isolation is impossible to the brute. An animal is conscious of itself as suffering, but the consciousness is perceptual. There is no separation of the self as an entity distinct from the suffering which is a mere accident thereof; no conception of a self which may suffer or not suffer, may act or may not act, may be connected with the body or may sever that connection. Just as there is a vast difference between the perception of an object as here and not there,

of an occurrence as now and not then, of a touch as due to a solid body; and the conception of space, time, and causation; so is there a vast difference between a perception of an injury as happening to one's self, and a conception of self as the actual or possible subject of painful consciousness. This difference is clearly seen by Mr. Mivart, who therefore speaks of the *consentience* of brutes as opposed to the *consciousness* of man. Consciousness he regards as conceptual; consentience as perceptual.*
And, as before stated, I should be disposed to accept his nomenclature, were it not for its philosophical implications. For Mr. Mivart regards the difference between consciousness and consentience as a difference *in kind*, whereas I regard it as a generic difference. I believe that consentience (perceptual consciousness) can pass and has passed into consciousness (conceptual consciousness); but Mr. Mivart believes that between the two there is a great gulf fixed, which no evolutionary process could possibly bridge or span.

The perceptual volition of animals, then, is a state of consciousness arising when, as the outcome of perception and emotion, motor-stimuli prompting to activity conflict with inhibitory stimuli restraining from activity. The animal chooses or yields to the stronger motive, and is conscious of choosing. But it cannot reflect upon its choice, and bother its head about free-will. This involves conceptual thought. When physiologists have solved the problem of inhibition, they will be in a position to consider that of volition. At present we cannot be said to know much about it from the physiological standpoint.

Still, as before indicated, the fact of inhibition is unquestionable and of the utmost importance. It has before been pointed out that through inhibition, through the suppression or postponement of action, there has been rendered possible that reverberation among the nervous processes in the brain which is the physiological concomitant of æsthetic and conceptual thought. We have just

---
* In the sense in which I have used the word; not as he uses it himself.

seen that, in association with inhibition, the faculty of volition has been developed. And we may now notice that the postponement or suppression of action is one of the criteria of intelligent as opposed to instinctive or impulsive activities. This is, however, subordinate to the criterion of novelty and individuality.

Granting, then, that an action is shown to be intelligent from the novelty of the adjustments involved, and from the individuality displayed in dealing with complex circumstances (instinctive adjustments being long-established and lacking in originality), we may say that the level of intelligence is indicated by the complexity of the adjustments; their precision; the rapidity with which they are made; the amount of prevision they display; and in their being such as to extract from the surrounding conditions the maximum of benefit.

Before closing this chapter, I will give a classification of involuntary and voluntary activities:—

	Initiation.	Motive.	Result.
A. Involuntary (automatic and reflex)	Sense-stimulus	Unconscious re-action of nerve-centres	Automatic or reflex act
B. Involuntary (habitual and instinctive)	Percept (perhaps lapsed)	Impulse (perhaps lapsed)	Involuntary activity
C. Voluntary (perceptual)	Percept	Appetence	Voluntary activity
D. Voluntary (conceptual)	Concept	Desire	Conduct

In the involuntary acts classed as automatic and reflex, the initiation and the result may be accompanied by consciousness, but the intermediate mental link which answers to the motive in higher activities is, I think, unconscious. In habitual and instinctive activities the consciousness of the percept and the impulse may in some cases have become evanescent, or, to use G. H. Lewes's phrase, have lapsed. In the case of some instincts, originating by the natural selection of unintelligent activities, the perceptual element may never have emerged, and the initiation may have been a mere sense-stimulus.

The division of voluntary activities into perceptual and conceptual follows on the principles adopted and developed in this work. As to the terminology employed, I agree with Mr. S. Alexander * that it is convenient to reserve the terms "desire" and "conduct" for use in the higher conceptual plane. Animals, I believe, are incapable of this higher desire and this higher conduct. It only remains to note that it is within the limits of the fourth class (of voluntary activities initiated by concepts) that morality takes its origin. Morality is a matter of ideals. Moral progress takes its origin in a state of dissatisfaction with one's present moral condition, and of desire to reach a higher standard. The man quite satisfied with himself has not within him this mainspring of progress. The chief determinant of the moral character of any individual is the *ideal self* he keeps steadily in view as the object of moral desire—the standard to be striven for, but never actually attained.

* "Moral Order and Progress."

CHAPTER XII.

MENTAL EVOLUTION.

The phrase "mental evolution" clearly implies the existence of somewhat concerning which evolution can be predicated; and the adjective "mental" further implies that this somewhat is that which we term "mind." What is this mind which is said to be evolved? And out of what has it been evolved? Can we say that matter, when it reaches the complexity of the grey cortex of the brain, becomes at last self-conscious? May we say that mind is evolved from matter, and that when the dance of molecules reaches a certain intensity and intricacy consciousness is developed? I conceive not.

"If a material element," says Mr. A. R. Wallace,* "or a combination of a thousand material elements in a molecule, are alike unconscious, it is impossible for us to believe that the mere addition of one, two, or a thousand other material elements to form a more complex molecule could in any way tend to produce a self-conscious existence. The things are radically distinct. To say that mind is a product or function of protoplasm, or of its molecular changes, is to use words to which we can attach no clear conception. You cannot have in the whole what does not exist in any of the parts; and those who argue thus should put forth a definite conception of matter, with clearly enunciated properties, and show that the necessary result of a certain complex arrangement of the elements or atoms of that matter will be the production of self-consciousness. There is no escape from this dilemma—either all matter is con-

* "Contributions to the Theory of Natural Selection," p. 365.

scious, or consciousness is something distinct from matter; and in the latter case, its presence in material forms is a proof of the existence of conscious beings, outside of and independent of what we term 'matter.'"

There is a central core of truth in Mr. Wallace's argument which I hold to be beyond question, though I completely dissent from the conclusion which he draws from it. I do not believe that the existence of conscious beings, outside of and independent of what we term "matter," is a tenable scientific hypothesis. In which case, Mr. Wallace will reply, "You are driven on to the other horn of the dilemma, and must hold the preposterous view that all matter is conscious."

Now, I venture to think that the use here of the word "conscious" is prejudicial to the fair consideration of the view which I hold in common with many others of far greater insight than I can lay claim to. And it seems to me that we cannot fairly discuss this question without the introduction of terms which, from their novelty, are devoid of the inevitable implications associated with "mind" and "consciousness" and their correlative adjectives. Such terms, therefore, I venture to suggest, not with a view to their general acceptance, but to enable me to set forth, without arousing at the outset antagonistic prejudice, that hypothesis which alone, as it seems to me, meets the conditions of the case.

According to the hypothesis that is known as *the monistic hypothesis*, the so-called connection between the molecular changes in the brain and the concomitant states of consciousness is assumed to be identity. Professor Huxley suggested the term "neuroses" for the molecular changes in the brain, and "psychoses" for the concomitant states of consciousness. According to materialism, psychosis is a product of neurosis; but according to monism, neither is psychosis a product of neurosis, nor is neurosis a product of psychosis, but *neurosis is psychosis*. They are identical. What an external observer might perceive as a neurosis of my brain, I should at the same moment be feeling as a psychosis. The neurosis is the outer or objective aspect; the psychosis is the inner or subjective aspect.

It is almost impossible to illustrate this assumption by any physical analogies. Perhaps the best is that of a curved surface. The convex side is quite different from the concave side. But we cannot say that the concavity is produced by the convexity, or that the convexity is caused by the concavity. The convex and the concave are simply different aspects of the same curved surface. So, too, are molecular brain-changes (neuroses) and the concomitant states of consciousness (psychoses) simply different aspects of the same waves on the troubled sea of being. Again, we may liken the brain-changes to spoken or written words, and the states of consciousness to the meaning which underlies them. The spoken word is, from the physical point of view, a mere shudder of sound in the air; but it is also, from the conceptual point of view, a fragment of analytic thought.

Now, we believe that the particular kind of molecular motion which we call neurosis, or brain-action, has been evolved. Evolved from what? From other and simpler modes of molecular motion. Complex neuroses have been evolved from less complex neuroses; these from simple neuroses; these, again, from organic modes of motion which can no longer be called neuroses at all; and these, once more, from modes of motion which can no longer be called organic. And from what have psychoses, or states of consciousness, been evolved? Complex psychoses have been evolved from less complex psychoses; these from simple psychoses; these, again, from—what? We are stopped for want of words to express our meaning. We believe that psychoses have been evolved. Evolved from what? From other and simpler modes of—something which answers on the subjective side to motion. We can hardly say "of consciousness;" for consciousness answers to a *particular* mode of motion called neurosis. So that unless we are prepared to say that all modes of motion are neuroses, we can hardly say that all modes of that which answers on the subjective side to motion are conscious. I shall venture, therefore, to coin a word * to meet my present need.

It is generally admitted that physical phenomena, including those which we call physiological, can be explained (or are explicable) in terms of energy. It is also generally admitted that consciousness is something distinct from, nay, belonging to a wholly different phenomenal order from, energy. And it is further generally admitted that consciousness is nevertheless in some way closely, if not indissolubly, associated with special manifestations of energy in the nerve-centres of the brain. Now, we call manifestations of energy "kinetic" manifestations, and we use the term "kinesis" for physical manifestations of this order. Similarly, we may call concomitant manifestations of the mental or conscious order "metakinetic," and may use the term "metakinesis" for all manifestations belonging to this phenomenal order. According to the monistic hypothesis, *every mode of kinesis has its concomitant mode of metakinesis, and when the kinetic manifestations assume the form of the molecular processes in the human brain, the metakinetic manifestations assume the form of human consciousness.* I am, therefore, not prepared to accept the horn of Mr. Wallace's dilemma in the form in which he states it. All matter is not conscious, because consciousness is the metakinetic concomitant of a highly specialized order of kinesis. But every kinesis has an associated metakinesis; and *parallel to the evolution of organic and neural kinesis there has been an evolution of metakinetic manifestations culminating in conscious thought.*

Paraphrasing the words of Professor Max Müller,† I say, "Like Descartes, like Spinoza, like Leibnitz, like Noiré, I require two orders of phenomena only, but I define them differently, namely, as kinesis and metakinesis.

---

* I consider that an apology is needed for the coinage of this and of two or three other words, such as "construct," "isolate," and "predominant." I can only say that in each case I endeavoured to avoid them, but found that I could not make my meaning clear, or bring out the point I wished to emphasize without them.

† "Science of Thought," pp. 286, 287.

According to these two attributes of the noumenal, philosophy has to do with two streams of evolution—the subjective and the objective. Neither of them can be said to be prior. . . . The two streams of evolution run parallel, or, more correctly, the two are one stream, looked at from two opposite shores." And again,* " Like Noiré, I would go hand-in-hand with Spinoza, and carry away with me this permanent truth, that metakinesis can never be the product of kinesis (materialism), nor kinesis the product of metakinesis (spiritualism), but that the two are inseparable, like two sides of one and the same substance."

According to this view, the two distinct phenomenal orders, the kinetic and the metakinetic, are distinct only as being different phenomenal manifestations of the same noumenal series. Matter, the unknown substance † of kinetic manifestations, disappears as unnecessary; spirit, the unknown substance of metakinetic manifestations, also disappears; both are merged in the unknown substance of being—unknown, that is to say, in itself and apart from its objective and subjective manifestations.

It will, no doubt, be objected that the final identity of neuroses and psychoses is an assumption. It is pure assumption, it will be said, that these molecular nervous processes, and those percepts and emotions which are their concomitants, are simply different aspects, outer and inner, objective and subjective, physiological and psychological, of the same noumenal series. This must fully and freely be admitted. Any and every explanation of the connection of mind and body is based on an assumption. The commonplace view of two distinct entities, a mind which can act on the body and a body which influences the mind, is a pure assumption. The philosophic view, that there are two entities, body and mind, that neither can act on the other, but that there is a pre-established harmony between the activities of the one and the activities of the other, is, again, a pure assumption. The materialistic view, that matter

---

* " Science of Thought," p. 279.

† I use " substance " here in its philosophical sense.

becomes at last self-conscious, is a pure assumption. The idealistic view, that the world of phenomena has no existence save as a fiction of my own mind, is, once more, a pure assumption. It is not a question of making or of not making an initial assumption; *that we must do in any case*. The question is—Which assumption yields the most consistent and harmonious results?

Again, an answer will, no doubt, be demanded by some people to the question—*How* does that which, objectively considered, is neurosis become subjectively felt as psychosis? Is not the identification of neurosis and psychosis a begging of the question, unless the *how*, the *modus operandi*, is explained? If, in the latter query, by "begging the question" the adoption of an initial assumption is meant, I have already answered it in the affirmative. To the direct question—How does the objective neurosis become conscious as a subjective psychosis?—while freely admitting that I do not know, I enter the protest that it is philosophically an illegitimate question; for an answer is impossible without transcending consciousness. An illustration will, perhaps, make my meaning clear. Suppose that a sentient being be enclosed within a sphere of opaque but translucent ground glass, into the substance of which there are wrought certain characters. Suppose that external to this there is another similar but larger sphere, similarly inscribed, and that a second sentient being is enclosed in the space between the two spheres. By an attentive study of the two spheres, this second sentient being arrives at the conclusion that the markings on the convex surface of the inner sphere answer to the markings on the concave surface of the outer sphere; and he is led to the conviction that what *he* sees as markings on the convex, the being within the sphere sees as markings on the concave. He is, however, perplexed by the question—How can this be? He is acquainted with a certain inner surface and a certain outer surface. He is led to correlate the markings of the one with the markings of the other. But the question how the two can have such different aspects is beyond his solution. Puzzle as he may, he can never solve it. It can only be solved (and how simple then the solution!) by a being *outside both spheres*, who can see what the enclosed being, "cabin'd, cribb'd, confined," could never see, namely, that the characters were wrought in the translucent glass of the spheres. By which parable, imperfect as it is, I would teach that we can never learn how kinetic manifestations have a metakinetic aspect without getting outside ourselves to view kinesis and metakinesis from an independent standpoint. Or, in the words of Sir W. R. Hamilton,[*] "How consciousness in general is possible; and how, in particular, the consciousness of self and the consciousness of something different from self are possible . . . these questions are equally unphilosophical, as they suppose the possibility of a faculty exterior to consciousness and conversant about its operations."

The only course open to us, then, in this difficult but important problem is to make certain assumptions, and see how far a consistent hypothesis may be based upon them. I make, therefore, the following assumptions: First, that there is a noumenal system of "things in themselves" of which all phenomena, whether kinetic or metakinetic, are manifestations. Secondly, that whenever in the curve of noumenal sequences kinetic manifestations (convexities) appear, there appear also concomitant metakinetic manifestations (concavities). Thirdly, that when kinetic manifestations assume the integrated and co-ordinated complexity of the nerve-processes in certain ganglia of the human brain, the metakinetic manifestations assume the integrated and co-ordinated complexity of human consciousness. Fourthly, that what is called "mental evolution" is the metakinetic aspect of what is called brain or interneural evolution.

It would require far more space than I can here command to deal adequately with these assumptions, and meet the objections which have been and are likely to be raised against them. I must content myself with drawing atten-

[*] Quoted in Professor Veitch's "Hamilton," p. 77.

tion to one or two which seem at once obvious and yet easily met.

It may be asked—What advantage has such a view over realistic materialism? Why not assume that neural processes, when they reach a certain complexity, give rise to or produce consciousness?

First of all, I think, the objection raised by Mr. Wallace, in the passage before quoted, to materialism is unanswerable. Secondly, realistic materialism ignores the fact that kinetic manifestations for us human-folk are phenomena of consciousness. To this we will return presently. Thirdly, realistic materialism, and any view which regards the physical series as one which is independent of the psychical accompaniments, and which regards consciousness as in any sense a by-product of neural processes, are open to an objection which was forcibly stated by the late Professor Herbert.* "It is clearly impossible," he says, "for those . . . who teach that consciousness is [a by-product and] never the cause of physical change, to dispute that the actions, words and gestures of every individual of the human race would have been exactly what they have been in the absence of mind; had mind been wanting [had the by-product never emerged], the same empires would have risen and fallen, the same battles would have been fought and won, the same literature, the same masterpieces of painting and music would have been produced, the same religious rites would have been performed, and the same indications of friendship and affection given. To this absurdity physical science [realistic materialism] stands committed." I believe that Professor Herbert's argument, of which this passage is a summary, is, as against realistic materialism, sound and unanswerable. Finally, as Professor Max Müller has well observed,† "Materialism may in one sense be said to be a grammatical blunder; it is a misapplication of a word which can be used in an oblique

---

* T. M. Herbert, "The Realistic Assumptions of Modern Science Examined," 2nd edit., p. 123.
† "Science of Thought," p. 571.

sense only, but which materialists use in the nominative. In another sense it is a logical blunder, because it rests on a confusion between the objective and the subjective. Matter can never be a subject, it can never know, because the name was framed to signify what is the object of our knowledge or what can be known." Materialism, then, for more than one sufficient reason, stands condemned.

It should be stated, however, that Professor Herbert seems to regard the monistic view I am advocating as committed to the absurdity indicated in the passage I have quoted. I am convinced that he was here in error. Indeed, he seems to have failed to see the full bearing of the monistic hypothesis; for while he combats it, he comes very near adopting it himself. With this, however, I have no concern. I have only to show that, on the assumptions above set down, we are not committed to the "absurdity" of supposing that intelligence and consciousness have had no influence on the course of events in organic evolution—that they have only felt the inevitable sequence of physical phenomena without in any way influencing it. According to the monistic hypothesis, kinesis and metakinesis are co-ordinate. The physiologist may explain all the activities of men and animals in terms of kinesis. The psychologist may explain all the thoughts and emotions of man in terms of metakinesis. They are studying the different phenomenal aspects of the same noumenal sequences. It is just as absurd to say that kinetic manifestations would have been the same in the absence of metakinesis, as to say that the metakinetic manifestations, the thoughts and emotions, would have been the same in the absence of kinesis. It is just as absurd to say that the physical series would have been the same in the absence of mind, as to say that the mental series would have been the same in the absence of bodily organization. For on this view consciousness is no mere by-product of neural processes, but is simply one aspect of them. You cannot abstract (except in thought and by analysis) metakinesis from kinesis; for when you have taken away the one, you have taken the other also. To speak of the organic activities being conceivably the same in the absence of consciousness, is like saying that the outer curve of a soap-bubble would be the same in the absence of the inner curve. Whatever hypothetical existences this statement may be true of, it assuredly is not true of soap-bubbles.

To pass on from this point to another, it is possible—I trust not probable, but still not impossible—that some one may say, "But how, on this view, can perception be accounted for? Granted that in the neural processes of the individual organism kinesis is accompanied by those metakinetic manifestations which we term 'consciousness,' how will this account for our perception of a distant object? Yonder scarlet geranium is a centre of kinetic manifestations; it is fifty yards and more away. How can I here, by any metakinetic process, perceive the kinesis that is going on out there?"

For one who can ask this question, I have written the chapter on "Mental Processes in Man," and have used the term "construct," in vain. In vain have I endeavoured to explain that the seat of all mental processes is somewhere within the brain; in vain have I indicated the nature of localization and outward projection; in vain have I reiterated that the object is a thing we construct through a (metakinetic) activity of the mind; in vain have I insisted that our knowledge is merely *symbolic* of the noumenal existence; and perhaps in vain shall I again endeavour to make my meaning clear.

When we say that we perceive an object, the mental process (perception) is the metakinetic equivalent of certain kinetic changes among the brain-molecules. The object, *as an object* (as a phenomenon or appearance), is there generated. As before stated, I assume the existence of a noumenal system of which the noumenal existence, symbolized as object, is a part. But what we term the object is a certain phase of metakinesis accompanying certain kinetic nerve-processes in the brain. In other words, phenomena are states of consciousness, and cannot, for the percipient, be anything else.

"It comes to this, then," an idealist will interpose: "states of consciousness are metakinetic; phenomena are states of consciousness; therefore phenomena are metakinetic. Your kinesis vanishes, and you are one with us, a pure idealist."

Before showing *wherein* I am not a pure idealist, let me state *why* I am not. For the pure idealist, phenomena being states of consciousness, *and nothing more*, the world around resolves itself into an individual dream. Were I to hold this view, this pen which I hold, this table at which I write, the spreading trees outside my window, my little sons whose merry voices I can hear in the garden, my very body and limbs, all are merely states of my own consciousness. This I am not prepared to accept. Do what I will, I cannot believe that such an interpretation of the facts is true.

For this reason I make my first assumption that there is a noumenal system of things in themselves, of which all phenomena, whether kinetic or metakinetic, are manifestations. I differ from the pure idealist in that I believe that phenomena, besides being states of consciousness, *have another, namely, a kinetic, aspect*. What are for me states of consciousness are for you neural processes in my brain. These are, again, for you states of consciousness; but still for some one else they are kinetic processes. And an ordinary extraneous object, like this table, is the phenomenal aspect to me of a noumenal existence; and since that noumenal existence appears to you also in like phenomenal guise, the table is an object for you as well as for me, and not only for us, but for all sentient beings similarly constituted. The world we live in is a world of phenomena; and it has a phenomenal reality every whit as valid as the noumenal reality which underlies it. And that phenomenal reality has two aspects—an inner aspect as metakinesis, and an outer aspect as kinesis.

I must not here further develop the manner in which the hypothesis of monism presents itself to my mind. I will only, before passing on to consider mental or metakinetic evolution, draw passing attention to two matters. We have seen that Professor Hering and Mr. Samuel Butler have suggested "organic memory" as a conception useful for the comprehension of embryonic reconstruction in development and other such matters (see p. 62). On the hypothesis of monism, this may be regarded as a kinetic manifestation of that which in memory rises to the metakinetic level of consciousness.

The other matter is of far wider import. Monism affords a consistent and comprehensible theory of the ego, or conscious self—that which endures amid the flux and reflux of our conscious states. The ego, or self, is that metakinetic unity which answers to, or is the inner aspect of, the kinetic unity of the organism.* Only here and there, in fleeting and changing series, does the metakinesis rise to the level of consciousness. But the metakinetic unity is as completely one, indivisible, and enduring, as is the physical organism which is its kinetic counterpart. No one questions that there is an enduring organism of which certain visible activities are occasional manifestations; no one who has adequately grasped the teachings of monism can question that the enduring ego, of which certain states of consciousness are occasional manifestations, is the metakinetic equivalent of the organic kinesis. This solution of a problem which baffles alike materialists and idealists is, as it seems to me, as satisfactory as it is simple.

And now let us pass on to consider the question of mental or metakinetic evolution. What, on the principles above laid down, can we be said to know or have learnt about it?

The inevitable isolation of the individual mind has long been recognized. "Such is the nature of spirit, or that which acts," says Bishop Berkeley, "that it cannot be

---

* Strictly speaking, of the brain; but since the brain has no organic independence of the body, it is best here to focus attention on the unity of the organism.

itself perceived, but only by the effects that it produceth." "Thinking things, as such," writes Kant, "can never occur in the outward phenomena; we can have no outward perception of their thoughts, consciousness, desires; for all this is the domain of the inward sense." How comes it, then, that there is nothing of which, practically speaking, we are more firmly convinced than that our neighbours have each a consciousness more or less similar to our own? Certain it is that no one can come into sensible contact with his brother's personality and essential spirit. My brother's soul can never stand to me in the relation of object. Subject he never can be to any but himself. What, then, is he—his metakinetic self, not his kinetic material body—to me? In Clifford's convenient phrase, he is an eject. And what is an eject? An eject is a more or less modified image of myself, that I see mirrored, as in a glass darkly, in the human-folk around me. Into every human brother I breathe the spirit of this eject, and he becomes henceforth to me a living soul. Or, if this mode of presentation does not meet with approval, I will say that an eject is that metakinetic unity I infer as identically associated with the organic and kinetic unity of my brother's living body. And I base the close metakinetic correspondence that I infer on the close kinetic correspondence that I observe. But since the only form or kind of metakinesis that I know is that of human self-conscious personality, it is certain that the metakinetic eject is an image of myself; it is and must be, in a word, anthropomorphic.

Too much stress can scarcely, I think, be laid on the human, nay, even the individual, nature of the eject. All other-mind I am bound to think of in terms of my own mind. The men and women I see around me are like curved mirrors, in which I see an altered reflection of my own mental features. By certain signs I may be able to infer in this or that human mirror graces or imperfections that I lack. But throughout my survey of human nature, every estimate of intellectual or moral elevation or degradation that I form must ever be measured in terms of my own subjective base-line. My conception of humanity must always be, not only anthropomorphic, but idiomorphic.

Once more, let it be remembered that the metakinesis that rises to the level of consciousness is that which forms the inner aspect of the neural kinesis of my brain or yours. For each of us, then, that metakinesis is the only possible metakinesis which we can know as such and at first-hand. And for the pure idealist it is the only metakinesis which he can know at all. Not so with us. We have assumed a noumenal system of "things in themselves," of which all phenomena, whether kinetic or metakinetic, are manifestations. We have assumed that kinesis cannot emerge into the light of being without casting its inseparable metakinetic shadow. We have assumed that when the kinetic manifestations assume the integrated and co-ordinated complexity of nerve-processes in certain ganglia of the human brain, the metakinetic manifestations assume the integrated and co-ordinated complexity of human consciousness. Human physiology is teaching us more clearly every day that all human activities are, physically speaking, the outcome of neural processes. Such neural processes are in us conscious. Therefore, granting our assumptions, the conclusion that my neighbour is a conscious self, just as I am, is not only legitimate, but (as we see from the daily conduct of men) inevitable. In other words, certain kinetic phenomena have for us inevitable metakinetic implications.

Now, when we pass from man to the lower animals, the metakinetic implications become progressively less inevitable and less forcible as the kinesis becomes more dissimilar from that which obtains in the human organism. The only metakinesis that we know directly is our own human consciousness. In terms of this we have to interpret all other forms of metakinesis.

It is unnecessary to go over again the ground that has already been covered in previous chapters, in which we have endeavoured to give some account of what seem to us the legitimate inferences concerning the mental processes in animals. The point on which I wish here to insist is that, outside ourselves, we can only know metakinesis in and through its correlative kinesis. Underlying kinetic evolution, we see that, on the hypothesis of monism, there must have been metakinetic evolution. But of this mental or metakinetic evolution we neither have nor can have independent evidence. Such evolution is the inevitable monistic corollary from kinetic evolution. More than this it is not and cannot be. And only on the monistic hypothesis, as it seems to me, is it admissible to believe in mental evolution,* properly so called.

But does not, it may be asked, the hypothesis of monism, if carried to its logical conclusion, involve the belief in a world-consciousness on the one hand, and a crystal-consciousness on the other? If, according to the hypothesis, every form of kinesis has also its metakinetic aspect, "must we not maintain," in the words of Mr. J. A. Symonds, "that the universe being in one rhythm, things less highly organized than man possess consciousness in the degree of their descent, less acute than man's? Must we not also surmise that ascending scales of existence, more highly organized, of whom we are at present ignorant, are endowed with consciousness superior to man's? Is it incredible that the globe on which we live is vastly more conscious of itself than we are of ourselves; and that the cells which compose our corporeal frame are gifted with a separate consciousness of a simpler kind than ours?" To such questions W. K. Clifford replied with an emphatic negative. "Unless we can show," he said, as interpreted by Mr. Romanes,† "in the disposition of the heavenly

---

* I ought not to pass over without notice the "psychological scale" which Mr. Romanes introduces in a table prefixed to "Mental Evolution in Animals." It would be unjust to criticize this too closely, for it is admittedly provisional and tentative. If such a scheme is to be framed, I would suggest that the various phyla of the animal kingdom be kept distinct. I question, however, whether any one can produce a scheme which any other independent observer will thoroughly endorse. And I am inclined to think that the wisest plan is to tabulate the kinetic manifestations which we can actually observe rather than the metakineses of which we can have no independent knowledge.

† *Contemporary Review*, July, 1886. See Clifford's "Lectures and Essays," vol. i. pp. 72 and 248; vol. ii. p. 67.

bodies some morphological resemblance to the structure of a human brain, we are precluded from rationally entertaining any probability that self-conscious volition belongs to the universe."

I conceive that both parties, opposed as they seem, are logically right; and I venture to think that the terms I have suggested will help us here. Mr. Symonds used the word "consciousness" to signify metakinesis in general; Clifford used it to signify that particular kind of metakinesis which in the human brain rises to the level of consciousness. Not only is it not inconceivable, but it is a logical necessity on the hypothesis of monism, that answering to the kinetic rhythm of the universe there is a metakinetic rhythm; but unless the gyrations of the spheres have some kinetic resemblance to the dance of molecules in the human brain, the metakinesis cannot be inferred to be similar to the consciousness of man.

Similarly, with regard to the supposed self-consciousness of the so-called social organism. Mr. Romanes, in his article on "The World as an Eject,"* leads up to his conception of a world-eject through the conception of a society-eject— an eject, he tells us, that, for aught that any one of its constituent personalities can prove to the contrary, may possess self-conscious personality of the most vivid character. Its constituent human minds may be born into it, and die out of it, as do the constituent cells of the human body; it may feel the throes of war and famine, rejoice in the comforts of peace and plenty; it may appreciate the growth of civilization in its passage from childhood to maturity.

This, of course, may be so; or it may not. Who can tell? But Clifford was on firm monistic ground when he maintained that, unless the kinesis be similar, we have no grounds for inferring similarity of metakinesis.

The study of kinesis leads us to recognize different kinds or modes of its manifestation. There is one mode of kinesis in the circling of the planets around the sun, another mode of kinesis in the orderly evolutions of a great

* *Contemporary Review*, July, 1886.

army, another mode in the throb of a great printing-press; there is one mode of kinesis in the quivering molecules of the intensely heated sun, another in the wire that flashes our thought to America, and yet another in the molecular vibrations of the human brain. All are of the same order, all are kinetic. But they differ so widely in mode that each requires separate, patient, and long-continued study. So is it, we may conclude, with metakinesis. There may be, nay, there must be, many modes. But our knowledge is confined to one mode—that in which the metakinesis assumes the form of human consciousness.

I have been led to discuss this matter in order further to indicate the inevitable limits of our knowledge of metakinetic evolution. Our conclusions may be thus summarized: First, we can know directly only one product of metakinetic evolution—that revealed in our own consciousness. Secondly, the process of metakinetic evolution must be reached, if reached at all, indirectly through a study of kinetic evolution. Thirdly, we have no right to infer a mode of metakinesis analogous to human consciousness, unless the mode of kinesis is analogous to that which is observed in neural processes. And, fourthly, the closer the kinetic resemblance we observe, the closer the metakinetic resemblance we may infer.

The last point we have to notice, and it is by no means an unimportant one, is that, just as the kinetic evolution of the organism must be studied in reference to its kinetic environment, so, too, must the metakinetic evolution of mind be studied in reference to its metakinetic or mental environment.

Of course, in ordinary speech, and even in careful scientific description, we are forced, if we would avoid pedantry, to skip backwards and forwards from the kinetic to the metakinetic. We speak of a kinetic cow giving rise to metakinetic fear, and this determining certain kinetic activities. Why we thus interpose a mental link in a physical series has already been explained. The physical cow we know, the physical activities we know, the physical neuroses we scarcely know at all. On the other hand, fear we have ourselves experienced, and know well. Hence we introduce the mental link that we know in place of the physical link of which we are ignorant. And there can be no harm in our doing so when we are working on the practical, and not the philosophical plane. But when we are striving to go deeper, and are employing that gift of analysis which is man's prerogative, in order to proceed to a higher and more complete synthesis,—then we must be careful to keep separate those processes which analysis discloses to be distinct. And I repeat that, on the philosophical plane of thought, we must remember that *metakineses are determined by other metakineses, and by them alone.*

The reader who has kept his head among these slippery places will at once see that this is and must be so; for, as we have already seen (p. 474), all phenomena are states of consciousness, whatever else they may also be. The cow, as a phenomenon, is a *construct*, a product of mental activity, and woven out of states of consciousness. For the pure idealist she is this and nothing more. But for us she is a real external entity, manifested through phenomenal kineses. Hence in ordinary speech we separate the kinetic cow from its metakinetic symbols in consciousness (the convex from the concave aspect), and call the former the cow itself, and the latter our idea of the cow. But, as before maintained, my idea of an object is for me the object. And this is now justified by our deeper analysis.

The physiologist, dealing with organic phenomena in terms of motion (kinesis), proclaims that the physical series is complete, that there is no necessity for the introduction of feeling which is at best but a by-product. The idealist, dealing with the processes of thought and emotion in terms of consciousness, proclaims that his series is complete—an external material universe is an unnecessary encumbrance. Each proclaims a half-truth; each sees that half of the truth which alone is visible from his special standpoint. Monism combines the two (and is, of course, scouted by both). It sees not only that the one series does not in any case interfere with the other, but that the conception of such an interference involves an impossibility and incongruity. As soon could one speak of the convexities of one side of a curved surface interfering with the corresponding concavities of the other side, as of the metakinetic series interfering with the kinetic series, which is its other aspect. But if the one cannot interfere with the other, neither can the one exist without the other. To apply the same analogy, as well might one speak of the convexities of a curved surface existing without the concavities of its other side, as of the kinetic phenomena of organic life as being conceivably the same in the absence of conscious intelligence.

Remembering, then, that just as the environment of kinetic phenomena is itself kinetic, with which consciousness can in no wise interfere, so is the environment of metakinetic phenomena, perception, thought, and emotion, itself metakinetic. Let us now proceed to consider some of the implications.

We have already seen that, in what we may regard as the earlier phases of organic and mental life, the series between stimulus and activity is a simple one, which may be kinetically represented thus—

$$\text{Stimulus} \rightarrow \text{neural processes} \rightarrow \text{motor-activities};$$

but that when inhibition is developed, there arises an alternative, thus—

$$\text{Stimulus} \rightarrow \text{neural processes} \begin{array}{c} \nearrow \text{motor-activities.} \\ \searrow \text{inhibition thereof.} \end{array}$$

And we further saw that, as a result of this inhibition, the entering stimuli, instead of, as it were, rapidly running out of the organism in motor-activities, set up a more and more complex series of diffused and reverberating neural processes in the brain or other central ganglia.

From the metakinetic view-point these diffused and reverberating neural processes in the brain culminate in consciousness as thought, æsthetic emotion, and the higher conceptual mental activities. Deeply as these influence conduct, they are, to a large extent, independent of conduct. A man's thoughts and æsthetic yearnings may be of the truest and purest; but in the moment of temptation and action, when stimuli crowding in run through rapidly to action, he falls away. His conduct belies his ideals. Nevertheless, the ideals were there, but too far away in the region of thought and abstract æsthetics to be operative in action.

Now, we may divide the metakinetic concomitants of neural processes into two categories: first, those which are intimately associated with neural processes directly leading to motor-activities; secondly, those which are, so to speak, floated off from these into the region of thought and æsthetic emotion, and which are therefore associated with neural processes only indirectly or remotely leading to motor-activities. Both have, of course, kinetic equivalents in neural processes, but the former are directly associated with activities and conduct, and the latter are not.

Let me exemplify. Interpretations of nature, theories, hypotheses, belong to the latter class. Their association with activities is in the main indirect. Whether we believe in materialism, idealism, or monism, our conduct is much the same. People got out of the way of falling stones, and guarded against being caught by the incoming tide, before science comprised both phenomena under the theory of gravitation. The conduct of human-folk was not much altered by the replacement of the geocentric by a heliocentric explanation of the solar system. It matters not much how a man explains the lightning's flash so long as he avoids being struck. The bird continues to soar quite irrespective of man's prolonged discussion of how it can be explained on mechanical principles. And in general the practical activities of mankind remain much the same (I do not say quite the same, for there are remote and indirect results of the greatest importance in the long run) whatever their particular theory of the universe may be.

Now, let us note the implication. We have said a good deal in earlier chapters about natural elimination and selection. To which category of neural kineses do they apply—to those associated with practical results; or to those associated with theoretical results (supposing these to obtain below the level of man); or to both? Clearly to those associated with practical results. It matters not what theories a lion, or an adder, or a spider hold (supposing, again, that they are capable of theorizing, which I doubt). Its practical activities determine whether it survives or not. So, too, with men, *so far as they are subject to natural elimination*. It matters not what may be the nature of their thoughts, their æsthetic yearnings, their ideals. According to their practical conduct, they are eliminated or escape elimination. In other words, elimination or natural selection applies only remotely or indirectly to the human race regarded as theorists, æsthetes, or interpreters of nature.

Before proceeding to indicate to what laws our theories and interpretations of nature and moral ideals *are* subject, we may note that there are sundry activities of man, the outcome of his conceptual thought and emotion, which are also, under the conditions of social life, to a large extent beyond the pale of elimination. I refer to the æsthetic activities—music, painting, sculpture, and the like; in a word, the activities associated with art, literature, and pure science. These, in the main, take rank alongside the ideas of which they are the outward expression. Natural selection, which deals with practical, life-preserving, and life-continuing activities, has little to say to them. They are neutral variations which, so far as elimination is concerned, are neither advantageous nor disadvantageous, and, therefore, remain unmolested.

We may, therefore, fully agree with Mr. Wallace, when he says,* "We conclude, then, that the present gigantic development of the mathematical faculty [as also of the musical and artistic faculties] is wholly unexplained by the theory of natural selection, and must be due to some altogether distinct cause." Nay, we may go further, and

* "Darwinism," p. 467.

say that it is only by misunderstanding the range of natural selection as an eliminator that any one could suppose that these faculties could be explained by that theory.

We must admit, then, that there are certain neural kineses which, from the fact that they are unassociated with life-preserving and life-continuing activities, are not subject to the law of elimination; and in the development of which natural selection cannot have been an essential factor. These, in their metakinetic aspect, are conceptual thoughts, emotions, and ideas. Remembering the distinction drawn in the chapter on "Organic Evolution" between *origin* and *guidance*, let us proceed to inquire, first, how these ideas have been guided to their present development; and, secondly, how we may suppose these special variations to have originated.

To understand their development, we must understand their environment. The environment of metakineses is, as we have already seen, constituted by other metakineses. What we have now to note is that the environment of conceptual ideas, as such, is constituted by other ideas. The immediate environment of an hypothesis is other hypotheses; of a moral ideal, other moral ideals; of an æsthetic thought, other æsthetic thoughts; of a religious conception, other religious conceptions. But not only are ideas environed by ideas of their own order; they are environed by ideas of other orders. Thus a scientific hypothesis or a moral ideal may be in harmony or conflict with religious conceptions, and its fate may be thereby determined; or a religious conception may be in harmony or conflict with psychological principles, and its acceptance or rejection thereby determined. So that we may say, in general, that *the environment of an idea is the system of ideas among which it is introduced.*

Of course, it must be clearly understood that it is with the individual mind that we are dealing. The scientific ideas, moral ideals, æsthetic standards, religious conceptions, of a tribe, nation, or other community, are simply representative, either of the general views of the majority of the individuals, or more frequently of a majority among a cultivated minority. In any case, we have seen that metakineses are and must be an individual matter. For each individual there is a separate ideal world.

Through certain activities, notably language spoken or written, men can symbolize to each other the ideas that are taking metakinetic shape in their own minds. All-important, however, as is this power of intercommunication by means of language, it does not a whit alter the fact that the idea and its environment have to work out their relations to each other separately in each individual mind. My neighbour may symbolize, through language, his ideas in such a form that similar ideas may be called up in my mind; but it is there that they have to make good their claim for acceptance in the environment of the system of ideas among which they are introduced.

Now, what is the guiding principle of the evolution and development of ideas in the world of their metakinetic environment? Is there any principle analogous to that of elimination which we have seen to be of such high importance in organic evolution? I believe that there is. *An idea is accepted or rejected according to its congruity or incongruity with the system of ideas among which it is introduced.* The process has, perhaps, closer analogy with elimination than with selection, inasmuch as it would seem to proceed by the rejection of the incongruous, leaving both the congruous and the neutral. An idea or hypothesis may be accepted, at any rate provisionally, so long as it is not in contradiction to the theories and beliefs already existing in the mind.

It may, however, be objected that this view is at variance with the familiar observation that there are many excellent people who hold and maintain theories which are exceedingly incongruous, which seem, indeed, to us mutually antagonistic. Yes, *to us*. Brought into the environment of *our* system of ideas, one or other of these antagonistic views would be eliminated through incongruity. Not so, however, with those who hold both. Amid the environment of a less logical and less coherent system of ideas, both can find admission, if not as congruous, still as neutral. A sense of their incongruity is not aroused.

But there are some people, it may be said, who consciously hold views which they admit to be incongruous; who base all their scientific reasonings on a continuity of causation, but who, nevertheless, believe in miraculous interruptions of that continuity. In this case, however, the incongruity is made congruous in a higher synthesis. They belie themselves when they suppose that they are holding incongruous views. Stated at length, what they admit is that miraculous interventions are incongruous, not for them, but for those whose whole system of thought is cast in another mould than theirs—for the materialist and the infidel.

I cannot discuss the matter further here. This is not the place to show, or attempt to show, how the evolution of systems of thought has caused, or is causing, certain ideas, such as that of slavery, religious persecution, the moral and physical degradation of our poor, to reach that degree of incongruity which we signify as abhorrent; or how that evolution has caused yet more primitive ideas to seem positively repulsive. Nor is it the place to show, or attempt to show, how the advance of scientific knowledge has been constantly accompanied by the elimination of incongruous conceptions. I must content myself with the brief indication I have given of the principle of elimination through incongruity as applied to ideas.

It may be said that such a principle does not account for the origin of the new congruous ideas, but only for the getting rid of old incongruous ideas. Quite true. But I have grievously failed in my exposition of natural selection through elimination if I have not made it evident that this objection (if that can be called an objection which, in truth, is none) lies also at the door of Darwin's generalization.[*]

---

[*] In both cases, the question to which an answer is suggested is not—What variations will arise? but—What variations will survive?

Now, from all that has been said in this chapter, it will be seen that, on the hypothesis of monism, we cannot regard organic and mental evolution as continuous the one into the other, but rather as parallel the one with the other—as the kinetic and metakinetic manifestations of the same process. Organic evolution is a matter of structure and activity. If the structure or the activity be not attuned to the environing conditions, it will be eliminated, those sufficiently well attuned surviving. Turning to the metakinetic aspect, we have seen that there are certain mental processes which are directly and closely associated with activities. Their evolution will be intimately associated with organic evolution. For if these processes lead to ill-attuned activities, the organism will be eliminated; and thus the evolution of well-attuned activities and their corresponding mental states will proceed side by side. We may, therefore, say, not incorrectly, that these lower phases of mental evolution are subject to the law of natural selection.

But when the neural processes which intervene between stimulus and activity become more complex and more roundabout; when, instead of being directly and closely associated with life-preserving activities, they are associated indirectly and remotely;—then they become, step by step, removed from their subjection to natural selection. And when, in man, the metakineses associated with these neural kineses assume the form of hypotheses, theories, interpretations of nature, moral ideals, and religious conceptions, these are, except in so far as they lead to activities which may conduce to elimination, no longer subject to the law of natural selection, unless we use this term in a somewhat metaphorical, or at least extended, sense. They are subject, as we have seen, to a new process of elimination through incongruity.

Similarly with that wide range of conduct in man which is the outcome of his conceptual life, and is removed from those merely life-preserving activities which are still, to some extent, under the influence of natural elimination.

Conduct is here modified in accordance with the conceptual system of which it is the outcome and outward expression. And this higher conduct is subject, not to elimination through natural selection, but to elimination through incongruity. Slavery would never have been abolished through natural selection; by this means the modest behaviour of a chaste woman could not have been developed. To natural selection neither the Factory Acts nor the artistic products in this year's Academy were due; by this process were determined neither the conduct of John Howard nor that of Florence Nightingale. Some evolutionists have done no little injury to the cause they have at heart by vainly attempting to defend the untenable position that natural selection has been a prime factor in the higher phases of human conduct. I believe that natural selection has had little or nothing to do with them as such. They are the outcome of conceptual ideas, and are subject to the same process of elimination through incongruity.

So soon as, in the course of mental evolution, the idea of slavery became incongruous, and in certain minds abhorrent and repulsive, steps were taken to check the conduct which was the outward expression of this idea. So, too, in other cases. The reformer must not, however, be too far in advance of his generation, if his reform is to be practically carried out. When his ideas are so "advanced" as to be incongruous with those of all but a very small minority of his contemporaries, even they are forced to confess that the nation is not yet ripe for the changes they contemplate.

No one will question that artistic products are the outcome of artistic ideas. In the slow and difficult progress of a new school of painting or of music, we see exemplified the rejection of the new ideas through their incongruity with the old-fashioned artistic systems. Only gradually do there grow up new generations for whom these new ideas are not incongruous. For them the old-fashioned systems become incongruous; and if the school becomes dominant, artistic products embodying the old ideas are eliminated through incongruity.

We are not all alike. Our mental systems are different. One artist will introduce into his canvas effects which, to the eye of another, will at once strike a jarring note of incongruity. To some minds the institution of slavery presents no incongruity. There are not wanting men for whom the degrading moral and physical conditions under which many of our poor are forced to live and work present little or no incongruity. To the Russian, English fidelity to the marriage vow is said to be as incongruous as, to an English woman, is the harem of an Eastern potentate.

In the higher phases of human conduct, then, the activities are subject to the law of the ideas of which they are the outcome—the law of elimination through incongruity.

I have said that natural selection has little or nothing to do with these higher phases of conduct. But has not human selection through preferential mating? I believe that it has; and I trust that it will have a still greater influence in the future. It is one of the noblest privileges of woman, for with her mainly lies the choice, that she may aid in raising humanity to a higher level. If once the idea of marrying for anything but pure affection could become utterly incongruous to woman's mental nature; and if once the idea of perpetuating any form of moral, intellectual, or physical deformity could become equally incongruous; the bettering of humanity, through the exclusion of the deformed in body and mind from any share in its continuance must inevitably follow. Here, again, ideas would determine conduct.

And what, we may now proceed to ask, is the physiological or kinetic aspect of this metakinetic process? The answer to this question involves the conception of what I would term "interneural evolution." Just as the environment of a conceptual idea is constituted by other conceptual ideas, so is the environment of its neural concomitant constituted by the other neural processes in the brain. Just as no idea can get itself accepted if it be in incongruity with the system of ideas among which it is introduced, so, too, can no neural process become established if it be not in harmony with the other neural processes of the cerebral hemispheres. The brain is a microcosm; its neural processes are interrelated; and *the environment of any neural process is constituted by other neural processes.*

A little consideration will show that this must be so; that it is only the physical or kinetic aspect of what is freely admitted when the mental or metakinetic aspect is under consideration. If it be admitted that states of consciousness are determined by other states of consciousness, and that states of consciousness are the concomitants of certain neural processes in the brain, it follows as a logical necessity that brain-neuroses, however originating, are determined in their evolution by other brain-neuroses; and that there has been a brain or interneural evolution, distinct from and yet intimately associated with the evolution of other bodily structures and activities. The more closely and directly brain-neuroses are associated with immediate activities, the more closely implicated is interneural evolution in the process of organic elimination through natural selection. But when long trains of neuroses take place in only remote and distant connection with other bodily activities, they are removed from the process of elimination through natural selection, and interneural evolution is allowed to proceed comparatively untrammelled.

I have already indicated my belief that abstraction (isolation), analysis, and conceptual ideas have been rendered possible through language, and are excellences unto which the lower animals do not attain. Hence I regard this comparatively untrammelled phase of interneural evolution as something essentially human, something which differentiates man from brute. And I would correlate man's greatly developed brain—inexplicable, I think, by natural selection alone—with this later and special phase of interneural evolution. Even in the lowest savage this brain-evolution has proceeded a long way. I am not fitted in this matter to offer an opinion which would carry much weight. But from all that I have read I gather that savages have in all cases elaborated a complex—often a highly complex—interpretation of nature and theory of things. The interpretation may seem *bizarre* and incongruous enough *to us*, full of fetishism and strange superstitions, but it is an interpretation; to the savage it presents no incongruity; to him the incongruity is in the oddly assorted beliefs of the missionary. His system of ideas is, in fact, one of the many possible systems to which mental evolution may give rise.

For what we call systems of thoughts, interpretations of nature, theories of things, are so many genera and species which have resulted from this later phase of metakinetic evolution. Our methods are at present too coarse, our powers too limited, to enable us to determine these species from their kinetic aspect. The brains of Kaffir and Boer, of ploughboy and merchant, of materialist and idealist, are too subtly wrought to enable us to trace the systems of kineses which were the concomitants of their scheme of beliefs. But we can learn something of the genera and species from their metakinetic aspect as symbolized through language and other bodily activities. They fall into certain groups, fetishistic, spiritualistic, materialistic, idealistic, monistic, and so on, and within these groups there are subdivisions. This is not the place to consider them or discuss their characteristics. What I wish to note about them is that, diverse as they seem and are, each is a coherent product of mental evolution. In each, all that is incongruous to itself has been or is being eliminated.

There are some people, however, who are surprised at the incongruity of interpretations of nature among each other. Fetishism, they say, has been proved to be utterly false. It constitutes a hideous and grotesque delirium. How can that which is utterly and completely false to nature have had a natural evolution? Now, for the *élite* of the Aryan race, whose systems of ideas have been moulded in accordance with the conceptions of modern science, no doubt the fetishism of the poor savage seems sufficiently incongruous and grotesque. So, too, does the system of ideas of the Right Rev. Bishop of —— appear no doubt, to the learned and eminent Professor ——, and *vice versâ*. And so, too, no doubt, does the system of ideas of the white man (who introduces firearms and firewater, and preaches the gospel of forgiveness and temperance) appear to the poor savage. Each in his degree wonders how this falsity, this incongruity, can have had a natural genesis. But in each case the falsity and the incongruity is not within the system itself, but between different systems.

Once more, I repeat that if the individual nature of the systems of ideas be not adequately grasped, the nature of mental evolution will not be apprehended. States of consciousness can only be determined by other states of consciousness; and states of consciousness are for the individual subject, and for him alone. Conceptual ideas are states of consciousness; and "falsity to nature" means, and can only mean, incongruity with the environing states of consciousness in the individual mind. For the savage there is no falsity to nature in his fetishism. The idea presents no incongruity with his system of ideas; no more incongruity than filed teeth, flattened head, or pierced nose do to his standard of beauty. It is with *our* system of ideas (*i.e.* mine or yours) that his fetishism is false and incongruous. The falsity or incongruity, I repeat, is not within the system itself, but between different systems.

It may still, however, be said—Only one interpretation of nature can be true; all others must be false. And the falsity is not merely incongruity with other ideas in other systems of thought or belief; it is falsity to the plain and obvious facts of nature.

We may freely admit that only one interpretation of nature can be true. But who is to determine which? Who can decide the question between monist and materialist? Who dare arbitrate between the bishop and the professor? The criterion of fitness in this case, as in others, is survival; and who can say what existing interpretation of nature (if any) shall outlive all its competitors? Who can say what will be the nature of the further evolution of any existing philosophical creed? The elimination of the false is a slow and gradual process; and many degenerate systems of ideas may linger on in the darker corners of the world of men. False or out of harmony as they seem to be with the higher phases of development; false or out of harmony as they would be with a different and more exalted environment; they are not false or out of harmony with the environment in the midst of which we find them; they are not false or out of harmony with "the plain and obvious facts of nature," as these exist for the ill-developed or savage mind.

The plain and obvious facts of nature, as interpreted by men of science in 1890, have simply no existence for the untutored or the savage intellect. For him they have not emerged into the light of consciousness. But while we cannot blame the savage for entertaining ideas which are false to facts which for him have no existence, we may none the less believe that his system of ideas is not among those which are destined to become predominant species. So far as we can judge, the winning species among systems of ideas and interpretations of nature are those in which the greatest number of ideas are fused into harmonious synthesis; in which all the ideas are congruous, few or none neutral; and in which the abstract or conceptual ideas, when brought into contact with concrete or perceptual states of consciousness, are found to be in harmony and congruity therewith.

There is one more question in this connection on which I must say a few words. How, it may be asked, has the world become peopled, for many primitive and savage folk, with a crowd of immaterial spiritual essences, so that it is scarcely too much to say that, for some of these peoples, everything has its double; and there is no material existence that has not its spiritual counterpart?

I would connect this almost universal tendency with the origin of abstract ideas (isolates) through language. When the named predominant gave rise to the isolate (see p. 374), it could scarcely fail that the primitive speakers and thinkers should tend to regard those qualities or properties which they could isolate in thought (conceptually) as also isolable in fact (perceptually). And we may well suppose, though this is, of course, hypothetical, that one of the earliest severances to be thus effected through isolation was the severance of mind and body. The first phenomena that the nascent reason would endeavour to explain would probably be those of daily life and almost hourly experience. Many familiar facts would seem to point to the temporary or permanent divorce of the part which is conscious and feels, from the part which is tangible and visible. During wakeful life the two are closely associated. The visible part, or body, is conscious. But during sleep, or under the influence of a heavy blow, the visible part, which before was conscious, is conscious no longer. The conscious part is, therefore, absent, but returns again after a while. On death the conscious part returns no more. The divorce of the two has become permanent.

And then comes in the confirmatory testimony of dreams. In dreams the savage has seen his enemy, though that enemy's body was far away. Here, then, is the spirit which has left the body during sleep. In dreams also the slain enemy or the dead chief appears. The spirit, permanently divorced from the body, still walks the earth in spirit-guise.

Many occurrences would seem like the fulfilled threats of dead enemies or the fulfilled promises of dead ancestors. How can these be explained? Are they not produced by the ghost of the departed enemy, by the spirit of the deceased ancestor? And if these spirits are still powerful to act, why not petition them to act in certain ways?

Probably primitive man would explain all activities anthropomorphically. What knows he of gravitation or the laws of the winds? He knows himself as agent, and attributes his activities to the immaterial spirit within him; for when this is absent during sleep or in death these activities cease. All acting things might, therefore, come to be regarded as dual in their nature—possessed of a sensible material bodily part, and an insensible active spiritual part. And thus the whole world might be peopled with living existences of the spiritual order.

Now, whether the fetishistic faith arose in some such way as this or not—and we can never know how it arose, but can only guess—there would be nothing in such primitive explanations which would violate the law of congruity. They would have, therefore, a perfectly natural genesis. The attempted interpolation at such a stage of primitive reason of any modern scientific conception would be futile. It would at once be rejected through incongruity.

The history of scientific conceptions seems to show that they were first adopted with regard to phenomena on the very horizon of thought—in regions, that is to say, most remote from the central citadel of the soul. Only gradually have they, little by little, encroached upon this centre; and the application of them to physiology and psychology is a matter of quite modern times. Even to-day only a minority, but an increasing minority, of thinkers are prepared indissolubly to unite the mind and body, so long divorced in thought, so completely united, as many of us believe, in their essential being.

I have now, I trust, illustrated at sufficient length the principle of elimination through incongruity in interneural and its associated metakinetic or mental evolution. This, however, like natural selection, is a matter of guidance; we have still to consider the question of origin.

In truth, we know too little on the subject to enable us to discuss it with much profit. From the kinetic or organic point of view, neural variations take their place among the other variations, the origin of which, as we have already found, is so hard to account for. There may be a tendency for neural vibrations to mutually influence each other (like two clocks placed side by side), and thus gradually to drag each other into one harmonious and congruous rhythm. But this, though not improbable, is purely hypothetical. There is the hypothesis of the inheritance of acquired variations, the increased congruity acquired by the parent being in some degree transmitted to the offspring. There is the view which Mr. Wallace adopts* with regard to the origin of accessory plumes, that such variations may be due to "a surplus of strength, vitality, and growth-power, which is able to expend itself in this way without injury," and not without profit. The development of the social habit, the mutual aid and protection thus afforded, may well have left a balance of the life-energy, previously employed in individual self-preservation, available for this purpose. And then there is always the hypothesis of favourable fortuitous variations to fall back upon.

On only one of these points do I propose to say a few words—that of the possible inheritance of acquired variations.

Let us restate the problem here for the sake of clearness. There is, according to the suggestion put forward in this chapter, an interneural evolution, leading to an harmonious development of the neuroses in the individual brain. But this special evolution of the brain is nowise independent of the more general evolution of the body. The human being, as an organism, is still subject to natural elimination and human selection. Elimination through the action of surrounding physical conditions, although it has played some part in the evolution of man, is not a factor of the first importance. Elimination through enemies is more important, but has not much bearing on the question at present before us—the evolution of the conceptual. Elimination by competition, again, though a factor of yet greater importance in human evolution, has, nevertheless, so far as individuals are concerned, but little bearing on our present question. Few are eliminated through the absence of the conceptual faculty. Natural elimination, then, is, as Mr. Wallace well pointed out, practically excluded in this matter. No doubt, in the struggle between tribes and

---

* "Darwinism," p. 293. It is strange that Mr. Wallace did not apply this view to the mathematical and artistic faculties discussed in his last chapter. It is true that such application tends to undermine the argument there developed. But Mr. Wallace is far too great and conscientious a thinker to be influenced by such a consideration.

nations, that community is most likely to be successful in which there is rational guidance. No doubt, during the earlier phases of the development of man on our islands, the elimination of the irrational was a factor in progress. But if we take the last three centuries of English history, I doubt whether it can be shown that there has been much elimination determined by the relative absence of conceptual ideas and emotions.

Human selection has been a much more important factor. Those individuals which showed the higher types of intellectual thought have been constantly selected. Riches, rank, and social position have been bestowed upon them. Of course, there have been exceptions; great intellects have been allowed to languish in their lifetime, and have only obtained recognition through their works after death. But every day there is less chance of a genius dying in a garret. And the best intellects, being thus selected and chosen out from among their fellow-men, form to some extent a distinct social class. Segregation is thus effected; and intermarriage takes place within this intellectual caste, with the result that the conditions are eminently favourable for the inheritance of intellectual qualities.

Now, is this process of selection of the intellectual, this segregation into a caste, and the inheritance of innate intellectual qualities sufficient to account for the facts of intellectual progress; or must we call in to our aid the inheritance of individual increments? I confess I cannot say. Direct and satisfactory evidence, one way or the other, is almost impossible to obtain.

Must we, then, leave the question undecided? I think we must so far as direct evidence is concerned. I may have a general belief that there has been some transmission of acquired increment of intellectual faculty. But unless I can substantiate it by definite facts, I cannot expect to convince any one who holds the opposite view. And definite facts of sufficient cogency I am unable to adduce. It is practically impossible to exclude the influence of human selection; and unless we can do this the followers of Dr. Weismann will not be satisfied.

Still, general belief—which means the net result of one's consideration of the subject—counts for something. We must remember the question is one of origin, and not of guidance. The guidance of human selection is unquestioned and unquestionable. But when we consider the intellectual progress of the last three centuries, and ask whether all this has originated in fortuitous brain-variations, which human selection has simply picked out from the total mass of available material, an affirmative answer seems to me a little difficult of acceptance. There seems to have been a definite tendency to vary in this particular direction, a general raising of the intellectual level, which is difficult to account for unless it be due to the persistent employment of the intellectual faculties.

To put the matter in another way. I do not think that, during the last three centuries, there has been a large amount of elimination of the unintellectual. Such elimination as there has been of this nature has probably been more than compensated by the slower rate of multiplication of the intellectual classes. Elimination, then, in this matter may be practically disregarded. But it is obvious that selection, without the removal or exclusion of the non-selected, does nothing to alter the *general level** with regard to the particular quality or faculty concerned. It is merely a classification of the individuals in order of merit in this particular respect. It is, in a word, a segregation-factor. It arranges the individuals in classes, but it does not alter the position of the mean around which they vary.

Let me explain by means of an analogous case. Fifty boys, who have been admitted to a public school, await examination in a class-room. They are at present unclassified, but there is a mean of ability among the whole

---

* If elimination of the unintellectual (not necessarily of the unintelligent) may be excluded, and if the unintellectual increase by natural generation more rapidly than the intellectual, the general level of intellectuality must, on Professor Weismann's principles, be steadily *falling*.

fifty. A week afterwards they are distributed in different forms. Some are selected for a higher form, others have to take a lower place. But though selection has classified the material, it has not altered the position of the mean of ability among the fifty boys. This can only be done by expelling a certain number or excluding them from the school.

Granted, therefore, that elimination is practically excluded, human selection can at most classify the individuals according to their intellectual faculties. It cannot raise the mean standard of intellectuality. If, therefore, this mean standard has been raised during the last three centuries, there has been a tendency to vary in this particular direction, which *may*,[*] to say the least of it, be due to the inheritance of individual increment.

I am, of course, aware that the matter is complicated by the increased and increasing diffusion of knowledge through the printing-press and by the extension and improvement of education. But education, to take that first, though it may raise the level of each generation, can have no cumulative effect. For the effects of education cannot, on Professor Weismann's hypothesis, be inherited. You may educate brain and muscle in the individual, but his heir will inherit no good or ill effects therefrom. Each generation goes back and starts from the old level. There is no summation of effect; or, if there is, it tells so far against Professor Weismann.

And with regard to the diffusion of knowledge, this, though it brings more grist to the intellectual mill, can have no effect in raising the mean standard of excellence in the mill itself. There is more to grind; but this does not improve the grinding apparatus; or, if it does, it tells so far against Professor Weismann's hypothesis. To vary the analogy, the diffusion of knowledge increases the store of available food; but it does not bring with it any additional power of digesting the food; or, if it does, it may be through inherited increments of mean digestive power.

[*] It *may* also, in part, be due to "organic combination."

It may, however, be maintained that there is no conclusive proof that the mean intellectual level of Englishmen to-day is any higher than it was in the days of the Tudors. If so, of course, my argument falls to the ground. I have no desire to dogmatize on the subject. I merely set down the reasons, such as they are, and for what they are worth, which lead me to entertain a general belief that the intellectual progress of Englishmen during the past three hundred years has been in part due to the inheritance of individually acquired faculty.

Mental evolution, then, is the metakinetic equivalent of interneural, or, in us vertebrates, brain-evolution. The brain forms a kinetic system in some sense independent of, and yet in constant touch with, the kinetic system of the world around. Its kineses, though they do not resemble, yet more or less accurately represent or symbolize, the kineses of the surrounding universe. As the kineses of the world around are interdependent and harmonious, so are the neural kineses of the brain interdependent and harmonious. And no modification of this kinesis which is out of harmony with the kinetic system already established in the brain can be incorporated with that existing system. Such attempted modification is eliminated through incongruity.

Associated with this brain-kinesis, and forming its inner aspect, is a metakinetic system in which the higher manifestations rise to the level of full consciousness; others form sub-conscious states; others are unconscious. But the whole form a coherent system answering to the coherent kinetic system.

Consciousness is thus associated only with the phenomena of that kinetic microcosm which we call the brain (or other interneural system). Obviously, therefore, it does not and cannot deal directly with anything outside the brain. Its knowledge is solely and entirely a knowledge of the representative occurrences of the interneural system. But out of these occurrences a surrounding world of phenomena is constructed in mental symbolism.

The brain itself, however, is part of the world of phenomena thus constructed in mental symbolism; and the world, therefore, dissolves in pure idealism, leaving only a fleeting series of states of consciousness, if we do not assume the existence of a system of "things in themselves" (noumena), of which kineses and metakineses are the phenomenal manifestations. Whether the "things in themselves" in any sense resemble their phenomenal manifestations, we cannot say. It is as difficult philosophically to conceive that they can as it is practically to conceive that they do not. And since, whether they do or do not, the world we live in is phenomenal; since it is to phenomena that we have to adapt our conduct; since it is with phenomena that all our thoughts and emotions have reference; since the world we construct in mental symbolism is the world in which we live and move and have our being; it is not only convenient, but logically justifiable, to call this world of phenomena the really existing world for us human-folk and other sentient organisms.

As in the kinetic interneural system, or brain, so, too, in the metakinetic system, no modification of the metakinesis which is out of harmony with the existing metakinesis can be incorporated therewith. Such attempted modification is eliminated through incongruity.

In the lower stages of mental evolution, those which belong to the perceptual sphere, where the neuroses are closely connected with the life-preserving activities of the organism, the survival or non-survival of the system of neuroses is largely dependent on the fitness of the associated activities to the conditions of life. But in the higher stages of mental evolution, those which belong to the conceptual sphere, the connection of certain brain-neuroses with life-preserving motor-activities becomes less close and direct. The corresponding ideas, thoughts, and emotions become floated off into a more abstract region. Here the system of ideas, as such, that is to say, so far as they are removed from life-preserving activities, is determined mainly by the law of congruity. But there are several such systems. There are, indeed, as many systems as there are minds; but these may be classified in several distinct groups, which we may liken to genera and species. These are the various interpretations of nature, theories of things, and the like; the systems of ideas, thoughts, conceptions, emotions, beliefs, which, as we say, belong to us, each and all, and which determine to which metakinetic species we belong. These are the highest products of mental evolution; and among them there is, so to speak, a struggle, if not for existence, at any rate for prevalence. Which shall eventually prevail—a spiritual interpretation of nature, a material interpretation, a monistic interpretation, or other, who shall say? But, so far as we can judge, the winning species among systems of ideas and interpretations of nature are likely to be those in which the greatest number of ideas are fused into harmonious synthesis; in which all the ideas are congruous; and in which the abstract or conceptual ideas, when brought into contact with concrete or perceptual states of consciousness, are found to be in harmony and congruity therewith.

INDEX.

Abstract ideas, 322, **363**
Acceleration, sense of, 269
Acceleration and retardation, 221
Achirus pellucidus, 83
Acquired characters, are they transmitted? 147; habits, are they inherited? 436; variations in the intellectual sphere, 497
Acræa, 203
Activities, organic basis of comparative psychology, 337; of animals, 415; voluntary and involuntary, classification of, 462
Adaptation, analogous, 117; modes of, 119; special, examples of, 179; to varying environment, 183
Advantage must be particular, 184; must be immediate and not prospective, 186; must be "available," 188, 211
Æschna, 289
Æsthetic preferences in insects and birds, 207; aspect of sensation, not primary, 243; motive not present to animal consciousness, 409
ALEXANDER, Mr. S., "Moral Order and Progress," 463
ALLEN, Mr. Grant, on evolution of flowers, 206; on pleasure and pain, 380
——, Mr. J. A., on colour and humidity, 164
Alternation of generations, 46
Amblyopsis spelæus, 271
American school of evolutionists, 221
Amœba, how it feeds, 5; reproduction of, 12, 38; diagram of, 12; protoplasmic functions of, 142
Amphibia, labyrinthodont, 288
Anabolism, constructive process, 32
Analysis (mental), 321
Ancon sheep, 226
ANDERSON, Mr., on one-eared rabbits, 226
Anemone, sea, reproduction of, 41; marginal beads of, 298; discrimination by, 359
Anger and rage, 389
Animal life, nature of, 1; diversity of, 177
Animal Intelligence, differs generically from man's reason, 350

Animals, characteristics of, 1; divided into protozoa and metazoa, 15; and plants, their relation to food-stuffs, the atmosphere, and energy, 15; intelligent not rational, 373; capacities for pleasure and pain, 391
Animistic ideas of savages, how developed, 494
Anisognathus, 226
Anomia, 265
Ant, sauba, of South America, 213; sense of taste in, 253; sense of smell in, 258; auditory organ of, 267; intelligence of, 357; activities often described as instinctive, 425; neuter insects, 440; Siamese, 449
Antagonism, advantages of, 394
Antennæ of insects, modifications of, 178; of emperor moth, 199; organ of hearing in, 267; modified hairs of, 297
Antennulæ of crayfish, 259
Anthophora, 438
Anticipation, 327
Antlers of deer in illustration of growth, 28
Aphides, absence of fertilization in reproduction, 44
Appetence and aversion, 343, 384
Apus, 46
Aquatic organisms, respiration in, 4; sense of smell in, 256
ARAGO, M., observation on turnspit dog, 404
Arctic hare and fox, 84; animals, colours of, 165; fox, cunning of, 368
ARGYLE, Duke of, on humming-birds, 110
Artemia salina and *milhausenii*, 164
Artistic faculties and natural selection, 484; products, evolution of, 489
Association a tendency to integration, 183; perceptual and mimicry, 202; and recognition marks, 203
Ateles, 210
Atmosphere, relations of animals and plants to, 15
Attacus, 179
Attention, 312
Attidæ, 208
Auditory apparatus in man, 262

Aurelia, life-cycle of, 45
Australian mammals and others convergent, 117
Automatic action, 415
Available advantage, 188, 211
Aversion and appetence, **343**, 384

Baboon, experiments with, 352
Bacilli attacked by leucocyles, 439
Bacillus violaceus, 80
BAILEY, Mr. E. H. S., on taste, 251
BALBIANI on *Chironomus*, 137
Balistes, 179
BARRETT, Mr. W. F., on sensitive-flame experiment, 298
Barrier, geographical, 99; time, in physiological isolation, 105
BARRINGTON, The Hon. Daines, on song of linnet, 454
BATESON, Mr. W., on lateral line, 252; on fishes hunting by scent, 256; on smell in shrimps, etc., 260; on hearing in fishes, 264; on hearing in *Anomia*, 265; on sight in fishes, 286; on rockling and sole, 352; on fascination in fishes, 388
Bats, tabulated measurements of wing-bones of, 65-73; wings, fortuitous variations in, 235; experiment with, 247
Beauty, standard of, 206; sense of, 467
Beaver, change of habit in, 445
BECCARI on gardener bower bird, 408
BECKER, Alexander, on variations in the balance of life, 112
Bees, divergent development of, 58; cuckoo, 90; latency in, 228; sense of taste, 253; sense of smell, 257; smell-hollows, 259; eyes and eyelets of, 289; intelligence of, 357; colour preferences in, 408; homing faculty in, 428; neuter insects, 440
Beetles of Madeira, 81; stag-, variability of male, 180; observations on dung-, 368
Begging in dogs, 345
BERKELEY, Bishop, quoted, 475
BERT, M. Paul, limits of sensibility to light, 296

BIDIE, Mr. George, anecdote of cat, 370
BINET, M., "Psychic Life of Micro-organisms," 360
Birds, influence of food-yolk on development of, 56; divergence among, 97; breeding area of comparatively restricted, 101; humming, Duke of Argyle on, 110; destruction of eggs of, 189; game-, white and black crossed, 225; taste in, 251; smell in, 256; hearing in, 264; sight in, 284; colour-vision in, 285; gardener bower, 408; humming, nests of, 408; perfect instincts of præcoces, 424; love antics of satin bower, 450; nests of, 453; song of, 454
BLOCHMANN on the development of the drone, 153
Blood, circulation of, 22
Body as distinguished from reproductive cells, 131
BOLL and KÜHNE, Messrs., on retinal purple, 216
BOLTON, Miss Caroline, on the bat, 247
Bombus muscorum, 90; *lapidarius*, 91
Bombyx quercus, 258
Bower bird, 408, 450
Brain, 31; decreased, of rabbits and ducks, 171; a microcosm, 491
BREHM's, Thierleben, quotation from, 405
Brine shrimp, modified by salinity of water, 164
BROOKS, Prof. W. K., his modification of pangenesis, 134; on the greater variability of the male, 237
BROWN, Prof. Crum, on sense of acceleration, 270
BROWNE, Sir J. Crichton, on ducks, 171
Budding, reproduction by, 42; in relation to heredity, 128
Bull, "Favourite," prepotent, 227; reversion in, 229
BUNYAN, John, on gateways of knowledge, 311
BUTLER, Mr. Samuel, on organic memory, 62, 475
Butterfly, protective resemblance in, 86; mimicry in, 87

Camel, wounded, 392
Canary, crested, 225; nest building of, 453
Capon, taking to sitting, 228
Capuchin monkey, Miss Romanes's observation on, 357; sympathy in, 397
CARLYLE, quoted, 331, 335
Carp at Potsdam, 265
CARTER, Dr. Brudenell, quoted, 285
Caste, idea of, in dog, 400
Cat, effect of African climate on, 164; defining its percept, 339; communication, 345; intelligence of, 370; and mouse, 399; punishing kitten, 400
Caterpillars, protective resemblance in, 82

Cattle of Falkland Islands, 203
Causation, 327
Cell, diagram of animal, 10; controlled explosions in, 31
Cessation of selection, effects of, 172
Chætodon, 83
Chætogaster limnæi, reproduction of, 42
Chaffinch, nest of New Zealand, 454
Chameleon, 296
Chance, 236
Change of conditions, 183
Characters, specific, 110
CHARBONNIER, Mr. Henry, measurements of bats, 63
CHATROCK, Mr. A. P., his experiments on colour-vision, 286; letter to, on dog and picture, 341
CHESHIRE, Mr., on smell-hollows in bees, 259
Chickens' aversion to protected caterpillars, 352; perfectly instinctive activities, 424
Chironomus, reproductive cells of, 137
Choice, 458
Circulation of the blood, 22
Classification, 323
CLIFFORD, W. K., on human consciousness, 341; on the eject, 476; on "world-consciousness," 479
Clover and bees, 113
Clytus arietis, 87
Cockchafer, smell-hollows of, 259
COCKERELL, Mr., on variations in snails, 75; on effects of moisture, 239
Cockroach, diagram of trachea or air-tubes of, 3; sense of taste in, 253; sense of smell in, 258
Cocoon, collective, 429
Colobus, 210
Colour, protective resemblance in, 82; warning of inedibility, 82; dependent on humidity, 164; direct action of climate on, 164; development of, 202; blindness, 273, 279; phenomena of, 278
Combination, organic, hypothesis of, 150, 246
Communication in dogs, 345; in bees, 358
Compensation of growth, 155
Competition, elimination through, 89
Concept, 325, 328
Conception, 325
Conceptual conduct and evolution, 488
Condor, rate of increase of, 57
Conduct, 463; influence of thought and æsthetics on, 483; conceptual, and natural selection, 488
Congruity, principle of, 486
Conjugation in protozoa, 39; of ovum and sperm-cell, 42
Consciousness, 32; and consentience, 326, 362; as a criterion of instinct, 432
Consentience, 326, 362

Construct and construction (mental), 312; three stages of, 324; inevitable nature of, 332; in mammals, 338
Continuity of reproductive cells, 131; germ-plasm, 134; cellular, 142; in mental development, 373
Convergence, phenomena of, 117
Co-ordinants, 303
COPE, Prof., on the effects of use, 210; and HYATT, Prof., on retardation and acceleration, 221
Correlated variation, 59, 216
Coati, organ of, 263
Coryne, Prof. Weismann on, 139
COUCH, Mr., on goldfinch song, 454
Crab, protective resemblance in, 87; hermit, 195; habit of decking itself, 457
Crayfish, smell in, 259; auditory organ of, 266
Crossing, effect on reversion, 230
Cruelty in cat, objective, 400
Crustacea, eyes of, 292
Ctenomys, 194
Cuckoo, the name onomatopoetic, 322; habits intelligent, 436; ejecting young birds, 437
Curiosity in prong-horn, 339
Cuttlefish, eyes of, 293
Cyclas, 265
Cycloptera speculata, locust resembling leaf, 86

DALLINGER, Dr., his temperature-experiments on monads, 147
Damais, 203
Daphnids, absence of fertilization in reproduction of, 45; colour-vision in, 292, 296; leucocytes of, 439
DARWIN, Charles. Natural selection and the struggle for existence, 77; divides the principle of selection into three kinds, 78; on selection of flowers and fruits by insects, 93; on sexual selection, 94; on prevention of free crossing in breeding, 99; on differential fertility, 104; on London rats, 106; on Galapagos archipelago, 109; on diverse adaptation, 111; on the influence of old maids on clover crops, 113; on the influence of parent on offspring, 122; on the co-ordinating power of her organization, 125; hypothesis of pangenesis, 131; on fur of arctic animals, 165; changes of structure attributed to use and disuse, 171; on blindness of tuco-tuco, 194; on the principle of economy, 194; on sexual selection, 198; on preferential mating, 204; on evolution of flowers, 205; on co-ordinated variations in the elk, 213; on acceleration, 222; on ancon sheep, 226; on prepotency, 227; on reversion, 229; on the effects of crossing, 230; on fortuitous variation, 236; on the subordination of the conditions to the organism,

236; on the greater variability of male, 237; on attention in monkeys, 342; on brain of ant, 358; on gestures of anger and rage, 389; on pleasures and pains of animals, 394; on bravery of a monkey, 396; on Abyssinian baboons, 405; on sense of humour in the dog, 406; on neuter insects, 440; on selection of oxen, 441; on acquisition of fear of man by birds, 443; on satin bower bird, 450

Death, natural introduction of, 188, 193
Deceit in dogs, 400
Degeneration, 183
Desert animals, inconspicuousness of, 89
DESCARTES on pineal gland, 288
Desire, 460, 463
Destruction, indiscriminate, as opposed to elimination, 76
Development of organisms distinct from growth, 6; reproduction and, 36; is differential growth, 49; of a vertebrate, diagrammatic account of, 51; comparative, of some vertebrates, 220
DE VRIES, 132, 159
Differentiation in protozoa, 40; in metazoa, 41; during development, 49; of reproductive cells, 143; and integration, 183; of tissues, 232
Difflugia, 360
Dimorphism in larvæ, 187
Discrimination in the sense of touch, 245; hearing, 262; sight, 275; its fundamental nature, 338; in sea-anemone, 359
Disease, elimination by, 80
Display, 207
Disuse, panmixia and, 189; negative and not positive, 196; use and, 209
Divergence among birds, illustrated from Wallace, 87; through diverse adaptation, 111
DIXON, Mr. Charles, effects of climate on the colours of birds, 164; on chaffinch nests, 454
Dog, effect of Indian climate on, 164, 167; greyhounds in Mexico, 167; sense of smell in, 255, 338; vague percept of, 359; and the feelings of other animals, 340; and pictures, 341; powers of communication, 344; swimming rivers, 365; cleverness of, 367; sympathy in, 397; idea of caste, deceit, 400; endurance of pain, 402; sense of justice in, 404; punishing pup, 405; sense of humour in, 406; swimming a deferred instinct in, 423; turning round to make a couch, 444
Dog-fish, sense of smell in, 257
Domestication, variations effected by, 171, 215; crossing and reversion, 230
Doris tuberculata, 84
Dreaming, 341; and the animistic hypothesis, 495

Dromia vulgaris, 457
Drones developed from unfertilized ova, 45; second polar cell extruded, 153
DUBOIS, M., on *Proteus*, 294
Ducks, Sir J. Crichton Browne on, 171; Dr. Rae on instinctive wildness of, 435
Duration of life, 186

Eagle, sclerotic plates of, 437
Ear, 263
Earthworm, respiration in, 4, 24; regeneration of lost parts, 41; sensitive to light, 293; outward projection in, 359
EATON, Rev. A. E., on insects of Kerguelen Island, 81
Ecitons, 427
Economy, principle of, 194
Education of ants, 428; of young animals, 455
Egg and hen, problem of, 130
Egg-cell and sperm-cell, diagram of, 13; conditions which determine production of, 60
Eggs, influence of food-yolk on mode of development of, 56; destruction of birds, 189
Ego, or self, 475
EIMER, Prof., on inhabitants of Nile valley, 165; on *Helix hortensis*, 226; on instinct, 436; on differential dread in birds, 444
Eject, meaning of, 476
Elaboration, 183
Elephant, rate of increase of, 57; intelligence of, 363, 369; use of tools by, 370; vindictiveness in, 401
Elimination, as opposed to selection, 79; its three modes, 80; as a factor in the origin of instinct, 447; of ideas through incongruity, 486; as applied to the intellectual faculties, 497
Embryology negatives preformation, 50
Emotions exemplified, 382; the expression of, 385; three orders of, 391; in vertebrata, 395
Encystment, 38, 49
Ends and means, 371
Energy, relations of animals and plants to, 16
Ennomos tiliaria, caterpillar, protective resemblance of, 85
Environment, direct effects of on the organism, 163; changes of, in relation to the organism, 183; as effects of direct or indirect?, 233; instances of effects of, 238
Equus, 118
Eristalis tenax, 87
Ethics in animals, 413
Euplœa, 203
Evolution of older writers, 50; and revolution, 119; organic, 177; meaning of term, 182; mental, 464; organic and mental not continuous, 488; interneural, 490
Excrement of birds, resemblance of spider to, 90
Excretion, an essential life-process, 3, 29

Expectation, 327
Experience dependent on memory, 395
Expression of the emotions, 385
Eye, structure of in man, 274; in mole, 284; pineal, 287; in insects, 288; facetted, 289; in crustacea, 292; in molluscs, 292; four types of, 294

FABRE, M., on *Sitaris*, 439
Facetted eye, 289
Factors of phenomena, laws of, 61
Falkland Islands, cattle of, 102; birds of, 443
Fear, dread and terror, 387; instinct of, 443
Feelings of animals, 8, 378
Female. *See* Sex-differentiation.
Female and male insects, differences between, 179; vigour expended on offspring, 236
Fertilization, nature of, 42; absent in parthenogenetic forms, 44
Fertility, differential, Darwin and Romanes on, 104; of hybrids, 105
Fetishism, its natural genesis, 492
FISCHER, Dr. Emil, on smell, 254
Fish, respiration in, 24; protective resemblance in, 83; amount of food-yolk in eggs of, 220; skate and turbot compared, 220; sense of taste in, 252; sense of smell in, 256; sense of hearing in, 264; sense of sight in, 286; fascination in, 388; love-antics of, 450
FISK, Rev. G. H. R., on sympathy in cat, 397
Fission, a process of cell-division, 37; in protozoa, 38; in metazoa, 41
Flight, instinctive nature of, 425
FLOURENS, M., on function of semicircular canals, 269
Flowers and fruits, selection of, 93; evolved through insect agency, 206
Folliculina, 360
Food-stuffs, relations of animals and plants to, 15; nature of and digestion of, 25
Food-yolk, influence of, on development, 55; the result of parental sacrifice, 57
FORBES, H. O., on Javan spiders, 90
FOREL, M., on taste of ants, 253; on vision of daphnids, 296; on happy family of ants, 428
Form-characteristics of animals, 2
Fortuitous variation, 235
Fosterage and protection, 219; result of female self-sacrifice, 238
FOTHERGILL, Mr., on dogs swimming rivers, 364
Fowl, variations in, attributed by Darwin to use, 171; crossing of, 227, 230
Fox, cunning of, 366
FRANCIS, Mr. H. A., 90

FRITSCH, Dr., Fig. of skull of *Melancrpeton*, 288
Frog, development of, 6; arrest of life in, 21; respiration in, 24; fishing, or angler-fish, 91; modified development of, 214; effects of simple stimulus on, 395
Fruits and flowers, selection of, 93

GABET, Messrs. HUC and, on Llama cow, 333
Galapagos Archipelago, species and varieties in, 99; climate of, 109
Gallus bankiva, 230
GALTON, Mr. Francis, on the coloration of the zebra, 84; his modification of pangenesis, 135; numerical estimate of inheritance, 150, 192; his investigations on twins, 169; on blended characters, 225; on the steps of evolution, 227
Ganglia, 31
Gannet, rate of increase of, 57
Gas-engine, analogy of, 30
GAUTIER, Théophile, his cat, 264
GEDDES, Prof. Patrick, and THOMSON, J. A., on anabolism and katabolism, 44; quoted, 50, 137, 237
Gemmules, pangenetic, 131
Generations, alternation of, 46
Generic idea, 326
Geographical barriers a means of segregation, 99
Geological changes, influence on natural selection, 113
Germ-plasm, continuity of, 138; convenience of, 140
Gills of mussel, 4; as respiratory organs, 24
Giraffe, co-ordinated variations in, 212
Glacial epoch, effects of, 113
Gland, pineal, 288
Goldfinch, song of, 454
GOLDSCHNEIDER, on temperature-sense, 249
GOULD, Dr., on humming-birds' nests, 408
GRABER, Dr., on colour-sensitiveness of earthworm, 293
GRANT, Mr. G. L., on New Zealand sparrows, 445
Grasshopper, auditory organ of, 266
Gregarina, reproduction in, 38
GRENACHER, Dr., experiment on moth's eye, 290
Grouse, white plumage in, due to reversion, 229
GROVE, Sir W. R., on antagonism, 394
Growth of organisms, 5; illustration of a deer's antler, 28; law of, after mutilation, 126
Guidance distinguished from origin, 242
Guillemot, eggs of, 410
GULICK, Rev. J. T., on landshells of Sandwich Islands, 109; on tendency to divergence, 151
GUPPY, Mr., on crab of Solomon Islands, 87

Habits of animals, 415
Habitual activities, 420; sense of satisfaction in performance of, 421
HAECKEL, Prof., plastidules of, 125; theory of perigenesis, 159
Halictus cylindricus, 90
HAMERTON, Mr. P. G., on the ignorance of animals, 333
HAMILTON, Sir Wm., quoted, 470
HANCOCK, Mr. John, on instinct of cuckoo, 437
HASSE, E., on bumble-bees, 259
HAUSER, on cockchafer, 259
HAYCROFT, Mr. J. B., on taste, 250
Hearing, sense of, 261
Heliconia, 203
Helix, nemoralis and *hortensis*, variation of, 75, 217, 226, 239
HELMHOLTZ, Von, on colour, 277; on local signs of retina, 308
Hen and egg, problem of, 130
HENSEN, on shrimps, 266
HERBERT, Prof. T. M., quoted, 471
HERDMAN, Prof., on sea-slug (*Doris*), 84; his modification of pangenesis, 135; on warning coloration in nudibranchs, 252
Heredity, an organic application of the law of persistence, 62; and the origin of variations, 122; in protozoa, 123; and regeneration of lost parts, 124; failure of, 192; and instinct, 435
HERING, Edward, on organic memory, 62, 475
HERON, Sir R., on crossing rabbits, 225
HERSCHELL, Sir John, on colour, 277
HERTWIG, Richard, observations on Infusoria, 39
HICKS, on Capricorn beetle, 267
HICKS' organ, 267
HICKSON, Dr., Fig. of eye of fly, 290
Hipparion, 118
Hippopotamus, instinctive activities in, 423
HOLLAND, Sir Henry, on inheritance, 223
Homing faculty of bees, 428
Horse, two different evolutions of, 118; effects of use on digits of, 210; sense of pain in, 392
HOWES, Prof., antennule of crayfish, 259
HUBER, Pierre, on smell in bees, 257; judgment and instinct, 452
HUC and GABET, Messrs., on Llama cow, 333
HUGGINS, Dr., his dog Kepler, 396
Humming-birds, 110
Humour, sense of, in dog, 406
HUXLEY, T. H., on limitation of variations, 151; on neurosis and psychosis, 456
HYATT, Prof., on acceleration and retardation, 221
Hybrids, fertility of, 105
Hydra, reproduction of, 14, 41; diagram of, 43; artificial division of, 124; budding in, 128; sexual reproduction of, 129
Hydra tuba, and medusa of aurelia, 45
Hydroids, development of, 46; Weismann on, 139
Hymenoptera, antennary structures of, 297; instincts of social, 441, 448

Ichneumon fly, instinct of, 430
Ichthyosaurus, pineal eye of, 288
Icteridæ, 454
Idea of an object, 313
Ideas, conceptual, their environment, 485; the law of their evolution, 486
Idealism, 474
Ignorance of animals, 333
Image, inverted in retina, 311
Imagination, constructive, 325
Imitation as a factor in habit or instinct, 443, 453
Immortality of protozoa, 12
Incongruity, elimination by, 486
Increase, law of, 58
Incubation, instinct of, 434
Individuality, a tendency to differentiation, 183
Inference, conscious and unconscious, 328; in animals, 361
Infertility of isolated forms, 108
Infusoria, reproduction in, 39
Inheritance, exclusive, a means of isolation, 104; of variations, 223; of acquired habits, 435; of acquired increments of intellectual faculty, 497
Inhibition, 385; as a condition of volition, 459
Innate capacity, 422; its importance, 429
Insects, tracheal respiration of, 3, 24; wingless, of Madeira, 81; of Kerguelen Island, 81; mimicry and protective resemblance in, 85, 88; segregation by colour, 101; antennæ of, 178; mouth-organs of, 179; and the evolution of flowers, 206; sense of touch in, 245; taste in, 253; smell in, 257; hearing in, 266; sight in, 288; perceptual powers of, 357; neuter, 440
Instinct and available advantage, 211; consideration of, 415; perfect, imperfect, and incomplete, 422; deferred, 423; blind prevision in, 429; gratification in performance of, 430; consciousness and, 432; primary and secondary, 434; three factors in the origin of, 447; as influenced by intelligence, 452; by imitation, 453; by education, 455; as distinguished from intelligence, 457
Instinctive emotion, 390, 395
Integration and differentiation, 183
Intellectual development, 486
Intelligence involved in selection, 95; distinguished from reason, 330, 365; lapsed, 435 involved in instinct, 446; as influencing instinct, 452; criteria of, 456

Index.

Interbreeding and intercrossing, 97
Interneural evolution, 490
Interpretations of nature, genera and species of, 492
Isle of Man, tortoiseshell butterfly of, 81
Isolates, 322, 384
Isolation, organic, or segregation, 99; mental, or abstraction, 322

JAEGER, Dr., on crossing of pigs, 230
JENKINS, Mr. H. L., on the elephant, 363
Judgment, 330

Kallima paralecta, leaf-butterfly, 86
KANT, quoted, 476
Katabolism, a disruption or explosive process, 32
Kea, of New Zealand, 446
Keimplasma. See Germ-plasm
Kentish plover, 83, 217
Kepler, Dr. Huggins's dog, 396
Kerguelen Island, wingless insects of, 81
Kinesis, 467
Kingfishers, 446
KIRBY and SPENCE, localization of smell in insects, 258; on hearing in a moth, 267; on instinct of ichneumon fly, 430
Kittens, instinctive antipathy to dog, 396
KLEIN, Mr. S., on *Bombyx quercus*, 258
KÜHNE, Messrs. BOLL and, on retinal purple, 276

Labyrinthodont amphibia, pineal eye in, 288
LAMONT, on reindeer, 392
LANE, Dr. Arbuthnot, on influence of certain trades on structure, 169
LANGLEY, Prof., on ætherial vibrations, 299
Language, 322; the instrument of analysis, 349; its origin and effects, 374
LANKESTER, Prof. E. Ray, his description of perigenesis, 159; on blind cave-fish, 194
Lapsing of intelligence, 435
LARDEN, W., on the Rhea, 89; on instinct in a snakelet, 424
Larmarckian school, 209
Larvæ, dimorphism in, 187
Latency, phenomena of, 227
Lateral line of fishes, 252
Leaf-butterfly, 86
LEE, Mr. Arthur, on communication in cat, 345
Leptalis, 87
LEROY, on abstract notion of danger in fox, 348
Leucocytes, rôle of, 439
LEWES, G. H., 437, 462
LEYDIG, on antennule of crayfish, 259
Life, duration of, due to natural selection, 186

Life-area, expansion and contraction of, 114
Limits of vision, 281; of sensation, 299
Limnæus truncatulus, 48
LINCECUM, Dr., on habits of Texan ants, 425
Linnet, song of, 454
Lion, observation on, 400
Liver-fluke, life-history of, 47
Local signs, 308
Localization, 307; in animals, 338; in medusa, 359
LOCKE, on difference between man and brute, 349
Logos makes man human, 375
LONBIERE, on instincts of Siamese ants, 449
LOTZE, quoted, 379
LUBBOCK, Sir J., "Senses of Animals," 246; sense of smell in ants, 258; auditory organ of ant, 267; on Hicks's organ, 267; on colour-sense in dog, 283; in insects, 291; in Daphnia, 292; on limits of colour-vision, 296; on antennary structures in hymenoptera, 297; on power of communication in dog, 345; in ants, 358; on colour preferences in bees, 407; on instinct of play and sympathy in ants, 414; on homing faculty in bees, 428; on sitaris, 439
Lucanus cervus, 180
Ludicrous, sense of, in dog, 406
LUMSDEN, Sir Harry, on partridges, 398
LYELL, the necessary precursor of Darwin, 121

MACH, Prof., on Macula acustica, 271
Machetes pugnax, 110, 178
MACKENNAL, Mr. Alexander, observation on a cat, 405
MACLAGAN, Miss Nellie, on sympathetic action in dog, 398
Madeira, wingless insects of, 81
Male. See Sex-differentiation
Male and female insects, differences between, 179; greater variability in, 237; vigour and vitality of, in secondary sexual characters, 237
MALLE, Dureau de la, on starling, 455
Mammals, respiration in, 21; early nutrition of, the result of parental sacrifice, 57; convergence in, 117; sense of smell in, 255; hearing in, 263; sight in, 283; perceptions of, 338
Man, elimination by physical circumstances, 81; alternation of good and bad times, 117; reversion in, 229
MANN, Mrs., on sympathetic action of dog, 397; anecdotes of dogs, 406
Mantis, protective and aggressive resemblance in, 90
Marsupials of Australia, 117
MARTINEAU, Dr., on wants, 382
Materialism, 464, 471

Mathematical faculty and natural selection, 484
MAUPAS, M., observations on infusoria, 39
MAYER, on mosquito, 267
McCOOK, Dr., sense of smell in ants, 258; habits of Texan ants, 425
McCOSH, Dr., quoted, 391
Means and ends, 371
Medusa, 46; sense of hearing in, 265; eyes of, 293; localization by, 359
Melanerpeton, 288
MELDOLA, Prof. R., 239
Memory, the revival of past impressions, 304; organic, Butler and Hering on, 62
Mental evolution, 464
MERCIER, Dr. Charles, on the criteria of intelligence, 456
MERRIFIELD, Mr., experiments on moths, 238
Metabolism, 32
Metakinesis, 467
Metamorphosis and transformation, 7
Metaphysics, 15
Metazoa, 15
Methona, 87
MIALL, Prof., Fig. of touch-hair of an insect, 248
Mice, white and grey, crossed, 225
Microbes, elimination among, 80
Micrococcus prodigiosus, 81
Microstomum lineare, reproduction in, 42
Mimicry, 87; as evidence of perceptual association, 202, 351
Mind, out of what evolved? 464
Mineral crystals, analogy of, 240
MITCHELL, James, his delicate sense of smell, 255
MIVART, Prof. St. George, on Saturnia, 163; on commonsense realism, 318; on ideas, etc., 328; on "practical intelligence," 362; on man and brute, 374; on consciousness and consentience, 461
Modifiability of individual organism, 163
Modifications of antennæ and mouth-organs of insects, 178
Mole, eye of, 284
Mollusks, variety of, 178; sense of smell in, 260; hearing in, 265; sight in, 292
Monads, reproduction of, 38; temperature experiments with, 147
Mongrelization, 168
Monistic hypothesis, 465
Monkey, *ateles* and *colobus* digits of, 210; examining marsupial pouch, 340; attention in, 342; capuchin, intelligence of, 367
Monospora biscuspidata, 439
MOORE, Mr. Thomas, on hybrids between Amherst and golden pheasants, 166
Mosaic vision, 291
Mouth-organs of insects, 179
Muciparous canals of fishes, 298
MÜLLER, Prof. Max, "Science of Thought," 325; on percepts,

375; on language and thought, 376; paraphrased, 467; on materialism, 471
Murex, 292
Mus rex and *imperator*, 106
Musical and artistic faculty, 484
Mussel, freshwater, gills of, 4; olfactory organ of, 260
Mutilation, law of growth after, 126; not the best kind of evidence of transmitted modifications, 162

NAEGELI, 159
NAISH, Mr. John G., on the cockatoo, 354
Natural selection, variation and, 61; two modes, elimination and selection proper, 79; and the effects of use and disuse, 174; not to be used as a magic formula, 184; and instinct, 445; and human thought, 484
Nerves, briefly described, 246; afferent and efferent, 303
Nestor notabilis, 446
Nests of bower-bird **and humming-bird**, 408; **instinctive building of, 453**
NETTLESHIP, Mr., on a lion, 400
Neural **processes, environment of, 491**
Neuroses and psychoses, 465
Neuter insects, 440
New Zealand sparrow, 445; parrot, 446; chaffinch, 454
NICHOLS, on taste, 251
Noctule, 66
NOIRÉ, on concepts, 325
Nomada solidaginis, 90
NORRIS, Mr. W. E., quoted, 420
Noumena, or "things in themselves," 470
Nucleus of animal cell, **10; as** controlling formative process in, 124
Nutrition **in illustration of the process of life, 25**

Object, nature of, 313, 437
Ocelli in insects, 288
Oecodoma cephalotes, 213
Onchidium, 293
Optogram, 278
Organic combination, **hypothesis** of, 156, 240
Organic evolution, **177; as basis** of comparative **psychology, 336**
Organic growth, 5
Organism, unity of, as regards **body** and germ, 161; relation of, **to** environment, 183
Organization, co-ordinating power of, 125; of bodily and mental activities, 419
Origin, distinguished from guidance, 242
Origin of species, 242
Origin of organic variations, 231; of metakinetic or mental variations, 496
Ornithoptera, 179
Otoliths, 265, 271
OWEN, Sir Richard, suggested germinal continuity, 135

Oyster-embryo set free early, 56; variation of Mediterranean, 164

Pachyrhyncus orbifex, **87**
Pagurus prideauxii, **457**
Pain, massive and acute, 379; capacities of animals for, 391
Pangenesis, 182
Panmixia and disuse, 189
Papilionidæ, 202
Paradise, birds of, 202
Paranucleus in protozoa, 39
Paramœcium, reproduction in, 39
Parasites, how they feed, 5
Parental sacrifice in birds and mammals, 57; its limits, 186
Parrot, intelligence of, 353
Parthenogenetic forms, **no** second polar cell in, 153; **the** drone an exception, 153
Parus palustris, 164
PEAL, Mr S., on **use of tools by** elephant, 370
PECKHAM, Mr. G. W., on love-antics of a spider, 208, 450
Pecten, 293
Pelagic animals, colours of, 83
PENZOLDT, Dr., on **smell, 254**
Percept, 325, 326
Perception, 311, 324; in animals, 339
Perceptual association, 202
Perigenesis of the plastidule, 159
Peripatus, 142
Persistence, law of, **61**
Pheasant, hybrids between Amherst and golden, 106; golden, hen with cock's plumage, 228
Phenogámi, 223
Phenomenal **nature of object,** 315, 320, 331
Photographic psychology, **320,** 326
Phrynocephalus mystaceus, 90
Physiological isolation, 104
Physiological and psychological activities, 304; series, 386, 417
PICTON, Mrs. E., on Skye terrier, 398
Pigeons, correlated variations in, 216; silky fantail prepotent, 227
Pigs, intestines of, 171; crossing of, 226, 230
Pike, teeth of, 437
Pineal gland, 196, 288
Pipistrelle, wing of, 64
Pipits as illustrating divergence, 97
Pitch, musical, 261
Plasm, 10
Plasmogen, 10
Platyglossus, 83
Play, instinct of, 450
Pleasure and the special senses, 243; massive and acute, 379; capacities of animals for, 391
Plecotus auritus, 68
Plesiosaurus, pineal eye of, 288
PLOSS, Herr, on sex-differentiation in man, 59
Plover, Kentish, 83, 217
Polar cells, extrusion of, 51; and variation, 153
Postponement of action, 385
POULTON, Mr. E. B., on colours of animals, 84; on *Phrynoce-*

phalus mystaceus, 90; on caterpillars and chrysalids, 165; dimorphism in larva, 187; observations on edibility of caterpillars, 212; "Theories of Heredity," quotation from, 214; on the eating of unpalatable insects, 445
Predominant defined, **349**; and language, 374
Preferential mating, a means of segregation, 102; and sexual selection, 197
Preformation and evolution of older writers, 50
Prepotency, 227
Presentations of sense, 318
Previous sire, effect of, 168
Prevision as a criterion of **intelligence,** 457
Principles, mechanical, **368**
Process **of** life, 20
Progress, or continuous adaptation, 119; adaptation to more complex circumstances, 183
Pronghorn, curiosity in, 339
Proposition, 329
Protective resemblance and mimicry, 82; general resemblance, **83**; variable resemblance, 84; special resemblance, 86; to another organism, 87; coloration, a means of segregation, 101
Protection, fosterage and, 219
Proteus, **sensitive** to light, 294
Protista, 15
Protohippus, **118**
Protophyta, 15
Protoplasm, 10
Protozoa, nature of, 15; transmission of acquired faculty in, 147; origin of metazoan variations in, 156; psychology of, 360
Psithyrus rupestris, 90
Psychological and physiological activities, 304; series, 386, 417
Psychoses and neuroses, 465
Ptarmigan, on colour of, **185**

Rabbit, brain of, 171; Angora **crossed, 225**; one-eared, 226; **deprived of** long lip-hairs, 247; *papilla foliata* of, 250; effects of superabundant food on, 394
RAE, Dr., on dogs swimming rivers, 364; on "abstract reasoning" in the fox, 366; on wild and tame ducklings, 435
Rage and anger, 389
RAMSAY, Dr. Wm., on smell, 255
Rats of Solomon Islands, 100; **of the** London Docks, 106; at **South** Kensington, 115
RAYLEIGH, Lord, on colour-blending, 283; **on** sensitive-flame experiments, 298;
Reality, meaning of term, 314
Reason distinguished from intelligence, 330, 365; as defined by Mr. Romanes, 372
Recepts, 326, 368
Recognition-marks, 103; involve perception, 351
Reconstructs and reconstruction (mental), 318

Reflex action, 415; and instinct, 422
Regeneration of lost parts, 41; in relation to heredity, 124; law of growth concerned in, 126
Reindeer wounded, 392; change of habit in, 445
Remnants or vestiges, 198
Reproduction, nature of, 13; and development, 36; in the protozoa, 38; in the metazoa, 41; by budding, 42; sexual, 42; peculiar modifications of, 45; developmental, 143
Reproductive cells, continuity of, 131
Resemblance, protective, 82; aggressive, 80
Respiration an essential life-process, 3; in illustration of the process of life, 21
Retardation and acceleration, 221
Retina of man, 274; of birds, 284
Retinal purple, 276
Revenge, 401
Reversion, 191
Revolution and evolution, 119
Rhea, neck resembling snake, 88
Rhinolophus ferri-equinum, hipposideros, 65
Rhyme-association in parrot, 356
Ribot, M., on attention, 343
Richardson, Mr. Charles, on railway servants killed by train, 386
Riley, Prof., on *Phengodini*, 223
Romanes, Prof. G. J., on physiological isolation, 104; on the cessation of selection, 190; on the failure of heredity, 192; on the reversal of selection, 193; on sense of smell in dog, on colour-sense in chimpanzee, 283; on ideas, 326; on dog cowed by noise, 340; on abstract ideas in animals, 348; on parrot, 353; on localization and discrimination, 359; examples of animal intelligence considered, 362; on abstract ideas in the capuchin, 368; definition of reason, 372; on strange attachments in birds, 396; on some emotions in animals, 400; on endurance of pain in dogs and wolves, 402; on sense of humour in dog, 407; on indefinite morality in animals, 413; definition of instinct, 422; on education of ant, 428; on homing faculty of bees, 428; on consciousness and instinct, 432; summary of his conclusions on instinct, 434; on instincts of Siamese ants, 449; his psychological scale, 478; on the world as an eject, 479
Rotation, sense of, 269
Rotifers, absence of fertilization in reproduction, 45
Roux, on extirpation of cleavage-cell of frog's egg, 214
Rowell, G. A., on "Beneficent Distribution of Pain," 392
Ruffs, variability of males, 110, 178

Russell, Mr. W. J., on smell in the dog, 255

Saitis pulex, 450
Salinity of water, effects of, on brine-shrimp, 164
Salmon, new variety of, in Tasmania, 89
Saturnia, modification of, by changed food, 163; *carpini* (emperor moth), 258
Savages, fetishistic belief in, 494
Schaur, Mr., observations on a terrier, 405
Schmankewitsch on *Artemia*, 164
Sclater, Mr. W. L., on mimicry in an insect, 88
Sedgwick, Mr. Adam, on development of peripatus, 142
Seebohm, Mr. H., on birds' eggs, 410
Segregation, 99
Selection, as compared with elimination, 70; illustrated, 92; artificial, 172; cessation of, 190; reversal of, 193; sexual, or preferential mating, 197, 452; as a factor in the origin of instinct, 447; as applied to the intellectual faculties, 486
Selenia, illunaria, and *illustraria*, 238
Self, the, or ego, 475
Self-consciousness, 460
Semicircular canals, 262, 269
Senility, introduction of, 184
Sensation defined, 305, 324
Sense-feelings of animals, 393
Senses of animals, 243; organic and muscular, 244; touch, 245; temperature-sense, 249; taste, 250; smell, 253; hearing, 261; sight, 272; contact and telæsthetic, 249; problematical, 267
Sensibility, 385; variations of, 440
Sensitive, special use of the term, 9
Sensitiveness and sensibility, 385
Sentiments, 361; in animals, 403
Sex-differentiation, 56
Sexual union of ovum and sperm a source of variations, 149; characters, secondary, 197; selection, 197
Shame in monkey, 402
Sheep, Youatt on, quoted, 455
Shells, land, of Sandwich Islands, 99
Shipp, Captain, experiment on an elephant, 401
Sight, sense of, 272
Sitaris, instinct of, 438
Skertchley, Mr. S. B. J., on leaf-butterfly, 86
Slave-making ants, 425
Smell, sense of, 253
Smerinthus ocellatus, 165
Smith, Mr. G. Munro, on elimination among microbes, 80
Snail, variations in banding of shells, 75; sense of smell in, 260; auditory sac, 265; eye of, 292; *spicula amoris* of, 459
Snakes, mimicry in, 88
Snipe, drumming of, 448

Sollas, Dr. W. J., on regeneration of tentacle in snail, 127
Somatic, or body-cells, 193
Sommering, Fig. of semicircular canals, 270
Spalanzani, his experiments on bats, 248
Spalding, Douglas, on instinctive emotions, 385; on perfect instincts of chicken, 424; on deferred instinct in swallow, 425
Sparrows in New Zealand, 445
Specific characters, utility of, 110; constancy of, 111
Spencer, Mr. Baldwin, Fig. of pineal eye, 288
Spencer, Mr. Herbert, law associated with his name, 37; physiological units, 125, 153; on lap-dogs, 195; on the Irish elk and giraffe, 212; on diminution in ear-muscles, 215; definition of pleasure and pain, 381; on æsthetics, 412; on instinct and reflex action, 422
Sperm-cell and egg-cell, 13; conditions which determine production of, 60
Sphex, instinct of, 429, 456
Spiders, hunting, mimicry in, 89; Javan, Mr. H. O. Forbes on, 90; love-antics of, 208, 450; ocelli of, 289
Spinoza, quoted, 61, 379, 460
Sponges, reproduction of, 41, 42
Spongilla, reproduction of, 46
Spore-formation, reproduction by, 38
Squirrel of Sarepta, 113
Stag-beetles, variation in males of, 180
Star-fish, embryo set free early, 56
Starling, modified song of, 455
St. John, observations on a retriever, 400
Stenorhynchus, 457
Sterility, how developed, 198
Stewart, Mr. Duncan, on sympathy in cat, 398
Stimuli, 302
Strange, Mr., on love-antics of satin bower-bird, 450
Striped ancestor of Equidæ, 230
Struggle for existence, 79; variations in the intensity of, 112
Sturge, Miss Mildred, on the parrot, 355
Stylonichia, observations of M. Maupas on, 39
Sully, Mr. James, on concepts, 325; on propositions, 329; on judgment and reason, 336; on emotion, 396; on æsthetic sense of beauty, 411
Sutton, Mr. Bland, on hen pheasant like the male, 228; on the action of leucocytes, 439
Swallow and swift, convergence in, 117; cliff, of United States, 445
Swayne, Mr. S. H., on the elephant, 369
Symbolic nature of mental products, 314
Symonds, Mr. J. A., on "world-consciousness," 478
Sympathy in animals, 397

Tameness, instinctive, 435
TANNER, Miss Agnes, on a thrush, 398
Tasmanian salmon, 99
Taste, standard of, 95, 295; sense of, 250
Teeth of pike, 437
Temperature-sense, 249
Terror, 387
Thaumalea picta and *amherstiæ*, 106
Thekla, instinct of, 430
"Things in themselves," or **noumena, 470**
THOMAS, Mr. Oldfield, on rats of Solomon Islands, 100
THOMSON, Mr. J. A., Prof. Patrick GEDDES, and, on anabolism and katabolism, 44; quoted, 50, 137, 237; his "History and Theory of Heredity," 35
Thought, 482
Thrush, bearing in, 264; sympathy in, 398
THUNBERG on young hippopotamus, 423
Tissues of the body, 20
TOOKE, Mr. Hammond, on egg-eating snake, 88
Tools, use of, by animals, 370
Touch, sense of, 245
Transformation and metamorphosis, 7
Transparency of some marine organisms, 83
TREAT, Mrs., her experiments on caterpillars, 59
Tricks, 355
Trionyx, 181
Trochus, 292
Tuco-tuco, 194
TURNER, Sir Wm., on New Guinea natives, 169
Turkey, instinctive emotion in the, 395
Twins, Mr. Galton's investigations on, 169
TYLOR, Alfred, on coloration in animals and plants, 201

Udders, enlarged, of cows, 215
Ultra-violet rays, 296
Unicellular organism. *See* Protozoa
Unity of organism, 161, 234
Use and disuse, 146, 209
Utility of specific characters, 110

Vanessa urticæ, 165
Varanus benegalensis, 288
Variation, correlated, 59; and natural selection, 61; tabulated by A. R. Wallace, 63; in wing-bones of bats, 63; advantageous, neutral, and disadvantageous, 95; in climatal and geographical conditions, 112; secular, in climate and life area, 113; effect of good times and hard times on, 114; heredity and the origin of, 122; a source of, in use and disuse, 146; sexual union, a mode of origin of, 149; in definite directions, 151; produced by extrusion of second polar cell, 153; protozoan origin of, 156; due to the action of environment, 163; to the effects of use and disuse, 163; to domestication, 171; in male stag-beetles, 180; in mating preferences, 205; co-ordinated in Irish "elk" and giraffe, 212; nature of, 216; in amount of developmental capital, 221; inheritance of, 223; origin of, 231; limitations of, 232; fortuitous, in bat's wing, 235; definite direction of, 238; in limits of colour-vision, 281; in habits and instincts, 445, 456; in mental evolution, 496
Vertebrata, diagrammatic account of development of, 51
VERWORN, Dr., on protozoa, 440
Vespertilio mystacinus, 70
Vesperugo leisleri, 65
Vesperugo noctula, 67
Vesperugo pipistrellus, 69
Vigour and vitality, application of, in male, 237; in female, 238
Vindictiveness, 401
Vision, 272; mosaic, 291
Volition, 459
Volucella bombylans, 99
Voluntary and involuntary activities, 416
Vorticella, 38

WAELCHLI, Dr., on colour-globules in birds, 284
WALLACE, Mr. A. R., tabulations of variations, 63; on tortoiseshell butterfly of Isle of Man, 81; on protective colours in fishes, 83; on divergence among birds, 97; on recognition-marks, 102; on papilionidæ of Celebes; 165; on the dull colours of hen birds, 199; on origin of secondary sexual characters, 200; and A. Tylor on physiological guidance, 201; on preferential mating, 203; on reversion in grouse, 229; on migration in birds, 428; on nest-building in birds, 453; on the song of birds, 455; on materialism, 464; on mathematical and artistic faculties, 484, 497
WALKER, R., on reversion in bull, 229
WARD, Mr. J. Clifton, on dog, 345

Warning-coloration, 82; involves perception, 351
WARREN, Mr. Robert Hall, a dog anecdote, 344
Wasp, use of antennæ, 291
Waste and repair essential life processes, 8
Water, changes of salinity in, 164
Water-ouzel, 446
WATERTON, Charles, 256
WATSON, "Reasoning Power of Animals," 369
WEBB, Dr., his operation on an elephant, 369
WEBER, on musical discrimination, 309; on muscular sensation in eye, 310
WEIR, Mr. Jenner, on nest-building in birds, 453
WEISMANN, Dr., on continuity of germ-plasm, 138; on distinctness of germ-plasm from body-plasm, 140; on meaning of second polar cell, 153; on protozoan origin of variations, 156; on the introduction of senility and death, 184; on the distinction of birds' eggs, 189; on the effects of panmixia, 190; on acceleration, 222; his views applied to instinct, 438; the intellectual faculties, 497
WESTLAKE, Miss Mabel, on the parrot, 353
Whiskered bat, 70
White, in arctic forms, 165; Mr. Poulton on production of, 202; in grouse, instance of reversion, 229
Wildness of birds, instinctive, 435
WILL, F., on taste in bees, 253
WILSON, Sir Charles W., on wounded camels, 392
WILSON, Edward, measurement of bats, 63
Wing-bones of bats, measurement of, in illustration of variation, 63
Words, "understanding" of, by animals, 347
Wrasse, keenness of vision of, 287

Xiphocera, 178

YOUATT "On Sheep," 455
YOUNG, Thomas, his colour-vision theory, 277
YUNG, his experiments on tadpoles, 59

Zebra, inconspicuousness of, in dusk, 84
Zuyder Zee, new variety of herrings in, 99

www.ingramcontent.com/pod-product-compliance
Lightning Source LLC
Chambersburg PA
CBHW051156300426
44116CB00006B/324